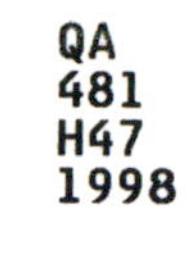
QA
481
H47
1998

D0215564

This Book is Due...
JUL 26 2007
OCT 04 2007
This Book is Due...

A Mathematical History
of the
Golden Number

A Mathematical History of the Golden Number

Roger Herz-Fischler

'Αχρον χαὶ μέσον λόγον εὐθεῖα τετμῆσθαι λέγεται, ὅταν
ᾖ ὡς ἡ ὅλη πρὸς τὸ μεῖζον τμῆμα, οὕτως τὸ μεῖζον
πρὸς τὸ ἔλαττον. —Euclid, *Elements*, VI,def.3
[Euclid–Heiberg, II, 72]

Non me pare, excelso Duca, in più suoi infiniti effetti al presente estenderme, peroché la carta non supliria al negro a esprimerli tutti.... —Paccioli, *Divina proportione*
[Paccioli, 1509, Chap. XXIII]

DOVER PUBLICATIONS, INC.
Mineola, New York

Published in Canada by General Publishing Company, Ltd., 30 Lesmill Road, Don Mills, Toronto, Ontario.
Published in the United Kingdom by Constable and Company, Ltd., 3 The Lanchesters, 162–164 Fulham Palace Road, London W6 9ER.

Bibliographical Note

This Dover edition, first published in 1998, is an unabridged republication of *A Mathematical History of Division in Extreme and Mean Ratio*, originally published by Wilfrid Laurier University Press, Ontario, Canada, 1987.
The Dover edition incorporates a slight rearrangement of the original pagination to accommodate a new preface and a section of "Corrections and Additions," both prepared specially for this edition by the author.

Library of Congress Cataloging-in-Publication Data

Herz-Fischler, Roger, 1940–
[Mathematical history of division in extreme and mean ratio]
A mathematical history of the golden number / Roger Herz-Fischler.
p. cm.
Originally published: A mathematical history of division in extreme and mean ratio. Waterloo, Ont., Canada : Wilfrid Laurier University Press, c1987.
"Incorporates . . . a new preface and a section of 'Corrections and additions,' both prepared specially for this edition by the author"—T.p. verso.
Includes bibliographical references.
ISBN 0-486-40007-7 (pbk.)
1. Ratio and proportion—History. I. Title.
QA481.H47 1998
512.7—dc21 97-52729
CIP

Manufactured in the United States of America
Dover Publications, Inc., 31 East 2nd Street, Mineola, N.Y. 11501

Pour

Eliane, Mychèle et Seline

To all those past and present whose work and studies resulted in this book and to those whose aid, direct or indirect, enabled me to complete it.

A MATHEMATICAL HISTORY OF DIVISION IN EXTREME AND MEAN RATIO—"THE GOLDEN NUMBER"

[Publisher's Note: This is the author's preface to the original edition, published under the title *A Mathematical History of Division in Extreme and Mean Ratio.* That title reference—as well as the abbreviation DEMR—has been retained in the body of the text.]

The first unequivocal appearance of DEMR (division in extreme and mean ratio—"the golden number") occurs in the *Elements* of Euclid. But when and how did this concept arise? *A Mathematical History of Division in Extreme and Mean Ratio* is the first work to make a complete and in-depth study of this question as well as of all aspects of the historical development (from the origins to 1800) of a concept which has played an important role in the development of mathematics and evoked much commentary. A detailed analysis of the role of DEMR in the *Elements* and of the historical implications is followed by a discussion of other mathematical topics and of proposals by modern commentators concerning the relationship of these concepts to DEMR. Succeeding chapters discuss the Pythagoreans, examples of the pentagram before −400, other historical theories, and the writings of the pre-Euclidean mathematicians. The author then gives his own views on the origin, early development, and chronology of the concept of DEMR, many of which go against what is often assumed to be true in the literature.

The second half of the book traces the development after the time of Euclid, through the later Greek period, the Arabic world, India, and into Europe. The emphasis throughout is on a clear but rigorous presentation of the work of each author in the context of the mathematics of the time and on the transmission of results and concepts.

This work will be of interest not only to mathematicians and historians of science, but also to classicists, archaeologists, and to those interested in the transmission of ideas.

TABLE OF CONTENTS

PREFACE TO THE DOVER EDITION

'L'histoire, oserais-je dire, et sans aucune intention de paradoxe, c'est ce qu'il y a de plus vivant; le passé, c'est ce qu'il y a de plus présent."—Lionel Groulx [cited in *Le Courrier du patrimoine*, automne 1997, 15]

The reception given to my study of division in extreme and mean ratio has been most pleasing to me, particularly statements to the effect that this book has been a useful one. I hope that this new edition will make this fascinating topic available to a wider audience. As Appendix II shows, division in extreme and mean ratio has attracted mathematicians throughout the ages.

Reviews of a book constitute a type of appendix, for in addition to the general comments, favourable or unfavourable, they include material which was omitted by the author. The following is a list of all the reviews that are known to me.

Artmann, B. 1989. *Mathematische Semesterberichte* 36, pp. 141–42.

Bidwell, J. 1992. *Ancient Philosophy* 12, pp. 248–50.

Crawford, C. 1987. *Canadian Book Review Annual*, pp. 331.

Fraser, C. 1988. *Mathematical Reviews*, review 88:01006.

Grattan-Guinness, I. 1989. *British Journal for the History of Science* 22, pp. 84–5.

Gyula, M. 1989. *Centaurus* 32, pp. 244–45.

Høyrup, J. 1990. *Historia Mathematica* 17, pp. 175–78.

Perol, C. 1988. *Bulletin de l'association des professeurs de mathématiques de l'enseignement public*, n°666, pp. 644–45.

Pour, R. 1989. *Choice* April 1988, p. 1277; May 1989, p. 1470.

Pottage, J. 1989. *The Mathematical Gazette* 73, pp. 265–67.

Unguru, S. 1989. *Isis* 80, pp. 298–99.

As I note on page xii, my examination of theorem XIV** of section 24 turned into a very long and extremely complex study. Because of the length, and the interruption in the flow of the material that this would have created, I decided to publish this study separately:

"Theorem XIV** of the First 'Supplement' to the *Elements*", *Archives Internationales d'Histoire des Sciences* 38, 1988, pp. 3–66.

A glance at my comparison of Pappus and Book XIV and the "genealogy of XIV**", [1988, 9, 44], should convince the reader that there is much work left to be done in the field of ancient Greek and Arabic mathematics. The mere existence of XIV** and the "ratio lemma" suggests that we are far from completely understanding the mathematical thought process of the Greek mathematicians; see [1988, fn. 4]. I also hope to see the disappearance of the unfortunate attitude—see [1988, fn. 2, 8, 11, 50, 57, 67, 77]—of some authors who assume that if a result does not appear in a "pure" Greek manuscript, then it is either a later development or else it was "obvious" to the Greek mathematicians.

As Ivor Grattan-Guinness pointed out in his review, there were two other books published in the 1980s that dealt with a specific aspect of Greek mathematics: Wilbur Knorr's *The Ancient Tradition of Geometric Problems* and David Fowler's *The Mathematics of Plato's Academy*. Unfortunately the world of the history of mathematics lost one of its outstanding members this year with the death of Professor Knorr. My relationship to him was limited to his being—as I later found out when he sent me some additional material—the referee of my first article, [Fischler, 1979a]. However his forthright criticism and commentaries at that point were certainly influential, not only in the rewriting of that article, but also on my later approach to Greek mathematics. On another note, I was pleased to learn that David Fowler is presently preparing a second edition of his book.

I announced in the preface that I would be publishing another book dealing with the non-mathematical history of the golden number. Both before and after I worked on *Division in Extreme and Mean Ratio*, I did research on this topic; see under "Fischler" in the bibliography. The last chapter of that book was going to deal with the various theories of the shape of the Great Pyramid, in particular those involving the golden number; see [Fischler, 1979b]. However as happened with *Division in Extreme and Mean Ratio*, I let my research go where the sources and my curiosity led, without considering such matters as the finishing date. Thus the planned last chapter turned into a separate book, *The Shape of the Great Pyramid / A Historical, Sociological, Philosophical and Analytical Study*. This work will be published by Wilfrid Laurier University Press in 1998 or 1999.

The other book, tentatively entitled *Golden Numberism*, is perhaps sixty percent complete. However experience has taught me that it is unwise to predict either the length, the date of completion, or the final contents of a book before it is sent to the editor. An aperçu can be found in:

"The Golden number, and Division in Extreme and Mean Ratio" in *Companion Encylopedia of the History and Philosophy of the Mathematical Sciences*, I. Grattan-Guinness ed., London, Routledge, 1994, 1576–1584.

Those who wish to see how intricate such studies can be, and how a knowledge of the history of mathematics is most useful for resolving certain enigmas in the field of art history, may consult my article:

"Le Nombre d'or en France de 1896 à 1927", *La Revue de l'art*, 1997, n° 4.

My thanks go out again to Sandra Woolfrey and the Wilfrid Laurier University Press for their confidence in my work. I also thank the reviewers of my book for their comments. Finally I would like to dedicate this reprint of my book to the memory of Ivor Bulmer-Thomas. He was a fine scholar, but in addition he was a humanist and a charitable person. In the first category we have his *Selections Illustrating the History of Greek Mathematics*; in the second, one of his last works, "The Star of Bethlehem—A New Explanation—Stationary Point of a Planet" (*Q. J. Royal Astromical Society* 33(1992), pp. 363–74). Finally his interest in good causes was evident in his taking almost total responsibility for the British organization, Friends of Friendless Churches.

Roger Herz-Fischler

Ottawa, Ontario, Canada

November 11, 1997

FOREWORD

The story of how this history came to be written is perhaps not without interest. In 1972, when I was still contentedly proving theorems in abstract probability theory [Fischler, 1974, 1976], I was approached by the then chairman of the Department of Mathematics and asked to take over a course for first-year architecture students at Carleton University. There had been some discontentment on the part of the students, and since the Mathematics Department did not want to lose control of the course, the order came through to "keep them happy." I decided that the best way to keep the students content was to keep myself content by talking about things that interested me. In particular, having heard various things about the so-called "golden number," I decided that I would read about it and use some of the material in the classroom. Eventually, as sometimes happens (cf. Bulliet [1975, v]), I decided to make a detailed investigation of some of the claims concerning the supposed non-mathematical manifestations of the "golden number" (see, for example, Fischler [1979b,c; 1981a,b]). As a result of these investigations I decided to write a book dealing with the findings of my research in which, for completeness, it seemed appropriate to say something about the purely mathematical history.

At first it seemed as if this mathematical history would be fairly short and straightforward. This opinion was based on a preliminary reading not only of parts of the *Elements*, but also of some of the standard histories of Greek mathematics. However, two things soon became clear: the early Greek aspect was not as clear-cut as it was often made out to be and the historical aspects that needed to be considered neither started nor ended with the early Greeks.

While it turned out that the later history of DEMR had essentially never been dealt with, at least in a unified form, the writings on the early history suffered from a number of defects. On the one hand, the literature turned out to be surprisingly large, but consisted to a large extent of scattered writings in which the authors were ignorant of or ignored the writings of others. On the other hand, the writings were often based on the slimmest of evidence; on vague references in the classical literature; on a priori assumptions about the state of mathematics at certain periods; or on out and out speculation. I also noticed that no one had made an in-depth study of the role of DEMR in the *Elements*. Finally, the literature suffered from those deficiencies which unfortunately are not limited to the present case: references to historical material which was difficult to obtain and/or available only in the original language (Babylonian, Greek, Latin, Arabic); obscure bibliographical references; as well as incorrect translations, incorrect inferences from quotations, and misrepresentation of the mathematical process actually involved in the original.

Because of all this I decided to make a separate, complete and detailed study of the mathematical history of DEMR that I hoped would not suffer from these defects. This book is the result of my research. My studies on the non-mathematical history of the "golden number" will appear in another book. The two aspects are completely separate from one another except that in several instances (see, for example, Sections 12,vi; 16,D,iii) historical commentary by certain authors has been influenced by writings on the non-mathematical aspects and conversely some of the non-mathematical claims have been based on some supposed historical truths.

This book was written without the benefit of any nearby colleagues who were interested in the history of mathematics, but having had an example set by my wife (see E. Herz Fischler [1977]), I did not feel that that was necessarily a handicap. From a distance I received valuable comments and material from David Fowler, whom I had never met until he incredibly deduced that I would show up at the Institut d'Esthétique et des Sciences de l'Art in Paris on the first Monday of September 1982.

I wish to express my appreciation to the many scholars, librarians, and others who provided me with references, material, and helpful suggestions.

Foremost among the difficulties I faced were linguistic ones. I had the choice of being incomplete, or waiting until my next lifetime to learn Greek, Latin, Akkadian, Sumerian, Arabic, Italian, etc., or finding people who would help me. I preferred the latter option. I state this not as an apology for my lack of knowledge, nor as an excuse for the linguistic errors which may be found, but, rather, so that my readers will not be deceived. As a word of encouragement to others in the same position, I note that my fluency in French combined with a dictionary and calculator and a disregard for the fine points of Latin and Renaissance Italian enabled me to understand the mathematics of the texts of Section 31 on my own. Of all those who helped me with linguistic matters, my special thanks go to Len Curchin, presently of the Classics Department of the University of Waterloo, a true scholar's scholar who fortunately for me was doing his doctoral work at the time I was writing the first part of the book. Not only did he read Latin, including the Latin of medieval and Renaissance manuscripts, and Greek, but also Akkadian and Sumerian (see, for example, Curchin [1977; 1979a,b; 1980]). Furthermore, he seemed to have an inexhaustible knowledge of the ancient world and came up with such items as Quotation 5 of Section 11,A. All the translations from Latin and Greek sources not attributed to specific texts—except for a few from Chapter IX—are due to him. The discussion of Scholium 73 to II,11 in Section 11,C is based on his analysis of the text; in addition, he made many suggestions pertaining to the discussion of Theaetetus in Section 18 as well as in various other spots (see also Curchin and Fischler [1981] and Curchin and Herz-Fischler [1985]).

As the colophon at the end of Section 32 indicates, I completed this study in July 1982. I tried to be complete in my coverage of the literature up to that point, but in the course of final revisions I was only able to include some of the works that have appeared since then or older ones which I subsequently came across. The only major revision concerns Theorem XIV,** which is discussed in Section 24,A. I had already come to the conclusion that this theorem was in one of the early Greek manuscripts and had discussed this briefly in an appendix to Section 1. I had no inkling, however, of the complexities of the problem that my research over the last two and a half years has revealed (see Herz-Fischler [1985]). To have included this material, complete with linguistic, mathematical, and alphabetical analyses, would have presented many problems and perhaps distorted this book. I have thus only mentioned the matter and will publish the full study elsewhere.

As well, I express my thanks to the interlibrary loans staff at Carleton University for their patience and never-ending search for all those books and articles that I needed; the referees of the manuscript for their time and perceptive comments; Walter R. Powell of the CAD Canada Group of Ottawa for his incredibly rapid production of the drawings using the Canadian ACDS computer graphics system; and finally to the staff of Wilfrid Laurier University Press who acted as "sages-femmes" in transforming my manuscript into a book.

This book has been published with the help of a grant from the Canadian Federation for the Humanities, using funds provided by the Social Sciences and Humanities Research Council of Canada.

A GUIDE FOR READERS

The purely technical aspects of this book are the result of training and tastes, on the one hand, and a reaction against certain aspects of the historical literature, on the other. While I have tried to make the book self-contained and internally comprehensible from a strictly mathematical viewpoint, I have also insisted on being very detailed as far as references are concerned. I have followed the lead of almost all mathematical as well as of many other scientific and historical journals (for example, *Historia Mathematica*) by giving author and date references directly in the text. Furthermore, since I felt that everything worth saying about the mathematical history of DEMR should appear in the text and the rest should be left out of an already large work, I have avoided using notes and footnotes, and as a result I must apologize for any possible prejudice to the future well-being of any authors concerned (see Leviant [1973, 99]). Many of the bibliographical references include background material or material indirectly related to the main topic. Similarly, I have often noted—sometimes in the bibliography alone—various works which, a priori, might be related to this study, but which examination has shown are not. I hope this will spare future scholars some effort.

I have included another bibliographical feature which I often wished had been included by other authors. No library has all the books needed for a study such as this, and rare indeed is the serious researcher in a broad field who can function without an effective interlibrary loan system. Thus I have noted the location of all books that I obtained on interlibrary loans, as well as those at Carleton University.

A second bibliographical feature is the use of the bibliography as a partial index. Since this is in a sense a book about books, whether they are original texts or commentaries, it seemed desirable to indicate directly with each entry in the bibliography where the work is referred to in the text. This feature, together with a detailed table of contents (which also serves as a chronological chart) and a list of quotations, makes a general index superfluous.

A. Internal Organization

Sections are numbered consecutively with arabic numerals from the beginning to the end of the book; for example, Section 15.

Subsections are indicated by uppercase letters and start again in each section; for example, Section 20,C.

Sub-subsections are indicated by small roman numerals and start again in each subsection; for example, Section 20,C,i.

Quotations are indicated in the form of "Q.5" and start again in each chapter. If no chapter is indicated, it is the present one that is being referred to.

Equations are indicated in the form of (12) and start again in each chapter. If no chapter is mentioned, it is the present one that is being referred to.

Figures are numbered consecutively in each chapter; for example, Figure II-1.

B. Bibliographical Details

The form of the bibliographical entry used depends upon the nature of the work. (An asterisk following an entry indicates that I did not actually consult the work.)

i. Original books and articles:

Diels, H. 1934. *Die Fragmente der Vorsokratiker*, Bd. 1. Edited by W. Kranz. 10th ed. Berlin: Weidmann, 1961.

The date 1934 following the author's name corresponds to the original edition, if known; the date 1961 corresponds to the edition I consulted.

ii. Articles from the *Dictionary of Scientific Biography* (*DSB*) and the *Oxford Classical Dictionary* (*OCD*):

Vogel, K.–*DSB* 1. "Diophantus of Alexandria." *DSB*, IV, pp. 110-19.

The number following *DSB* distinguishes one of several articles; this one appears in Volume IV by Vogel in *DSB*. Full bibliographical details for *DSB* and *OCD* are given under those entries.

iii. Editions and translations of works:

Plato–Fowler 2. *Greater Hippias* in *Plato*, vol. 6, pp. 333-424. Translated by H. Fowler. London: Heineman, 1953.

Here the 2 following Plato–Fowler distinguishes one of several editions of Plato by Fowler.

Square brackets are used in the text for references; the text references are the same as the bibliographical entries except that the author's initial is omitted in the text. Thus, "[Plato–Fowler 2, 408]" or "Diels [1934, 58] comments. . . ." If there is a further immediately obvious reference, then a shortened form is sometimes used; for example, [Diels, 62] or [p. 62].

Because many editions of Euclid, and other works, are constantly referred to, I simply write "in the Euclid–Frajese edition" in the text.

For quotations, the first reference, always bracketed alone, indicates the source used. If this source gives an English translation, then this is what has been used word for word unless otherwise indicated. If the source gives a non-English quotation, then it is this quotation that has been translated. The following bracket contains other sources and/or translations.

C. *Abbreviations*

DEMR Division in extreme and mean ratio.

I note, for the reader who might object to seeing this abbreviation several thousand times, not only that it was helpful in preserving my sanity and that of Wilfrid Laurier University Press, but that it also continues the tradition found in a medieval edition of Euclid [Euclid–Adelard III, fol. 334v = p. 664].

"Quiquid accidit uni linee divise secundum *p.h.m.* [*&*]. *d.e.* (i.e., proportionem habentum medium et duo extrema) omni linee similiter divise probatur accidere." ("Whatever happens to one line divided according to EMR is proved to happen to every line likewise divided"; cf. Section 24,A.)

EMR Extreme and mean ratio.

D. *Symbols*

To use symbols in rendering mathematics that was written out is to invite criticism which is often justified. This is particularly true of Euclid, and especially those parts involving ratio, proportion, numbers, etc., because the *Elements* is far from being clear on these matters and also because it uses various terms loosely and not always consistently. But to avoid the use of symbols would have created great difficulties, and so I have employed them and given explanations of any difficulties that may arise. Further, I have sometimes used the same symbol to mean two different things. Mathematical discourse, as ordinary discourse, relies on a certain initial primary ambiguity to avoid very cumbersome statements. This primary ambiguity is eliminated when the context of the statement is taken into account. As is the case with Euclid's *Elements*, the reader should have no difficulty deciding in each instance which of the two possibilities is meant.

AB The segment whose endpoints are A and B or the length of this segment.

$S(AB)$ The square, in the sense of the geometrical figure, whose sides have the same length as the line segment AB or the area of the square.

$R(AB,CD)$ The rectangle, in the sense of the geometrical figure, two of whose sides are equal in length to the line segments AB and the other two to CD or the area of the rectangle.

For the designation of rectangles and squares, I have followed the lead of Euclid–Frajese in using the initials of these words in the language in which I was writing rather than using the O and T notation of Dijksterhuis as employed, for example, by Mueller [1981, 56 fn. 59].

$R(AB,CD) = S(EF)$ The areas of the rectangle and the square are equal.

I = II The geometrical figures labelled I and II in the accompanying figure have equal area.

$=$ Used for numerical equality of areas, angles, chords, and arcs.

$\doteq$ Approximately.

$\approx$ Used to indicate similarity of geometrical objects.

$\cong$ Used to indicate congruency of geometrical objects.

$\equiv$ Used to indicate that the two geometrical quantities are really one and the same even though the labels may differ.

$A:B = C:D$ The numerical ratio of the magnitudes of A and B is the same as that of the ratio of the magnitudes C and D. See my comments at the beginning of this list and also the comments on *Elements* V in Section 1.

ΔABC Triangle ABC.

$\measuredangle ABC$ Angle ABC.

r Radius of a circle or its length.

d Diameter of a circle or its length.

D Diameter of a sphere or its length.

a_n Side of the regular n-gon or its length. Where several polygons are mentioned, it is understood that they are all inscribed in the same circle.

A_n Area of a regular n-gon.

d_5 Diagonal of a pentagon or its length.

e_n Edge of the regular polyhedron with n faces or its length.

S_n Surface area of the regular polyhedron with n faces.

V_n Volume of the regular polyhedron with n faces.

$R(x)$ Square root of x. This fine symbol from the Italian Renaissance has many advantages over the usual "Johnny-come-lately"; see Cajori [1928, 366]. Let the doubting reader consider Francesca's answer to problems 10 and 15 of Section 31,C. Note how close in form and spirit $R(x)$ is to the $SQR(x)$ used in connection with programs for modern computers.

II,11; VI,def.3 This refers to the eleventh theorem of Book II and the third definition of Book VI of Euclid's *Elements*. The numbering corresponds to the critical edition [Euclid–Heiberg]. The numbering used in Euclid–Heath varies somewhat; see Mueller [1981, 317]. When the numbering of an older edition or manuscript is used, the modern numbering is also given for reference. A prime after a number (e.g., XIII,13′) indicates my own variation on a result. A bracket, e.g., XIII,16[a], is used to designate a sub-result that does not exist as a formal entity, but which I wish to emphasize.

E. Dates

−386 The 386th year before the year 1 of the Julian system, i.e., the "common era."

72 The 72nd year of the Julian or Gregorian (after 1582) systems.

vth century The fifth century before the first century of the Julian system.

2nd century The second century of the Julian system.

For a detailed discussion of these and other chronological systems, see Neugebauer [1975, 1061]. A question mark after a date indicates either that it is estimated or that it is disputed in the literature.

F. Quotations from Primary Sources

The numbering of the quotations starts again in each chapter. Only a short form of reference and an indication of the contents are given. For references to Euclid's *Elements*, see the indication following each theorem in Section 1.

Chapter II. Mathematical Topics

Q.1 Plato, *Republic*, 527A (squaring and applying); Section 5,D.

Q.2 Plato, *Meno*, 86E (applying area); Section 5,D.

Q.3 Plutarch, "Pleasant Life Impossible," 1094B (Pythagoras—right triangle or application of areas); Section 5,D.

Q.4 Plutarch, "Table Talk," 720A (Pythagoras—application of areas); Section 5,D.

Q.5 Proclus, *On Euclid I*, on I,44 (Pythagoreans—application of areas); Section 5,D.

Euclid, *Division of Figures*, Propositions 18, 19, 20, 26, 27,I; Section 5,H.

Euclid, *Data*, Propositions 84, 85, 58,def.3; Section 5,J.

Marinus, *Commentary on Euclid's 'Data'* ("What is datum?"); Section 5,J.

Q.6 Plato, *Republic*, 546B ("nuptial number"); Section 6.

Q.7,8 Proclus, *Commentary on Plato's 'Republic'*, chapters 23, 27 ("side and diagonal numbers"); Section 6.

Q.9 Aristotle, *Prior Analytics*, I, 23, 41a (incommensurability of the side and diagonal of a square); Section 7.

Q.10 Plato, *Theaetetus*, 147D (Theodorus—incommensurability); Section 7.

Q.11 Plato, *Laws*, 820C (incommensurability); Section 7.

Chapter IV. The Pythagoreans

Q.1 Aristotle, *Metaphysics*, 985B (Pythagoreans—mathematics); introduction.

Q.2 Lucian, "In Defense of a Slip of the Tongue in Greeting," 5 (pentagram—symbol of Pythagoreans); Section 11.

Q.3 Scholium to Q.2; Section 11.

Q.4 Scholium to Aristophanes, *Clouds*, 609 (pentagram—symbol of Pythagoreans); Section 11,A.

Q.5 Inscription from ɪvth century Athens; Section 11,A.

Q.6 Proclus, "Catalogue of Geometers"; Section 11,B.

Q.7 Iamblichus, *On Common Mathematical Knowledge*, 25 (Hippasus—dodecahedron); Section 11,B.

Q.8 Iamblichus, *On the Pythagorean Life*, 18 (Hippasus—dodecahedron); Section 11,B.

Q.9 Iamblichus, *On the Pythagorean Life*, 34 (Pythagorean—dodecahedron and incommensurability); Section 11,B.

Q.10 Scholium 1 to *Elements*, XIII (Platonic figures, Pythagoreans, Theaetetus); Section 11,B.

Q.11 Proclus, *On Euclid I*, on I,44 (Pythagoreans—application of areas); Section 5,D.

Q.12 Scholium 2 to *Elements*, IV; Section 11,C (Book IV is Pythagorean).

Q.13 Scholium 4 to *Elements*, IV; Section 11,C (Book IV is Pythagorean).

Q.14 Scholium 73 to *Elements*, II,11; Section 11,C (numerical version of II,11).

Chapter VI. The Classical Period: From Theodorus to Euclid

Q.1 Plato, *Euthydemus*, 290B (geometers—dialecticians); Section 16,A.
Isocrates, *Antidosis*, 265 (geometry as mental exercise); Section 16,A.

Q.2 "Text from Herculaneum" (Plato as director–problem-giver); Section 16,B.

Q.3 Proclus, "Catalogue of Geometers"; Section 16,B.

Q.4 Proclus, *On Euclid I* (method of "analysis"); Section 16,B.

Q.5 Aristotle, *On the Soul*, 414 (types of souls—series of figures); Section 16,C.

Q.6 Plato, *Phaedo*, 110B (earth—ball of 12 pieces); Section 16,D.

Q.7 Plato, *Timaeus*, 55C (dodecahedron—"the whole"); Section 16,D.

Q.8 Plato, *Timaeus*, 55C ("basic triangles"); Section 16,D.

Q.9 Plutarch, *Platonic Questions*, V, 10003 (decomposition of the pentagon); Section 16,D.

Q.10 Plutarch, "The Obsolescence of Oracles," 427 (decomposition of the pentagon); Section 16,D.

Q.11 Albinus, "Introduction to the Doctrine of Plato," 13 (decomposition of the pentagram); Section 16,D.

Q.12 Aetius, *Placita*, ii (Pythagoras—five solid figures); Section 16,D.

Q.13 Plato, *Republic*, 509 ("the divided line"); Section 16,D.

Q.14 Plato, *Timaeus*, 31B ("fairest of bonds"—proportion); Section 16,D.

Q.15 Plato, *Hippias Major*, 303B (individually irrational—collectively rational); Section 16,D.

Q.16 Diogenes Laetius, *Lives of Philosophers*, 24 (Leodamas—analysis); Section 17.

Q.17 *Suidas*, "Theaetetus"; Section 18,A.

Q.18 *Suidas*, "Theaetetus"; Section 18,A.

Q.19 Pappus, *Commentary*, on Euclid X (Theaetetus—irrational); Section 18,B.

Q.20 Plato, *Theaetetus*, 148 (incommensurability); Section 18,B.

Q.21 Iamblichus, *Arithmetical Speculations on Divine Things* (Speusippus—*On Pythagorean Numbers*); Section 19.

Q.22 Proclus, "Catalogue of Geometers" (the "section"); Section 20,A.

Q.23 Proclus, *On Euclid I* (arithmetic vs geometry); Section 20,A.

Q.24 Scholium 70 to *Elements*, II,11 (the "section"); Section 20,A.

Q.25 Proclus, *On Euclid I* (sections of solids); Section 20,A.

Q.26 Proclus, *On Euclid I* (section of conics); Section 20,A.

Q.27 Proclus, *On Euclid I* (uses of the word "section"); Section 20,C.

Q.28 Archytas, "Fragment D4" (arithmetic vs geometry); Section 22,C.

Q.29 Plato, *Republic* 525B (arithmetic vs geometry); Section 22,C.

Q.30 Aristotle, *Metaphysics* 1051a ("obvious" results); Section 22,C.

INTRODUCTION

The mathematical concept which is at the centre of all the discussions contained in this study is deceptively easy to define. Suppose that we have a line AB that we wish to divide at a point C. There are of course many ways to do this but the manner that interests us occurs when the division point C is such that, as far as the larger and smaller segments and the whole line are concerned, we have a constancy of the ratios involved in the sense that whole line:larger segment = larger segment:shorter segment, or

$$AB{:}AC = AC{:}CB. \qquad (1)$$

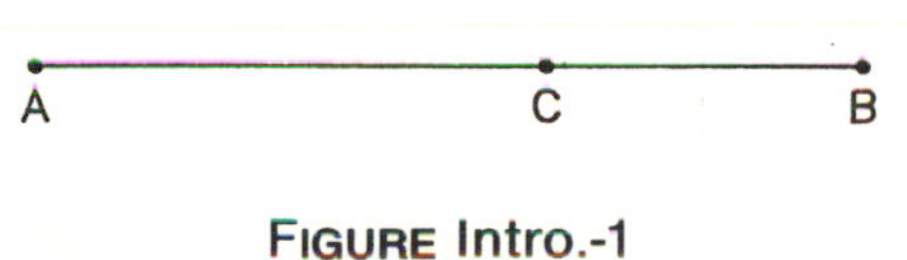

FIGURE Intro.-1

If the line has been divided in this way, then the terminology used is that the line has been divided in extreme and mean ratio (EMR). Indeed this is precisely the terminology of the third definition of Book VI of the *Elements* of Euclid (c. −300). Later on in Book VI (Theorem VI,30) we find described a manner of geometrically dividing a line in extreme and mean ratio.

If this were all there was to division and extreme ratio in Euclid, our story—at least the early part of it—would be a short one. To better understand the concept and historical problems, we must turn from Book VI which involves applications of the theory of proportions developed in Book V to Book II which involves a series of statements about squares, rectangles, and triangles. There in Theorem II,11, we find a construction which, while stated in terms of areas, also in effect defines the division of a line according to division in extreme and mean ratio (DEMR): "To cut a given straight line so that the area of the rectangle contained by the whole line and one of the segments is equal to the area of the square on the remaining segment."

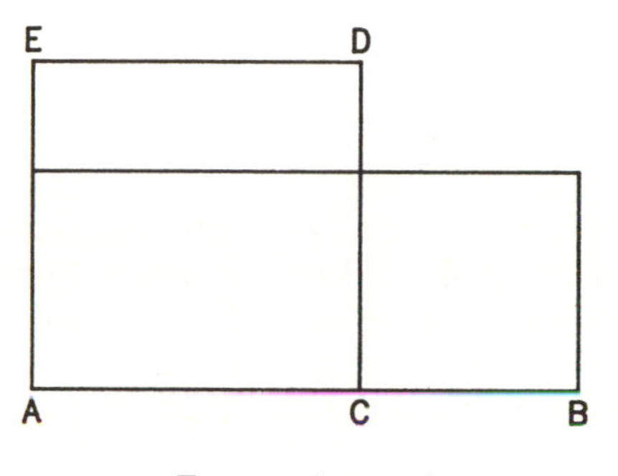

FIGURE Intro.-2

In terms of Figure Intro.-2, the requirements of the theorem state that the square $ACDE$ with sides equal to AC must have the same area as the rectangle with sides AB and CB. In modern notation, which in general will be avoided like the plague in this book, we can write this requirement as $AC \cdot AC = AB \cdot CB$. This in turn is equivalent to $AB{:}AC = AC{:}CB$ (i.e., the definition of DEMR).

Why are there two constructions (II,11 and VI,30) which lead to the same division point? One reason is that, as stated above, Book VI involves the theory of proportions which is only introduced in Book V and thus cannot be employed in Book II. But this is only begging the real question, which is why DEMR is introduced in the first place. Finding the complete answer to this question is in fact one of the major objectives of this book and, as will be seen, something which has been discussed either directly or indirectly to a great extent in the literature.

What can be answered here is the question: Where is the concept of DEMR used in the *Elements* itself? The first place is in Book IV—and thus must involve the area definition of II,11 rather than the proportion definition of VI,def.3—which has as its theme the construction of some regular polygons; Theorem IV,11 in particular is concerned with the construction of a regular pentagon.

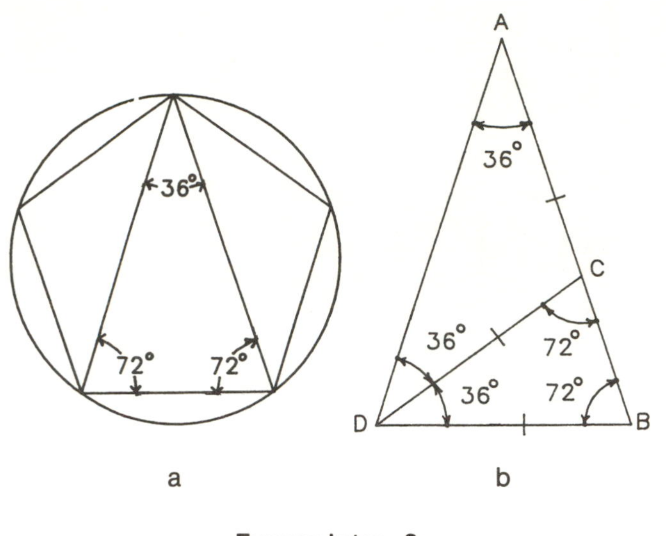

FIGURE Intro.-3

Since my aim in this introduction is to indicate briefly why DEMR enters into the construction of the pentagon, I will speak about actual angles and similar triangles. This, however, is not what we find in Euclid, where the actual sizes of the angle are never mentioned and the theory of similar triangles is only developed in Book VI. Indeed we have here a perfect example of why the study of the *Elements* is so fraught with pitfalls and so fascinating, for the reader is constantly tempted to redo the proofs or rephrase the statements in terms of later developments in the *Elements* or in mathematics and is forever wondering how and why all this came about. It is for this very reason that my study will begin in the first chapter with a presentation of the various results of the *Elements* followed by an exploration and an analysis which remains completely in the context of the Euclidean methods of each book.

If we now look at Figure Intro.-3a, we see that the regular pentagon will be determined by the $36°-72°-72°$ triangle formed by two adjacent diagonals. This observation is the basis of the construction of IV,11, while the triangle itself is constructed in IV,10. If we bisect one of the 72° base angles (Fig. Intro.-3b), then we see that the small triangle DCB thus formed will also be a $36°-72°-72°$ triangle. This in turn means that the sides of triangles ABD and DCB are proportional: $AB{:}DB = DB{:}BC$. But the triangles are isosceles, which means that $DB = DC = AC$. The proportion thus becomes $AD{:}AC = AC{:}BC$, which is precisely the requirement for line AB to be divided in EMR at C according to (1) above.

If we now examine the rest of the *Elements* we find that, aside from some rather straightforward uses of the pentagon in IV,12 and IV,16, neither DEMR nor the pentagon appears—except for the ratio definition and construction of Book VI—until Book XIII. There we find a series of results involving DEMR, properties of the pentagon—in particular the result of XIII,8, that the diagonals of the regular pentagon cut each other in EMR with the larger segment being the side of the pentagon—and then the construction of the icosahedron and dodecahedron, both of which involve the pentagon and DEMR.

Thus from a first brief glance we see that in the *Elements* at least the concept of DEMR is closely linked to the construction of the pentagon and related solids. But what does this tell us about the genesis of the concept of DEMR or the time at which it was first introduced? In fact the first twenty-two sections of this work are devoted to exploring these and related questions. To indicate briefly the nature of some of the difficulties, let us return to the area definition of DEMR found in Theorem 11 of Book II (Fig. Intro.-2). As mentioned earlier, Book II involves statements concerning areas of rectangles and squares. To indicate its nature more precisely let us look at Theorem II,6 which, along with the Pythagorean Theorem, is used to prove II,11.

In Figure Intro.-4 the given line AB, whose midpoint is C, is extended to an arbitrary point D. We draw the squares IV and V with sides CB and BD respectively, and these in turn determine the sides of the rectangles I, II, and III. Now think of cutting out these pieces and consider the big rectangle on top made up of I, II, and V. If we push II over to cover III, then push I over to replace II, and leave V where it is, the big rectangle has been changed into the big square, except for the square IV. This argument is really all there is to the proof of II,11, whose statement merely puts together what I have just said: area (rectangle with sides AD and DB) + area (square with side CB) = area (square with side CD).

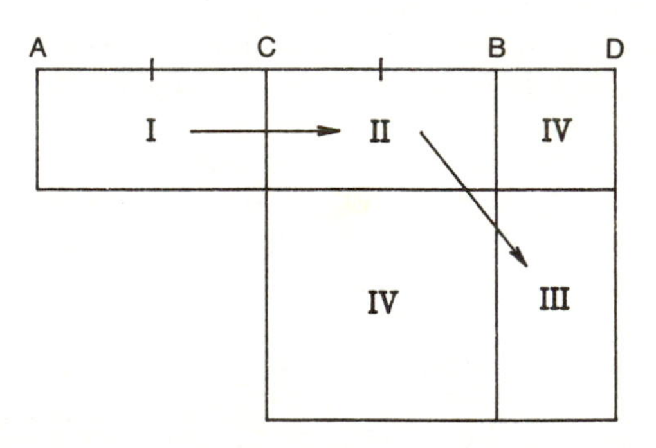

FIGURE Intro.-4

While all this looks simple and innocent enough, various authors have legitimately asked whether there is not more to this theorem than meets the eye. They have then proceeded to show that II,6 can be thought of in terms of algebraic identities, equations, and arithmetical concepts. The same is true of other results in

Book II and elsewhere in the *Elements*, and in particular of II,11 which, as we saw, essentially involves DEMR. Therefore, in order to investigate the origins of DEMR, all these various theories and their implications must be considered. Again we must look at suggestions that DEMR had an intuitive origin based on drawings of the pentagram. This in turn will lead us to archaeological examples and the mathematics of the Babylonians and Pythagoreans.

Our story will not end with Euclid's *Elements*, however, for the concept of DEMR continued to play an important role in later mathematics. One aspect of particular interest is the numerical value associated with DEMR. If we take an arbitrary line and divide it in EMR with the larger segment being taken as 1 and the whole line as x (Fig. Intro.-5), then the definition gives $x:1 = 1:x-1$ or $x^2-x-1=0$. The positive root of this quadratic equation is $(1+R(5))/2$ which is approximately 1.618. On several occasions I shall be delving into texts which involve numerical approximations, including some involving DEMR. In certain cases it will be a question of whether or not an author was in fact aware of such an approximation.

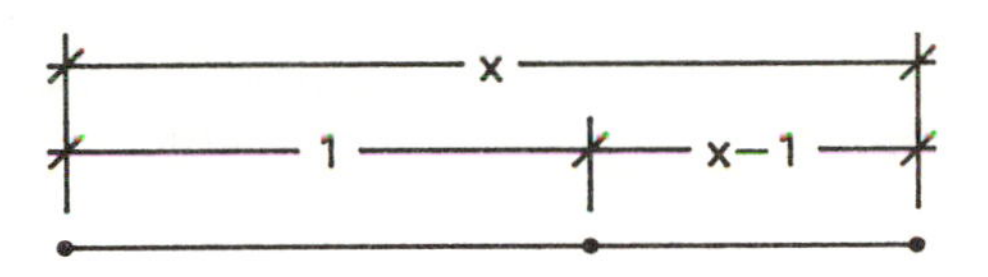

FIGURE Intro.-5

I will also be discussing the role played by DEMR in later Greek mathematics and in other places and times, in particular in the Arabic world and pre-modern Europe.

In brief, the purpose of this book is to present a complete history—from the origins to 1800—of all facets of the mathematical development of DEMR, including an English translation of all the relevant historical material, a discussion of all the various theories that have been brought forth concerning the development, and my own views concerning the early history.

There are many possible approaches to a topic like DEMR. Mine is based on the premise that the first step should be a detailed mathematical analysis of the extant results in the strict context of the works in which they are found. By avoiding any possible modern interpretation, I hope to ascertain the approach, spirit, and aim of the mathematician who presented the proof. In particular, in examining the *Elements* it is necessary first to consider the proofs in the strictly geometrical form in which they are presented, without considering possible external influences, mathematical or non-mathematical.

I wish to emphasize, however, that I am not in any way opposed to the use of other indications—historical, philological, or whatever. On the contrary, as will be evident to the reader, I have tried to cover the entire literature, to include all possible quotations, bits of evidence—linguistic, archaeological, and literary—and to bring out the salient point of each theory, no matter how incorrect I may personally have thought them to be. I have used non-mathematical arguments which I know will not sit well with everyone, but I bring these in later on, after finding out what the earliest text which unequivocally discusses DEMR has to tell us.

In view of this, Chapter I is entirely devoted to the transmitted Euclidean text and does not consider any extraneous mathematical or historical material. In Section 1 all the definitions and theorems from the *Elements* that are related or possibly related to the history of DEMR are brought together. There were several reasons for presenting this material rather than sending the reader to one of the standard texts of the *Elements* or historical commentaries. One was simple convenience, for the definitions and results are scattered throughout other material in the *Elements*. Among the other reasons, the principal reason is that I believed I would be more faithful to the strictly geometric spirit of Euclid by using an idiom which avoids the obscurities and indeed sometimes misleading phraseology of the literal translation (this problem is not confined to Euclid [cf. Orlinsky, 1970, 19]). For instance, on many occasions the statement of the theorem is difficult to understand; it is only when we read the beginning of the proof, where Euclid drops his pedantic phraseology and tells us with reference to a diagram exactly what he is going to prove, that we understand what the theorem is all about. Thus I have often presented the statement of the result of the theorem in a form that states explicitly what is given and what is to be proved. Another problem with the literal text is psychological. When modern readers see "the square on a line AB" it is very difficult for them to avoid putting in some algebraic connotation. It may very well be, as I will discuss in Section 5, that certain results had an algebraic origin, but it seems to me that we must avoid anything that, because of our mental conditioning, will automatically suggest this to us. Thus I have replaced all statements of the form "square on the line AB" by the symbolic statement $S(AB)$ and have followed a similar policy for rectangles.

Again for clarity, I have often used either four letters, instead of just two as in the text, or roman numerals to indicate rectangles. Similarly, I have not refrained from using a_5 instead of constantly writing such things as "Let AB be the side of the pentagon inscribed in the circle. . . ." I do not feel that such simplifications in readability are a modernization, algebrization, or simplification of the proofs themselves. On the contrary, they merely eliminate all the verbiage that stands between us and the real proof.

As far as the proofs themselves are concerned, I have varied my approach as befits the situation. For those key results—such as II,11, IV,10,11, and XIII,8—which will play an important role in my discussion, I have given the complete proofs as in Euclid, except that I have sometimes changed the internal order and omitted routine steps. For other results, I have occasionally given a sketch or several key points of the proof, and these are indicated by "sketch of proof" or "proof." On the other hand, for results such as XIII,16 and 17, and later on in Section 24, XIV,2, I was obliged to spend days rereading the proofs before I could truly say to my mathematical conscience "I understand" as opposed to "I followed the proof" or "I believe." Thus in these cases I have tried to give the reader an idea of what Euclid is doing in a "discussion" of the proof. In the case of XIII,13, I have placed together under the number XIII,13′ the lemma to the latter and a related result which is used but never explicitly stated. My purpose in my descriptions is not to replace Euclid, but rather—to paraphrase the poetic preface of Taisbak [1971]—to add a little light and dispel some loneliness.

Finally, since the various commentaries in the literature are not always clear enough or are sometimes prejudiced by certain a priori views, I have often explained the meaning of definitions and theorems.

Section 2 constitutes the key to understanding the early history, for here I analyze in detail exactly what Euclid is doing and what he is not doing in his treatment of DEMR. Euclid's *Elements* has too often been treated as a finished textbook or a logical system rather than what it really is, a work of mathematics. Because it is a work of mathematics, and despite its involving to an unknown extent the hand of one or more editors, the underlying process of mathematical creation, with all its brilliant moves and oversights, still shines through to a certain extent. Thus by investigating the Euclidean text in detail, as opposed to simply following various results step after step, we can form certain ideas about the origin of the concept of DEMR.

Many statements in the history of mathematics are based on supposed reconstructions of original or early proofs, and while reconstructions are not to be ruled out altogether, there are limits to their usefulness [cf. Molland, 1978]; in particular they must follow from an initial investigation of what is really known and observed. In view of this I have approached the historical problem by showing what Euclid could have done but did not do on several occasions while using the same techniques and results that we find in the related proofs of the *Elements*. From this approach I make certain deductions about the historical order of things. The key analysis is that of II,11, which represents an extended and somewhat modified version of an earlier article of mine [Fischler, 1979c]. My main conclusions run contrary to what usually is assumed in the literature.

In Chapter II, I consider those mathematical topics which either are associated in Euclid with DEMR or have been associated with it in the literature. Again because of the often unsatisfactory presentation in the literature, I have given my own explanation with emphasis on the meaning in the context of this book. While the other sections are essentially mathematical and textual, Section 5 consists of a reconsideration of a question which has been very controversial: whether or not certain results in Euclid, in particular in Book II, are a conscious geometrical formulation of certain algebraic results. Because of the extreme importance of this problem in relationship to the history of DEMR, I have re-examined the whole question from the beginning. Of particular interest should be the examination of a possible relationship of "application of areas" and "division of figures" in Section 5,H and the discussion of Euclid's *Data* in Section 5,J which is based on Herz-Fischler [1984].

After analyzing the text and considering mathematical aspects of the problem, I look at various historical documents and various theories that have been proposed. Because several authors have sought to connect the development of DEMR with pre-Greek mathematics as well as with various drawings, symbols, etc., in the shape of a pentagon, pentagram, or dodecahedron, I have made a study (the first of its kind to my knowledge) of all known examples from the vth century or earlier. The source of all these examples are shown on a map in Chapter III which also serves as a guide to the places mentioned throughout the book.

In Chapter IV we turn our attention to the Pythagoreans. First I present all the relevant material together, some for the first time in English translation. Next I consider the various theories that have appeared and which relate the Pythagoreans to the development of DEMR. Regarding the theories of others, I have tried in Chapter IV, and throughout the book, to bring forth the salient points and to offer where appropriate my comments and those of different authors. Where I have nothing to say or to add, I have preferred silence to sniping.

Chapter V discusses various theories that do not credit some part of the historical development to either a specific person or time. There is also a summary of discussions concerning XIII,1-5.

The classical period of Greek mathematics, which I associate with the time span from Theodorus to Euclid, is the subject of Chapter VI. All mathematicians of this period who have been associated in the literature with the development of DEMR are discussed. New points include comments on the life of Theaetetus in Section 18,A and, in Section 20,C, my views on the problem of the meaning of the word "section" in Proclus' "Catalogue of Geometers." A special problem was presented by Plato because of the sheer volume of the commentary on his mathematics, the obscurity of so many of his passages, and the gaps in our knowledge of the state of mathematics at the time. What I have attempted to present in Section 16, therefore, is both an overview of the general problem of the relationship of Plato to the mathematics of his time and a complete treatment of the literature dealing with Plato and extreme and mean ratio.

In Section 22 first I summarize the analysis of the Euclidean text made in Chapter I as well as of the various theories found in the literature that were discussed in the preceding sections, and then I present my own views on the origin, early development, and chronology of the concept of DEMR.

Chapter VII treats the post-Euclidean Greek period and such principal authors as Archimedes, Hero, Ptolemy, and Pappus. Of particular interest in this chapter are questions related to certain numerical values which we find in the works of these authors and the manner in which they were obtained. In some cases these texts have also been the centre of controversy. Section 24 deals with the so-called Book XIV of the *Elements* which has received short shrift in the literature. I enter into a detailed examination of the text and the various layers which are indicated both by the mathematics and the extant historical information, and on the basis of this I discuss questions of authorship and chronology. One of my claims is that one of the original versions of this Supplement to the *Elements* contained a result that is no longer in any of the extant Greek manuscripts.

Chapter VIII discusses the beginning of the diffusion of the concept of DEMR to the non-Greek world. In particular I focus on the very interesting material that has been preserved from the Arabic world.

The final chapter covers Europe from the Middle Ages to 1800. By closely analyzing the treatment of the pentagon, dodecahedron, and icosahedron by Fibonacci, Francesca, and Bombelli and comparing their different methods with those of Hero and Pappus, I wanted to indicate exactly what these mathematicians were really doing. Whereas most discussions in the literature of these and other authors try to paint broad pictures, I believe that it is necessary to compare the approaches of different mathematicians to a very narrow area, such as DEMR, in order to really understand how mathematics has developed.

Section 31,J discusses various numerical approximations to the ratio determined by DEMR, including a recently discovered marginal note from the early 16th century which indicates that the relationship between the so-called Fibonacci sequence and DEMR was known much earlier than Kepler, whose work is discussed in Section 32,A.

I conclude my study of DEMR with two appendices. The first looks at the terminology and names associated with this concept throughout the ages. The second appendix will, I believe, surprise many readers; it traces the admiration, to put it mildly, that DEMR has evoked from mathematicians from Campanus to Kepler.

CHAPTER I

THE EUCLIDEAN TEXT

Section 1. The Text

As I indicated in the Introduction, I felt that my study of DEMR must start off with a detailed analysis of the Euclidean text and that for various reasons it was appropriate to gather together all the relevant theorems here in Section 1. My reasons for the method of presenting the statements, proofs or sketches of proofs, and notations are explained in the Introduction and in the list of symbols which precedes it. I have omitted most of the congruence and similarity theorems as well as some other results which do not play an important conceptual role in the discussions.

The diagrams follow those of the critical edition of Euclid–Heiberg, except that the Greek letters have been replaced by their romanized versions as in Euclid–Heath. This has been done for ease of comparison even though great clarity could sometimes have been achieved (for example, in XIII,1,2) by relabelling as in the Euclid–Frajese edition. I have, however, followed the lead of this edition by identifying quadrilaterals by four letters instead of two, although sometimes regions are simply identified by roman numerals.

The results are followed by a partial list of those theorems of the *Elements* in which they are used. Sometimes just a reference to a book is given; a series of three dots indicates that the listing is not complete. In some cases I have also given a partial list of the principal results used in the proof. For all these lists I have followed, with some corrections and additions, the lists found in Euclid–Frajese. I have not methodically compared them with other lists and indications (such as those given in Euclid–Peyrard; Euclid–Heiberg; Euclid–Heath; Neuenschwander [1972; 1973]; Mueller [1981], although I have used these for various results. The interconnection between theorems will play an important role in the discussions of Section 2. In some cases I have added a few explanatory notes, none of which is related to DEMR. Where applicable I have indicated where in this book the theorem is discussed in relationship to DEMR.

BOOK I. This book contains some basic results concerning lines, triangles, and parallelograms.

THEOREM I,32. In any triangle an exterior angle is the sum of the two opposite interior angles. The sum of the interior angles is equal to two right angles.
Used in: IV,10; XIII,8,9,10,11,18 lemma . . .

THEOREM I,42 (a transformation of areas result). To construct a parallelogram one of whose angles is given and whose area is equal to the area of a given triangle.
Proof: See Section 4.
Used in: I,44,45.
Discussed in: Sections 4; 5,H.

THEOREM I,43 (complements). In the parallelogram *ABCD* the so-called complements—with respect to the diagonal—I and II have the same area.
Proof: See Section 3.
Used in: I,44; II,4(?),5,6,7,8; VI,27,28,29; X; XIII,1,2,3,4,5(?) . . .
Discussed in: Sections 2,D; 3; 4.
Note: See II,def.2 (gnomon); also Section 3.

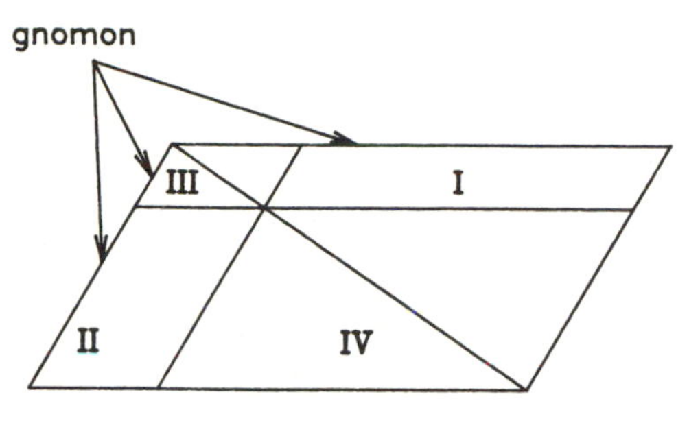

FIGURE I-1. I-43

THEOREM I,44 (an application of areas result). To construct (to apply) on a given line a parallelogram under the constraints that one angle is given and that the area is equal to the area of a given triangle.
Proof: See Section 4.
Used in: I,45; VI,25; X.
Discussed in: Sections 4; 5,C.

THEOREM I,45 (a transformation of areas result). To construct a parallelogram, one of whose angles is given, and whose area is equal to the area of a given rectilineal figure.
Proof: See Section 4.
Used in: II,14; VI,25; X (27 times).
Discussed in: Section 4.

THEOREM I,47 (Pythagorean Theorem). In a right-angle triangle with hypotenuse BC and legs AB and AC, $S(BC) = S(AB) + S(AC)$.
Used in: II,9,10,14; III,36; XIII,18 . . .

BOOK II. The origin, context, and meaning of the results of this book have been the subject of much controversy; I shall discuss these questions in detail in Section 5. In order to avoid any prejudgments, it is important to read these results in the first instance as simply dealing with the "geometry of areas," to use the terminology of Gow [1884, 190] and Szabó [1968, 226], that is, geometric statements concerning the areas of squares and rectangles.

THEOREM II,def.2. With respect to the parallelogram of Figure 1 both of the regions I + III + II and I + IV + II (i.e., the two complements with respect to the diagonal together with one of the parallelograms III or IV) are called a gnomon.
Used in: II,5,6; VI,27,28,29; XIII,1,2,3,4 . . .
Discussed in: Section 3.

THEOREM II,1. In the diagram let D and E be arbitrary division points of the line BC. Then $R(BG,BC) = R(BG,BD) + R(BG,DE) + R(BG,EC) \equiv \text{I} + \text{II} + \text{III}$.
Used in: Never used.
Discussed in: Section 5,A,G.

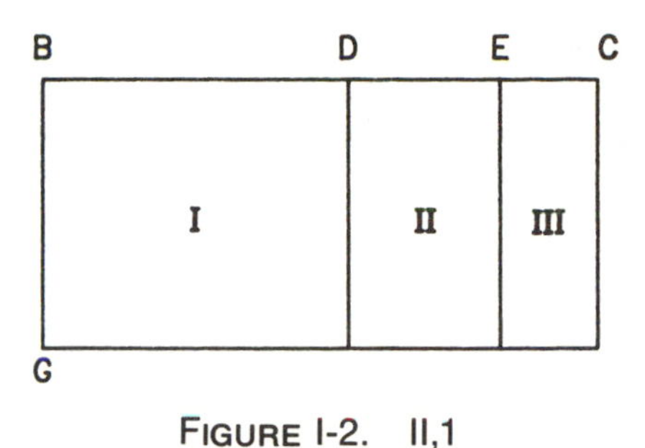

FIGURE I-2. II,1

THEOREM II,2. Let AB be any line and on it draw a square. If C is an arbitrary division point of AB, then $S(AB) = R(AB,BC) + R(AB,AC)$.

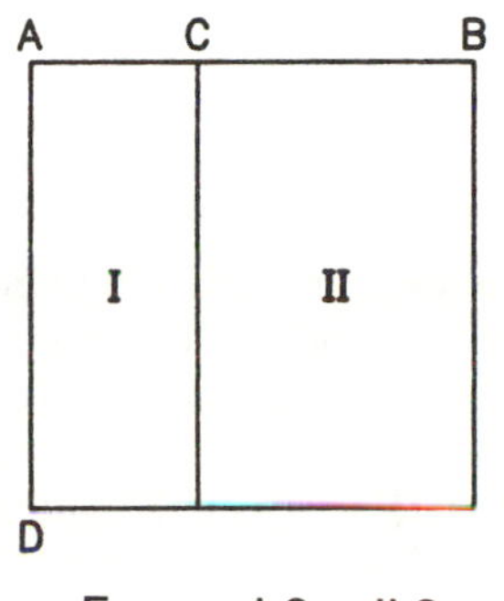

FIGURE I-3. II,2

Used in: XIII,10; implicitly in I,47.
Note: The proof does not use II,1.

THEOREM II,4. In the diagram let C be an arbitrary division point, then $S(AB) = S(AC) + S(CB) + 2 \cdot R(AC,CB) \equiv \text{I} + \text{I} + 2 \cdot \text{III}$.

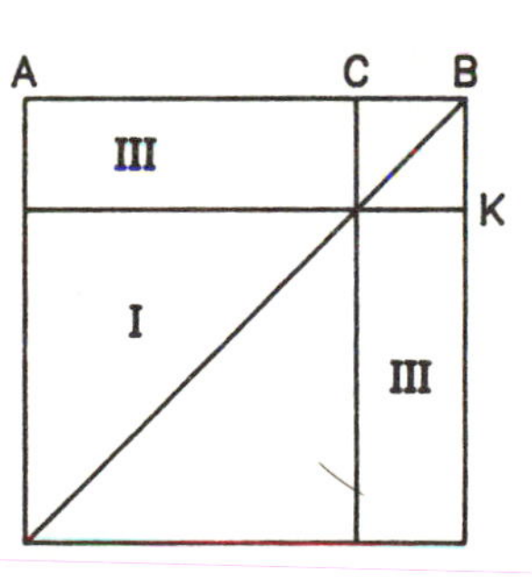

FIGURE I-4. II,4

Uses: I,43 (complements)(?).
Used in: II,12; IX; X; XIII,2 lemma (possibly not genuine).
Discussed in: Section 5,B.
Note: One usually presumes (e.g., Euclid–Frajese and Neuenschwander [1972]) that I,43 (complements) was what the mathematician had in mind when stating the equality of the rectangles marked III; however, the

word "complements" is not used. Since it is shown earlier in the proof that $CB = BK$, it is possible that the author had an extended version of I,35 in mind.

THEOREM II,5. Let AB be a line segment divided into two unequal segments AD and DB, and let C be the midpoint of AB. Then $R(AD,DB) + S(CD) = S(CB)$.

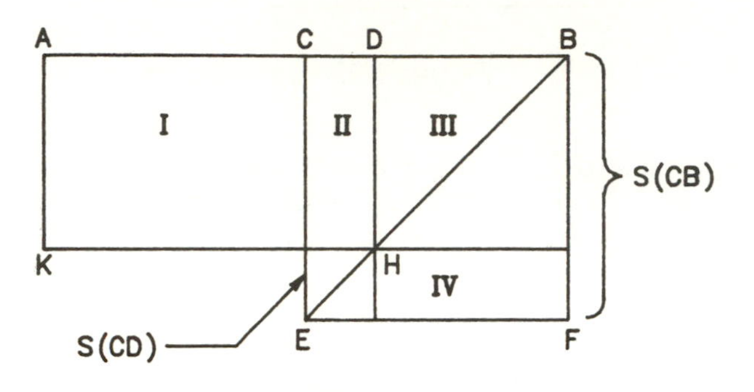

FIGURE I-5. II,5

Proof: Construct square $CBFE$ and draw line BE; this determines points H and K. We have that $R(AD,DB)$ $\equiv$ rectangle $ADHK \equiv$ I + II. But I = II + III, since C is the midpoint. Also by I,43, II = IV. Therefore $R(AD,DB) =$ (II + III) + (IV). This latter L-shaped region, the gnomon of II,def.2, is simply the difference of the two squares $S(CB)$ and $S(CD)$. Hence, by adding $S(CD)$ to both $R(AD,DB)$ and the gnomon, the result is proved.
Uses: I,43 (complements).
Used in: II,14; III,35; X,17, 41 lemma, 59 lemma.
Discussed in: Introduction; Sections 2; 5,B,E,J.

THEOREM II,6. Let AB be a straight line and C its midpoint. Add another straight line BD to AB. Then $R(AD,DB) + S(CB) = S(CD)$.

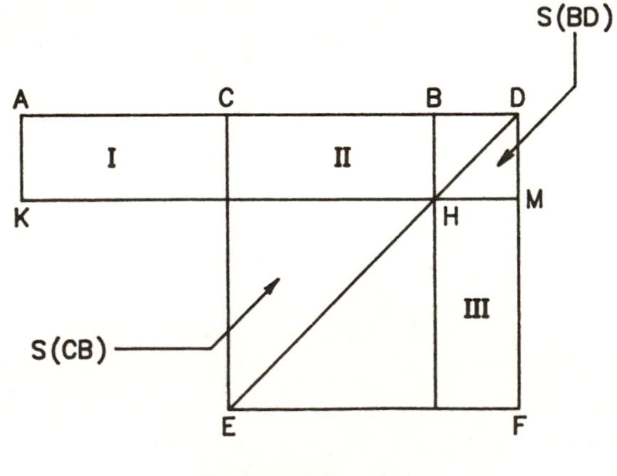

FIGURE I-6. II,6

Proof: Construct square $CDFE$ and draw line DE. This determines points H,K, and M. We have that $R(AD,DB) \equiv$ rectangle $ADMK \equiv$ I + II + $S(BD)$. But I = II, since C is the midpoint. Also, by I,43, II = III. Therefore $R(AD,DB) =$ II + $S(BD)$ + III. This L-shaped region, the gnomon (II,def.2), is simply the difference of the two squares $S(CD)$ and $S(CB)$. Hence, by adding $S(CB)$ to both $R(AD,BD)$ and the gnomon, the result is proved.
Uses: I,43 (complements).
Used in: II,11; III,36; X,28 lemmas 1,2.
Discussed in: Sections 2; 5,B,E,G.

THEOREM II,9. Let AB be a line with C as its midpoint. Let D be an arbitrary point of AB differing from C. Then $S(AD) + S(DB) = 2[S(AC) + S(CD)]$.

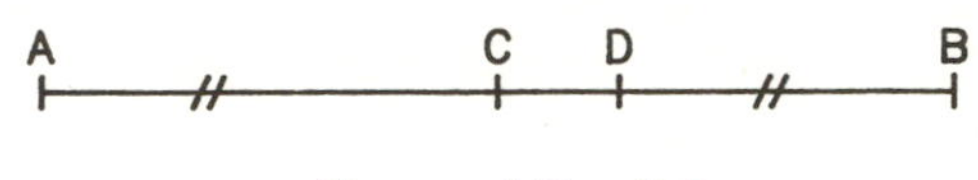

FIGURE I-7. II,9

Used: Never used.
Discussed in: Section 5,F.

THEOREM II,10. Let AB be a line with C as its midpoint. Let the line segment BD be added to AB. Then $S(AD) + S(DB) = 2[S(AC) + S(CD)]$.

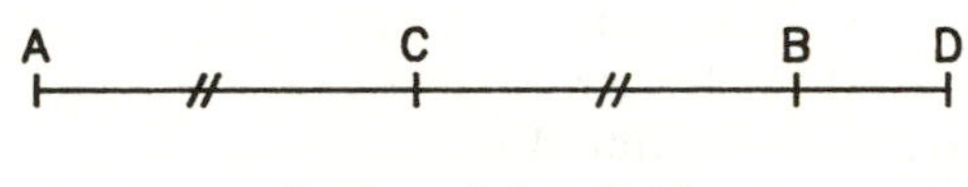

FIGURE I-8. II,10

Used in: Never used.
Discussed in: Section 5,F.

THEOREM II,11 (the area formulation of DEMR). To divide a line AB into two segments, a larger one AH and a smaller one HB so that $S(AH) = R(AB,BH)$.
Note: While the picture conjured up by the statement is that of Figure I-9a, the figure of the text is that of Figure I-9b.
Proof (Fig. I-9b): Construct square $ABDC$ and let E be the midpoint of AC. Obtain point F on the extension of EC so that $FE = EB$ and then construct square $AFGH$.

Since E is the midpoint of AC and AF is added to AC, II,6 tells us that $R(CF,FA) + S(AE) = S(EF) = S(BE)$. By the Pythagorean Theorem (I,47), $S(BE) = S(AE) + S(AB)$ so that when we subtract $S(AE)$ we obtain

(*) $$R(CF,FA) = S(AB).$$

In terms of the labelled areas, the equation (*) can be written I + III = II + III so that by subtracting III = $R(AH,HK)$ we have $S(AH) = R(BD,HB) = R(AB,HB)$. This shows that H is the desired division point.

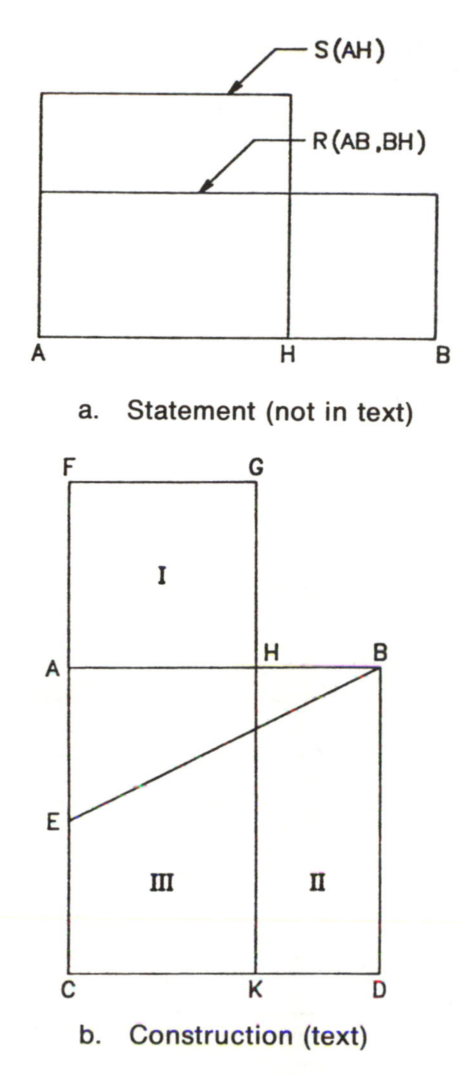

a. Statement (not in text)

b. Construction (text)

FIGURE I-9. II,11

Uses: I,47; II,6.
Used in: IV,10.
Discussed in: Sections 2,B,ii; 5,K,L . . .

THEOREM II,14 (a transformation of areas result). To construct a square whose area is the same as that of a given rectilineal figure.
Uses: I,45,47.
Used in: Book X (5 times).
Discussed in: Sections 4; 5,F; 22.

BOOK III. This book deals with the properties of circles.

THEOREM III,20 (Fig. I-10). Let B and C be two points on the circumference of a circle. Let $\measuredangle BOC$ be the central angle and let $\measuredangle BAC$, where A is any point on the circumference, be an inscribed angle. Then $\measuredangle BOC = 2 \cdot \measuredangle BAC$.
Used in: III,27; VI,33 . . .

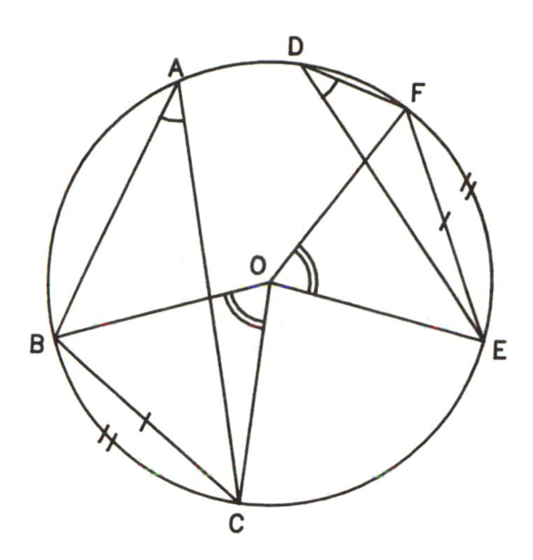

FIGURE I-10. III,20,26,27,28,29 (drawn on one circle)

THEOREM III,26 (Fig. I-10). Let BC and EF be arcs of equal circles. Suppose that central angles BOC and EOF are equal, then arcs BC and EF are equal. The same is true if the inscribed angles BAC and EDF are equal.
Used in: III,27,28; IV,11; XIII,10 . . .
Note: The statement says that the arcs will be equal if either the central or the inscribed angles are equal. In IV,11 it is equality of the inscribed angles that is used, whereas in XIII,10 it is equality of the central angles that is used. However, for some reason the proof of III,26 assumes that both the central and the inscribed angles are equal. This does not seem to have been observed in the various commentaries. See also the proof of XIII,8′ in Section 2,C.

THEOREM III,27 (Fig. I-10). With the notation of III,26, if the arcs are equal then the angles are equal.
Used in: III,29; IV,11; VI,33 . . .
Discussed in: Section 2.
Note: Compare with VI,33.

THEOREM III,28 (Fig. I-10). In two equal circles if chords CB and EF are equal then arc CB is equal to arc EF.
Used in: XIII,8 . . .

THEOREM III,29 (Fig. I-10). In two equal circles if arcs CB and EF are equal, then chords CB and EF are equal.
Used in: IV,11; XIII,10 . . .

THEOREM III,32 (Fig. I-11). Let BC be the arc of a circle and BF the tangent at point B. Let D be a point on that arc of the two arcs determined by B and C which is not included by $\measuredangle CBF$, then $\measuredangle BDC$ and $\measuredangle FBC$ are equal.

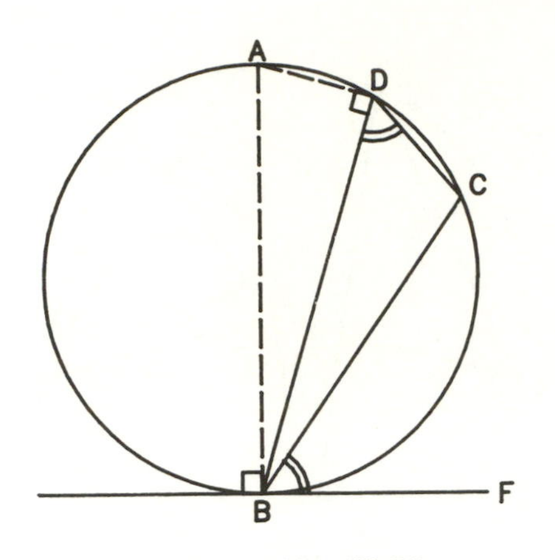

FIGURE I-11. III,32

Sketch of Proof: Draw the diameter AB so that $\measuredangle ABF = 90° = \measuredangle AOB$. Further, $\measuredangle BAD$ and $\measuredangle BCD$ add to 180° since their associated arcs correspond to the entire circumference. Combining these results—via subtractions and additions—with the theorems on the sum of the angles of a triangle and the angles along a line, the theorem is immediate.
Used in: IV,10 . . .

THEOREM III,35. Let AC and BC be two chords of a circle that meet at E, then $R(AE,EC) = R(DE,EB)$.

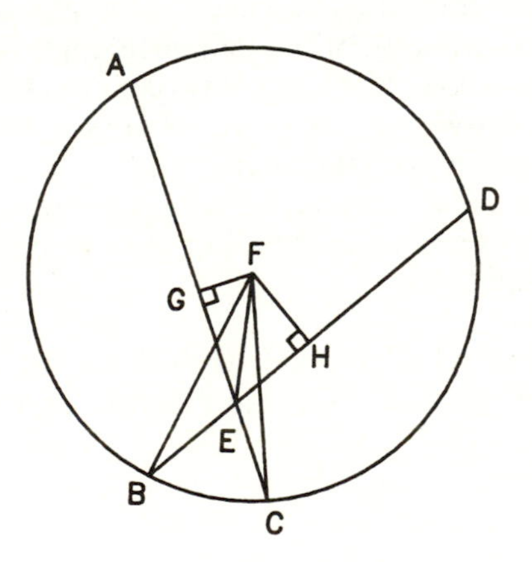

FIGURE I-12. III,35

Sketch of Proof: A perpendicular is dropped from the centre F and this divides AC into two parts. This brings II,5 into play which gives an expression for $R(AE,EC)$. This expression is related to FE and the radius FC via the Pythagorean Theorem (I,47). By working in the same way with BD, $R(DE,EB)$ is related to FE and FB. But FE is common and the radii FB and FC are equal, giving the desired equality.
Uses: I,47; II,5.
Used in: Never used.

THEOREM III,36. If from a point D outside a circle, a tangent line DB and a secant line DA are drawn, and if DA also cuts the circle at C, then $R(DA,DC) = S(DB)$.

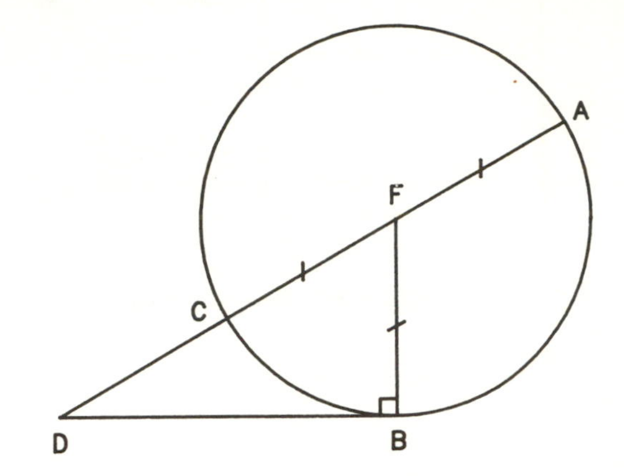

III,36,Case 1 (orientation changed to Case 2)

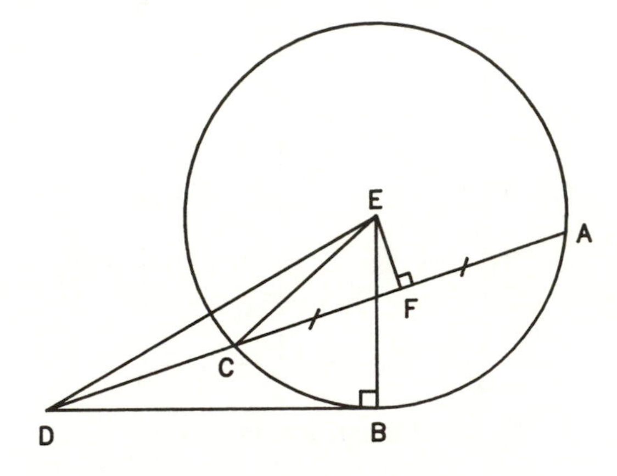

III,36,Case 2 (same orientation as text)

FIGURE I-13. III,36

Note: Euclid considers two cases: Case 1—when AC is a diameter; and Case 2—when AC is a chord. Case 2 does not use Case 1. In both, which are treated independently, there are two essential points to the proof: first, F is the midpoint of AC and so II,6 gives us a value for $R(AD,DC)$; and second, the right angle allows use of the Pythagorean Theorem.
Proof of Case 1: By II,6, $R(AD,DC) + S(FC) = S(FD)$. Applying the Pythagorean Theorem to the term $S(FD)$ and substituting FB for FC results in $R(AD,DC) + S(FB) = S(FD) = S(FB) + S(BD)$. Since $S(FB)$ is common to both sides, we have $R(AD,DC) = S(BD)$.
Proof of Case 2: By II,6 $R(AD,DC) + S(FC) = S(FD)$. The perpendicular EF is common to the two right triangles EFD and EFC; and so if we add $S(EF)$ to

both sides of the equality and use the Pythagorean Theorem, we obtain $R(AC,DC) + S(EC) = S(ED)$. Now substituting EB for EC and applying the Pythagorean Theorem once more, but this time to the right triangle EDB, we have $R(AD,DC) + S(EB) = S(DB) + S(EB)$. Since $S(EB)$ is common to both sides, we have $R(AD,DC) = S(BD)$.
Uses: I,47; II,6.
Used in: III,37.
Discussed in: Section 2.

THEOREM III,37 (the converse of III,36). Let D be a point outside a circle with centre F and let DA be a secant line which cuts the circle at C. Now suppose that B is a point on the circle such that the relationship $R(AD,DC) = S(DB)$ holds, then the line DB is in fact tangent to the circle at B.

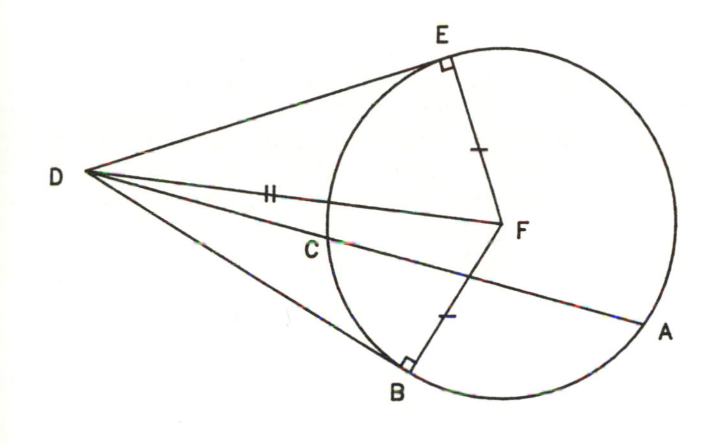

FIGURE I-14. III,37

Proof: Let DE be tangent to the circle then, by III,36 and the hypothesis, $S(DE) = R(AD,DC) = S(DB)$ so that $DE = DB$. Since $FE = BF$ and DF is common, we have that triangles DEF and DFB are congruent. This means that $\measuredangle DBF$ is also a right angle and thus DB is tangent.
Uses: III,36.
Used in: IV,10.
Discussed in: Section 2.

BOOK IV. This book gives the construction of various regular polygons.

THEOREM IV,2. To inscribe an arbitrary triangle in a circle.

THEOREM IV,6. To inscribe a square in a circle.

THEOREM IV,10. To construct an isosceles triangle with each base angle double the vertex angle; that is, the $72°-72°-36°$ triangle.

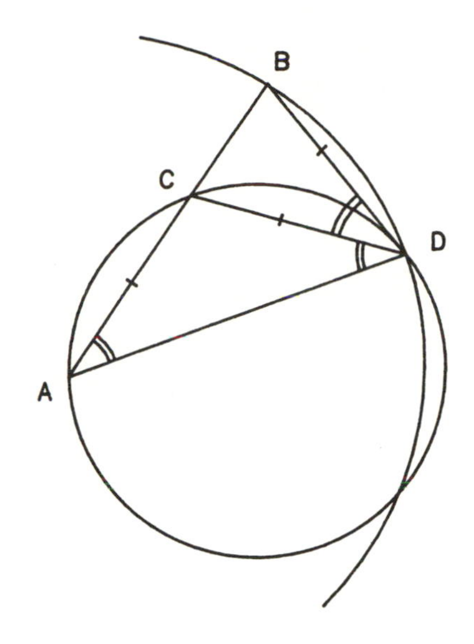

FIGURE I-15. IV,10

Proof: Take any line AB and draw a circle with A as centre and AB as radius. Now use II,11 to obtain point C with $R(AB,BC) = S(CA)$. Next pick point D on the circumference—this automatically makes $\triangle BAD$ isosceles with $AB = AD$ and $\measuredangle ABD = \measuredangle BDA$—such that $BD = AC$. Finally circumscribe a smaller circle about $\triangle ACD$. This construction has given us points A, C, B on the small circle and a point D outside the small circle such that $R(AB,BC) = S(BD)$. But this is precisely the set-up needed for III,37 and thus the line BD is actually tangent to the smaller circle. This in turn implies, via III,32, that $\measuredangle BOC = \measuredangle DAC$.

If we add $\measuredangle CDA$ to both sides, we have for the exterior angle BCD: $\measuredangle BCD = \measuredangle DAC + \measuredangle CDA = \measuredangle BDC + \measuredangle CDA = \measuredangle BDA = \measuredangle DBA \equiv \measuredangle CBD$. This means that $\triangle BCD$ too is isosceles with $DC = BD$. But by construction $BD = CA$ so that $DC = CA$, and thus $\triangle CDA$ is also isosceles with $\measuredangle CDA = \measuredangle DAC$. Putting all our information together, we have $\measuredangle BDA = \measuredangle DBA = \measuredangle BCD = \measuredangle CDA + \measuredangle DAC = 2 \cdot \measuredangle DAC = 2 \cdot \measuredangle DAB$.
Uses: II,11; III,32,37.
Used in: IV,11,12,16; XIII,16.
Discussed in: Section 2.

THEOREM IV,11. To inscribe a regular pentagon in a given circle.

Proof: Use IV,10 to inscribe the $72°-72°-36°$ triangle ACD. Bisect $\measuredangle ACD$ by line CE and bisect $\measuredangle CDA$ by line DB. Since the base angles of $\triangle ACD$ are twice the vertex angle CAD, all four of the angles ACE, ECD, BDA, and BDC will be equal to $\measuredangle CAD$. This in turn implies that the five arcs and chords around the circumference will be equal (III,26,27,29).
Uses: III,26,27,29; IV,10.

Used in: IV,12,16; XIII,16.
Discussed in: Section 2.

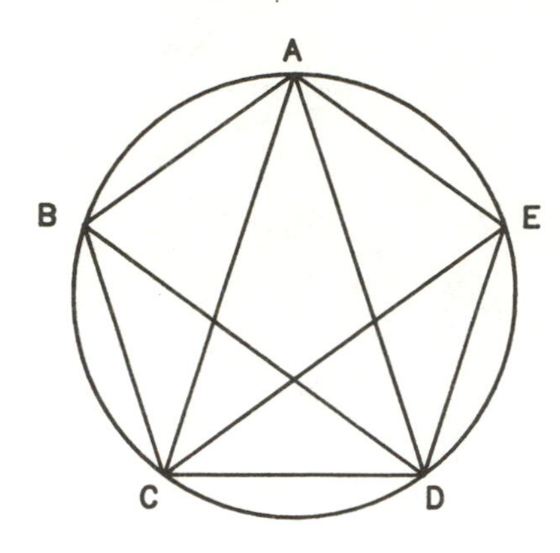

FIGURE I-16. IV,11

THEOREM IV,12. To circumscribe a regular pentagon about a given circle.
Uses: I,47 . . .
Used in: Never used.
Note: The proof uses the existence of an inscribed pentagon as in IV,11 but has no direct connection with DEMR.

THEOREM IV,13. To inscribe a circle in a given regular pentagon.
Used in: Never used.
Discussed in: Section 2,A.
Note: The proof does not involve DEMR, not even indirectly via IV,11. The pentagon is divided into ten $36°-54°-90°$ triangles as shown in Figure I-17.

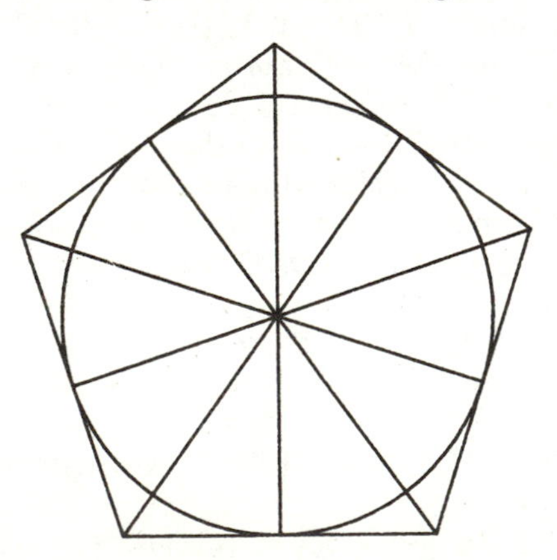

FIGURE I-17. IV,13

THEOREM IV,14. To circumscribe a circle about a given regular pentagon.
Used in: XIII,8,18 lemma.
Note: The proof does not involve DEMR, not even indirectly via IV,11.

THEOREM IV,15. To inscribe a regular hexagon in a circle.

THEOREM IV,16. To inscribe the regular pentadecagon in a given circle.
Proof: Inscribe an equilateral triangle, one side of which will cut off $1/3 = 5/15$ of the circumference. Next inscribe a regular pentagon, one side of which will cut off $1/5 = 3/15$ of the circumference. Thus we can construct an arc which is 2/15 of the circumference, and bisection in turn produces the required arc of 1/15 of the circumference.
Uses: IV,11.
Used in: Never used.
Note: The proof uses the existence of an inscribed pentagon as in IV,11 but has no direct connection with DEMR.

BOOK V. This book discusses the concepts of magnitude, ratio, and proportion. The concept of magnitude is not explicitly defined and there appears to be a great deal of vagueness concerning ratio and proportion as well as a certain ambiguity of language; see Sections 8; 13,ii. The various conceptual and historical problems are discussed in Mueller [1981, chap. 3] and Fowler [1979; 1982c; 1985].

THEOREM V,14. If $A:B = C:D$, then if A is greater (less, equal) than C, then B is greater (less, equal) than D.
Used in: V; XII; XIII,2,5,8,9,17.

BOOK VI. This book consists of applications of the theory of proportions developed in Book V.

THEOREM VI,def.3. A straight line is said to have been divided in EMR when the ratio of the whole line to the larger segment is the same as the ratio of the larger segment to the smaller segment.
Discussed in: Introduction; Sections 1-32; Appendices I,II (in other words, the whole book).

THEOREM VI,1. If two triangles (parallelograms) have the same altitude, then the areas of the triangles (parallelograms) have the same ratio as their bases.
Used in: XIII,1,2 . . .

THEOREM VI,3. In triangle ABC let AD be the angle bisector of $\angle BAC$. Then $BD:DC = BA:AC$. Conversely, if the proportion holds for a line AD, then that line is the angle bisector.
Sketch of Proof: Add the dotted lines and use similar triangles.
Used in: Never used.
Discussed in: Section 23,A.

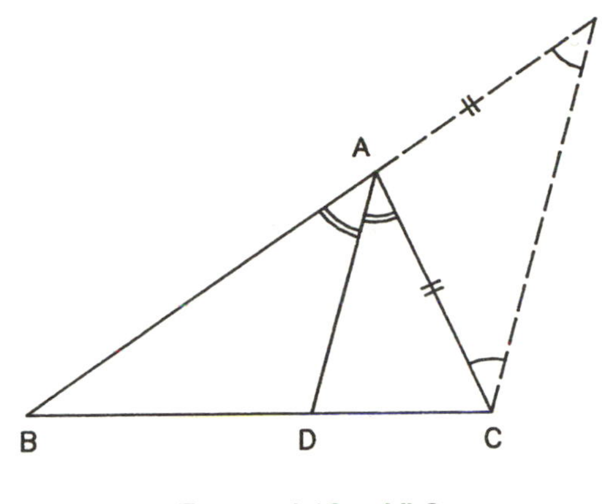

FIGURE I-18. VI,3

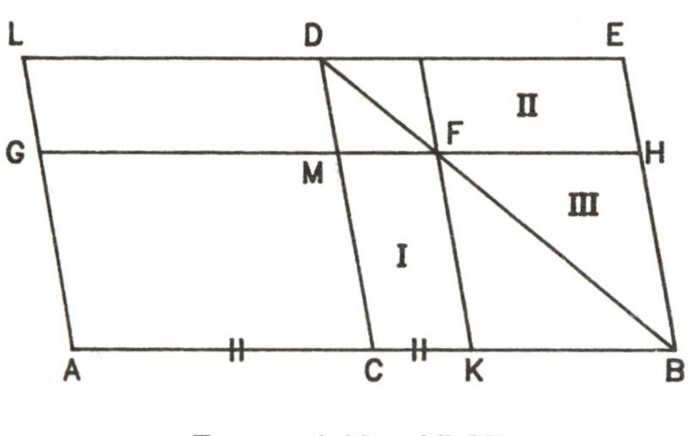

FIGURE I-19. VI,27

THEOREM VI,8. If in a right triangle we drop a perpendicular from the vertex corresponding to the right angle to the hypotenuse, then the two new right triangles will both be similar to the original. Similarity in turn implies various relationships between the sides and segments.

THEOREM VI,16. If lines A,B,C,D are proportional ($A:B = B:C$), then $R(A,D) = R(B,C)$ and conversely. The terms A and D are called the "extremes"; B and C, the "means."
Used in: VI,17; X.

THEOREM VI,17. If lines A,B,C are proportional (i.e., $A:B = B:C$), then $R(A,C) = S(B)$ and conversely. The terms A and C are called "extremes"; B is called the "mean."
Used in: XIII,1,2,3,4,5,6,10,13 . . .

THEOREM VI,25. To construct a rectilineal figure which is similar to a given rectilineal figure and which has the same area as another given rectilineal figure.
Uses: I,44,45.
Used in: VI,28,29.
Note: While the statement allows any shaped rectilineal figure, the proof assumes that it is a triangle that is involved and for this case uses I,44 (application of areas; no use is made of the angle mentioned in I,44). In case the given figure is not a triangle, one may use techniques similar to those used to prove I,45 from I,42,44 and VI,25, as it stands, from I,44,45.

THEOREM VI,27. Let AB be any line with C its midpoint and let $CBED$ be an arbitrary parallelogram. Let $KBHF$ (III) be a similar parallelogram and construct the figure shown. Then the area of parallelogram $ACDL$—which is the same as that of the parallelogram $CBED$—is greater than that of parallelogram $AKFG$. (The terminology used is that one has applied parallelograms $ACDL$ and $AKFG$ to line AB with respective "defects" $CBED$ and $KBHF$.)

Proof: I = II (I,43), therefore I + III = II + III. But since C is the midpoint of AB we have that parallelogram ($ACMG$) = parallelogram ($CBHM$) = I + III = II + III. Thus adding I to both sides, parallelogram ($AKFG$) = I + III + II (the gnomon). But the area of the gnomon is less than the area of parallelogram $CBED$ which in turn has the same area as parallelogram $ACDL$, again because C is the midpoint of AB. Thus the area of parallelogram $AKFG$ is less than that of parallelogram $ACDL$.
Uses: I,43 (complements).
Used in: VI,28.

THEOREM VI,28 (an application of areas result). It is required to perform the following construction. We are given a line AB, a parallelogram D and a rectilineal figure C. On AB we are to construct a parallelogram $ABRT$ so that the subparallelogram $ASQT$ has the same area as C, while the subparallelogram $SBRQ$ (II) is similar to the given parallelogram D. In order for this construction to be possible, a limitation must be put on the area of C. This condition is as follows: let E be the midpoint of AB and let parallelogram $EBFG$ be similar to parallelogram D. We assume that the area of parallelogram $EBFG$ is greater than or equal to that of C. (The terminology used is that one has applied to AB a parallelogram—$ASQT$—equal to the given rectilineal figure C and deficient by the parallelogram II similar to D.)
Sketch of Proof: By VI,27 the parallelogram $EBFG$ will have its area, which by hypothesis is greater than or equal to that of C, greater than that of the required parallelogram $ASQT$. In case the area of parallelogram $EBFG$ is equal to that of C we are done. In the case of inequality we are going to pick the subparallelogram $OQPG$ so that the remaining area—the gnomon = I + II + III—has its area equal to that of C; this parallelogram can be constructed via VI,25. Now consider the gnomon ≡ I + II + III. But II + III = IV, since E is the midpoint of AB, and further I = III, since they are complements (I,43). Thus IV + III = (III + II) + I ≡ gnomon = C. Since by construction, parallelogram $SBRQ$ will also be similar to D, all the requirements are satisfied.

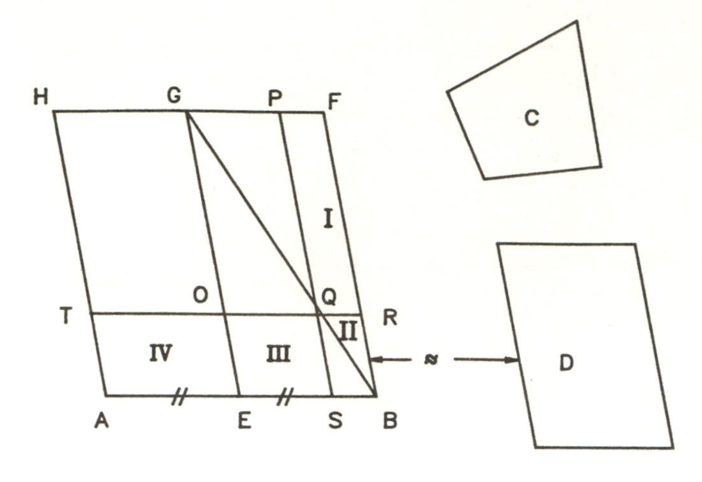

FIGURE I-20. VI,28

Uses: I,43 (complements); VI,25,27 . . .
Used in: X (14 times).
Discussed in: Section 2,I.

THEOREM VI,29 (an applications of areas result). It is required to perform the following construction. We are given a line AB, a parallelogram D, and a rectilinear figure C. On AB extended we are to construct a parallelogram $APOR$ whose area is the same as that of C and is such that the subparallelogram $BPOQ$ (II) is similar to D. (The terminology used is that one has "applied to AB a parallelogram—$APOR$—equal to the given rectilineal figure C and exceeding by the parallelogram II similar to D").

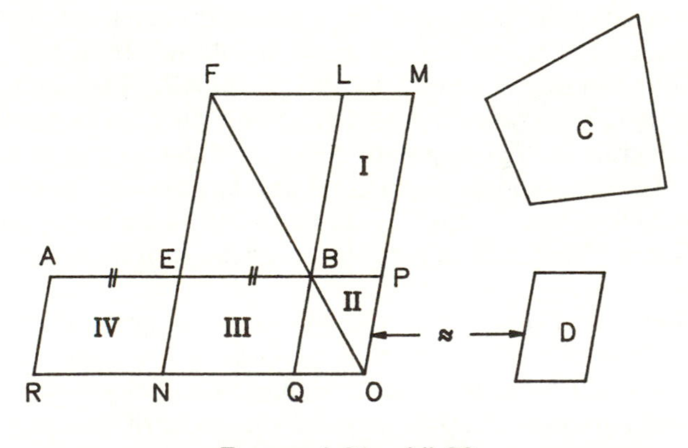

FIGURE I-21. VI,29

Sketch of Proof: Let E be the midpoint of AB. On EB construct parallelogram $EBLF$ similar to D. Now construct, using VI,25, the larger and similar parallelogram $NFMO$ whose area is the sum of the area of $EFLB$ and C. This means that the gnomon ≡ I + II + III has its area equal to that of C. But III = IV, since E is the midpoint of AB, and I = III, since these two regions are complements of a parallelogram (I,43). Thus IV + III + I equals C as required. Furthermore, by construction, parallelogram $QOPB$ is also similar to D.
Uses: I,43 (complements) . . .
Used in: VI,30.
Discussed in: Section 2,I.

THEOREM VI,30 (DEMR). To divide a line AB in extreme and mean ratio.

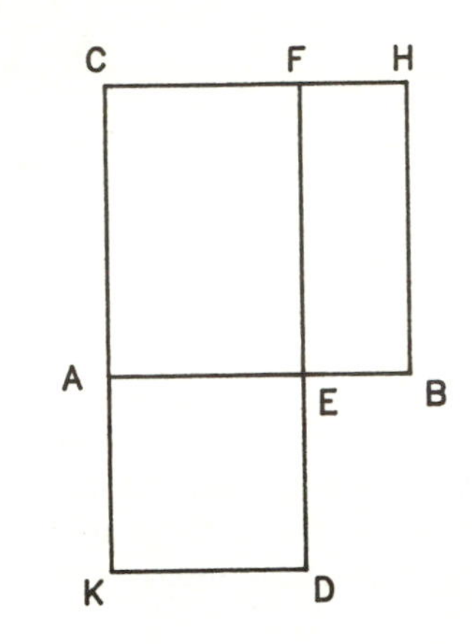

FIGURE I-22. VI,30

Proof: In VI,29 take the region C to be the square $ABHC$ and let the parallelogram D be a square. Then on AC (as opposed to the original line AB) we obtain rectangle $CFDK$ with area equal to that of the region C—square ($ABHC$)—and such that the "excess"—$AEDK$—is similar to D, a square.

If we now subtract rectangle $ACFE$ from both square $ABHC$ and rectangle $CFDK$ we have that $AEDK = EFHB$. This implies that $FE{:}ED = AE{:}EB$ and since $FE = AB$ and $ED = AE$ this gives $BA{:}AE = AE{:}EB$. Further, since BA is greater than AE the equality implies that AE in turn is greater than EB.
Uses: VI,29 . . .
Used in: XIII,1,2,3,4,5,6,8,9,10,11,16,17,18.

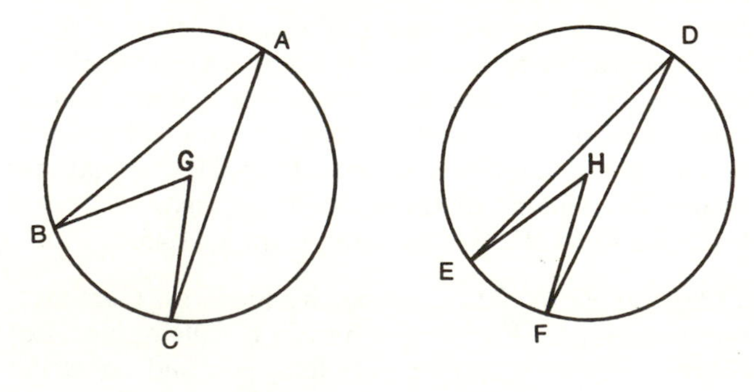

FIGURE I-23. VI,33

THEOREM VI,33. In the two circles shown arc BC: arc $EF = \angle BGC : \angle EHF = \angle BAC : \angle EDF$.
Used in: XIII,8,9,10.

Note: The proof uses III,27 to which the present result should be compared; see also the proof of XIII,8′ in Section 2,C.

BOOK VII. This is the first of three books that deal with the theory of numbers (integers).

THEOREM VII,1. The condition for two integers to be relatively prime (see VII,2).

THEOREM VII,2 (Euclidean Algorithm). We are given two integers which are not relatively prime—that is, they have a common divisor other than 1—and we are to find their greatest common divisor.
Proof: Since we are dealing with integers, the procedure can be described as follows. Let b_0 and b_1 be the integers with b_0 bigger than b_1. Suppose that b_0 goes into b_1 n_0 times with a remainder b_2, where b_2 is less than b_1. Then we can write $b_0 = n_0 b_1 + b_2$. Continuing in the same way with b_1 and b_2 we obtain the following set of statements:

$$b_0 = n_0 b_1 + b_2$$
$$b_1 = n_1 b_2 + b_3$$
$$\dots\dots\dots\dots\dots$$
$$b_j = n_j b_{j+1} + b_{j+2}$$
$$b_{j+1} = n_{j+1} \cdot b_{j+2}.$$

Since the b_i are decreasing, the sequence must stop with some last non-zero b_{j+2}. Working from the bottom equation up, we see that b_{j+2} divides all the preceding b_i and therefore b_0 and b_1. Conversely, starting at the top we see that any divisor of b_0 and b_1 must divide b_{j+2}. This shows that b_{j+2} is indeed the greatest common divisor. If b_{j+2} is 1, then b_0 and b_1 are relatively prime; this is the essence of VII,1.
Used in: VII.
Discussed in: Section 8.
Notes: (1) This so-called Euclidean Algorithm is discussed in most modern books on number theory; see, for example, Niven and Zuckerman [1960, 7]. (2) See the version for commensurable magnitudes in X,3.

BOOK X. This book deals with the construction and classification of incommensurable quantities; for a guide to its contents, see Taisbak [1982]. See also Mueller [1981]; Fowler [1983a; 1983b, 68]; and Section 7.

Book X is written from a geometric point of view and care must be taken not to think simply of the concepts in modern terms. Particular attention must be given to the difference between the dichotomy commensurable–incommensurable and the dichotomy rational (*rhētós*: "utterable," "expressible," "red")–irrational (*álogos*: "obscure," "unreasonable") (see the notes on definition 3).

If we have two comparable magnitudes A and B and if the length (area) of A is an integer multiple of the length (area) of B, then we say that B measures A. If we have a collection of comparable magnitudes $A,B,C\dots$, and there is a magnitude D which measures each member of the collection, then we say that D is a common measure of the magnitudes of the collection.

If A and B have a common measure, then any magnitude which is a common measure and which is such that no other common measure has a greater length (area) is called the greatest common measure.

THEOREM X,def.1. The magnitudes of a collection are said to be commensurable if they have a common measure. If there is no common magnitude then the magnitudes are called incommensurable.

THEOREM X,def.2. The straight lines $A,B,C\dots$ of a collection are called commensurable in square if $S(A)$, $S(B)$, $S(C)\dots$ are commensurable. If these squares are not commensurable then the straight lines are called incommensurable in square.

THEOREM X,def.3. Start with a fixed line segment F and call it rational. Then another line segment A is also called rational if either A is commensurable with F and/or if $S(A)$ is commensurable with $S(F)$. Otherwise A is called irrational....
Notes: (1) Euclid shows as part of X,9 that if $S(A)$ is not commensurable with $S(F)$ then A will not be commensurable with F. Thus instead of saying, "Otherwise A is called irrational," we could have said, "If $S(A)$ is not commensurable with $S(F)$ then A is called irrational." (2) In X,def.3 a line has, so to speak, two chances to be rational; even if the line A is not commensurable with the fixed line F, A will still be called rational if $S(A)$ is commensurable with $S(F)$. (3) Both the definitions of the dichotomies commensurable–incommensurable and rational–irrational involve "classes" of lines. The dichotomy rational–irrational is defined with respect to a fixed reference line and is defined for individual straight lines. On the other hand, the definition of the dichotomy commensurable–incommensurable involves two or more magnitudes and has nothing to do with a fixed line.

THEOREM X,def.4. Start with a fixed line segment F and call $S(F)$ rational. Then any plane figure A which is commensurable with $S(F)$ is also called rational. A plane figure A which is not commensurable with $S(F)$ is called irrational. Now if A is an irrational plane figure let B be a line segment such that $S(B) = A$, then the line B is also called irrational.

THEOREM X,1. Start out with two magnitudes A and B and suppose that A is greater than B. Now from the magnitude A subtract another magnitude C which is greater than one-half of A. Then from C subtract a magnitude D which is greater than one-half C, and keep

on repeating this process. Then eventually what will be left will have a magnitude of less than B.
Used in: X,2; XII.
Discussed in: Section 12,vii.

THEOREM X,2. Start with two magnitudes A and B and suppose that A is greater than B. Now subtract B from A as many times as possible and suppose that the magnitude that remains is C. Now subtract C from B as many times as possible and suppose that the magnitude that remains is D. Keep on repeating the process. Suppose that the magnitude that remains never measures the previous term that was subtracted. Then the original two magnitudes A and B will be incommensurable.
Idea of Proof: If there were a magnitude E that measured A and B, it would have to measure C. Then, since we now have that E measures both B and C, it must also measure D. In other words, at each stage E must measure whatever is left. But what is left will eventually be less than E and so we have a contradiction.
Used in: X,3.
Discussed in: Sections 8; 12,vii.

THEOREM X,3(4) (Euclidean Algorithm). We are given two (three) commensurable magnitudes A and B and we must find their greatest common measure.
Description of Method of X,3: Let A and B be two magnitudes with A greater than B. Keep on subtracting in the manner described in X,2. Eventually what is left, say F, measures the previous term in the process. This is the desired magnitude because by preceding back up the line we have that F must be a common measure of A and B. Furthermore, by starting at the beginning we have that any common measure of A and B must also be a common measure of F.
Used in: Never explicitly used.
Discussed in: Section 8.
Note: The corresponding result for integers is given in VII,2.

THEOREM X,5,6,7,8. Two magnitudes are commensurable if and only if the ratio of their lengths (areas) is the ratio of two integers. The two magnitudes will be incommensurable if and only if the ratio is not the ratio of two integers.
Used in: X.

THEOREM X,9. Squares on lines commensurable in length are in the ratio of squares of integers and conversely.

THEOREM X,21 (the irrational line called medial). Start with two rational straight lines AB and BC which are commensurable in square but which are not commensurable. Then $R(AB,BC)$ is irrational and if D is a line segment such that $S(D) = R(AB,BC)$ then D is irrational and is called medial.
Note: This is one of the three basic types of irrational line segments. The other two are the binomial (X,36) and the apotome (X,73).

THEOREM X,23, corollary. "From this it is manifest that an area commensurable with a medial area is medial."

THEOREM X,24. "The rectangle contained by the medial straight lines commensurable in length is medial."
Part of Proof: "For on AB let the square AD be described, therefore AD is medial."
Note: Euclid never explicitly defines the concept of medial area but from the corollary to X,23 and the proof of X,24 we can deduce that the term "medial area" applies to the area (that is, the numerical quantity associated with the region, as opposed to the region itself) of a square whose side is a medial line (X,21) or to the area of a rectangle such that the rectangle is commensurable with a square whose side is a medial line. Theorem 24 tells us that this is the case when the sides of the rectangle are medial and commensurable. As elsewhere there is a certain ambiguity of language.

THEOREM X,73. (The irrational line called apotome.) Start with a rational straight line AB. Let the segment BC be only commensurable in square with AB. Then the remaining segment AC is irrational and is called apotome.
Used in: X (28 times); XIII,6,11, 17[b].
Discussed in: Section 20,A.
Note: In XIII,6 we learn that each of the segments of a rational line divided in EMR is an apotome.

THEOREM X,76. (The irrational line called minor.) Start with a line AB. Let segment BC satisfy the following conditions: BC is incommensurable in square with AB; $S(AB + BC)$ is rational; the area of $R(AB,BC)$ is medial (see note on X,23,24). Then AC, the remainder when BC is subtracted from AB, is irrational and is called minor.
Used in: X; XIII,11, 16[b].
Note: In XIII,11 we learn that if we inscribe a pentagon in a circle with rational diameter then the side of the pentagon will be a minor.

BOOK XIII. The aim of this book is to describe the inscription in a sphere of the five regular solids—tetrahedron, octahedron, cube, icosahedron, dodecahedron. It is these last two constructions that involve DEMR. The book starts off with a group of propositions dealing with properties of DEMR.

THEOREM XIII,1. If a line is divided in EMR then $S(1/2 \text{ line} + \text{larger segment}) = 5 \cdot S(1/2 \text{line})$.
Proof: Let the line be AB and divide it at C in EMR with AC being the larger segment; extend BA to D so

that $AD = 1/2\,AB$. We must show that $S(CD) = 5\cdot S(AD)$. Draw the squares and rectangles as indicated. From VI,def.3 and VI,17 we have
(a) $\text{II} = R(AB,CB) = S(AC) = \text{I}$.
Since $AK = AB = 2\cdot AD = 2\cdot AH$ and $\text{IV} = \text{V}$ (I,43) we have
(b) $\text{III} = 2\cdot\text{IV}$ (by VI,1) $= \text{IV} + \text{V}$.
By adding (a) and (b) we thus obtain
(c) $S(AB) = \text{II} + \text{III} = \text{I} + \text{IV} + \text{V} \equiv$ the gnomon.
Using (c) together with $AB = 2\cdot AD$ we have
(d) gnomon $= S(AB) = 4\cdot S(AD)$.
Finally, adding $S(AD)$ to both sides of (d), $S(CD) =$ gnomon $+ S(AD) = 5\cdot S(AD)$.
Uses: I,43 (complements); VI,1,17 . . .
Used in: XIII,6,11.
Discussed in: Sections 2,D; 14.

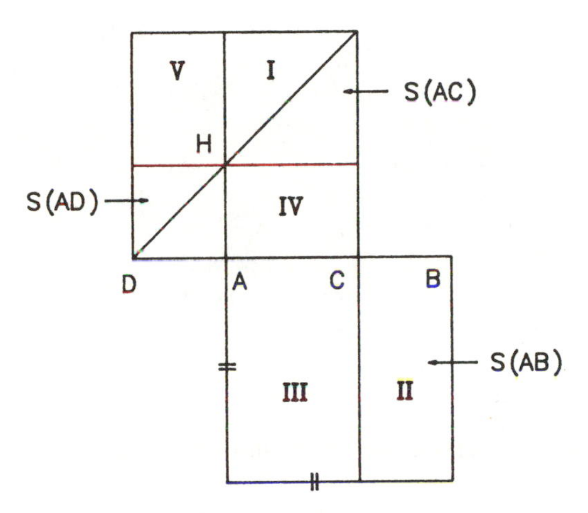

FIGURE I-24. XIII,1

THEOREM XIII,2. Let AB be a line, AD a line, AC a segment of it, and let $CD = 2\cdot AC$. Suppose that $S(AB) = 5\cdot S(AC)$. Then CB is the larger segment of CD when the latter is divided according to EMR.

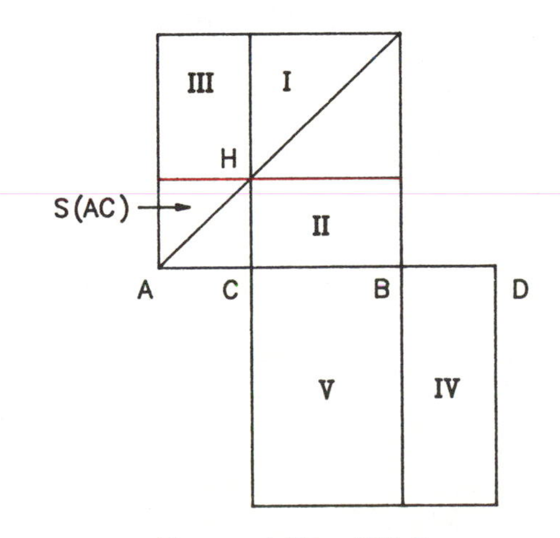

FIGURE I-25. XIII,2

Proof: Draw the squares and rectangles as indicated. We are given that $5\cdot S(AC) = S(AB) = \text{I} + \text{II} + \text{III} + S(AC) \equiv$ gnomon $+ S(AC)$. Subtracting $S(AC)$ from each side we obtain
(a) $4\cdot S(AC) \equiv$ gnomon $= \text{I} + \text{II} + \text{III}$.
Also, since $DC = 2\cdot AC$, we have $S(CD) = 4\cdot S(AC)$ and thus from (a) we have
(b) $\text{I} + \text{II} + \text{III} \equiv$ gnomon $= 4\cdot S(AC) = S(CD) = \text{IV} + \text{V}$.
Again, since $KC = CD = 2\cdot AC = CH$ and $\text{II} = \text{III}$ (I,43), we have
(c) $\text{II} + \text{III} = 2\cdot\text{II} = \text{V}$ (by VI,1).
Subtracting (c) from (b) we have $S(CB) \equiv \text{I} = \text{IV} \equiv R(CD,BD)$. Thus, by VI,17, $CD{:}CB = CB{:}BD$; and since DC is greater than CB (see lemma), CB must be greater than BD (V,14); that is, CB is indeed the larger segment.
Uses: I,43 (complements); V,14; VI,17; following lemma.
Used in: Never used.
Discussed in: Sections 2,D; 14.

THEOREM XIII,2 lemma. In the above, $2\cdot AC$—that is, DC—is greater than CB.
Uses: II,4.
Used in: XIII,2.
Discussed in: Sections 12,vii; 14.
Note: Heiberg [IV,254] considers this lemma to be not genuine.

THEOREM XIII,3. If a line is divided in EMR then $S(1/2$ larger segment + smaller segment$) = 5\cdot S(1/2$ larger segment$)$.

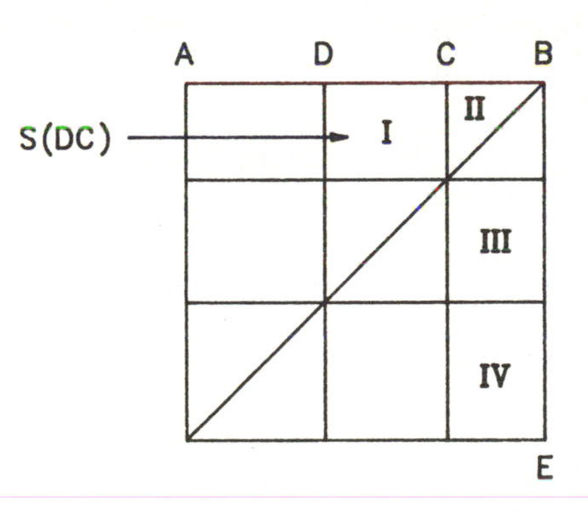

FIGURE I-26. XIII,3

Proof: Let the line be AB and divide it at C in EMR with AC being the larger segment. Let D be the midpoint of AC. Then we must show that $S(BD) = 5\cdot S(DC)$. Draw the squares and rectangles as indicated. Since $AC = 2\cdot DC$ we have
(a) $S(AC) = 4\cdot S(DC)$.
Because AB is divided in EMR at C, VI,17 gives
(b) $\text{II} + \text{III} + \text{IV} \equiv R(BE,CB) = R(AB,CB) = S(AC)$.

Thus combining (a) and (b), we have
(c) II + III + IV = 4 · $S(DC)$.
But $AD = DC$ and thus we obtain III = IV. Furthermore, I = III (by I,43), so that I = IV. If we add II + III to both sides of this latter equality and use (c) we obtain
(d) gnomon ≡ I + II + III = 4 · $S(DC)$.
Now add $S(DC)$ to both sides of (d) to finally obtain $S(BD)$ ≡ gnomon + $S(DC)$ = 5 · $S(DC)$.
Uses: I,43 (complements); VI,1,17.
Used in: XIII,16.
Discussed in: Sections 2,D; 12,vii; 14.
Note: The origin of this result is suggested by the proof of XIII,16; see the note on that result.

THEOREM XIII,4. If a line is divided in EMR then [S(line) + S(smaller segment)] = 3 · S(larger segment).

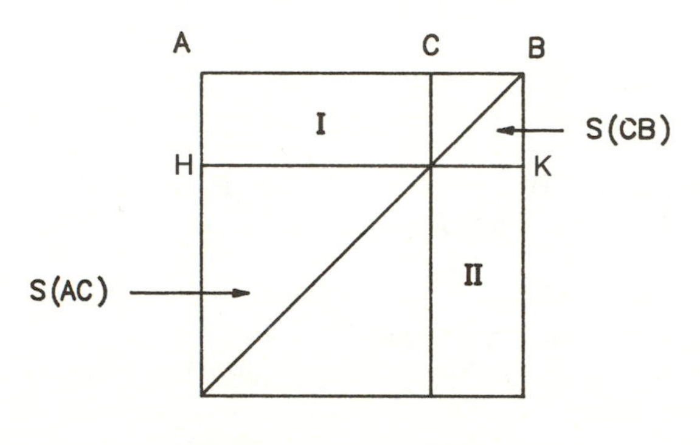

FIGURE I-27. XIII,4

Proof: Let the line be AB and divide it at C in EMR with AC being the larger segment. Then we must show that $S(AB) + S(BC) = 3 \cdot S(AC)$. Draw the squares and rectangles as indicated. By VI,def.3, and VI,17 we have
(a) rectangle $(ABKH) = R(AB,BC) = S(AC)$.
By I,43, I = II so that if we add $S(CB)$ to both sides we obtain rectangle $(ABKH)$ ≡ I + $S(CB)$ = II + $S(CB)$. Therefore
(b) gnomon + $S(CB)$ ≡ (I + $S(CB)$ + II) + $S(CB)$ = 2 · rectangle $(ABKH)$.
Now adding $S(AC)$ to both sides of (b) and using (a) and (b), we have $S(AB) + S(CB)$ ≡ (gnomon + $S(AC)$) + $S(CB)$ = 2 · rectangle $(ABKH) + S(AC) = 3 \cdot S(AC)$.
Uses: I,43 (complements); VI,17.
Used in: XIII,17.
Discussed in: Sections 2,D; 14.
Note: The origin of this result is suggested by the proof of XIII,17; see the note on that result.

THEOREM XIII,5. Let AB be a line which is divided in EMR at C, with AC being the larger segment. Extend BA to D with DA equal to AC. Then the line DB is divided in EMR at A, with AB being the larger segment.
Proof: Draw the squares and rectangles as indicated. By VI,def.3 and VI,17 we have I + II = $R(AB,BC) = S(AC)$. Also II + I = II + III and $S(DA) = S(AC)$ and so $S(DA)$ = III + II. Thus adding rectangle $(ABKH)$ to both sides $R(BD,DA)$ = rectangle $(DBKL) = S(DA)$ + rectangle $(ABKH)$ = III + II + rectangle $(ABKH) = S(AB)$. Thus by VI,17 $DB{:}BA = BA{:}AD$, and by V,14 EA is the larger segment.

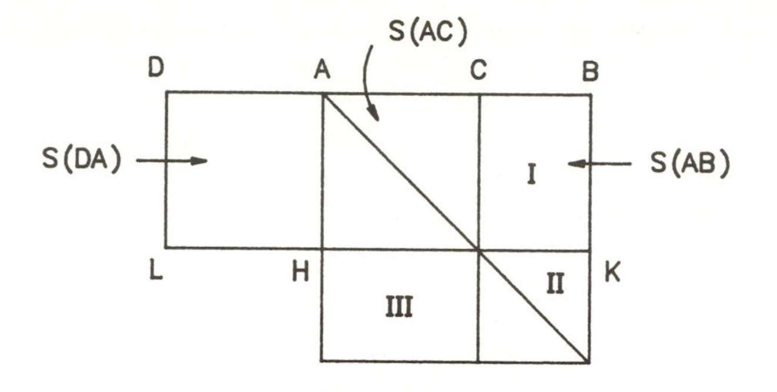

FIGURE I-28. XIII,5

Uses: V,14; VI,17.
Used in: XIII,17.
Discussed in: Sections 2; 12,vii; 14.
Notes: (1) It is not stated why I + II = III + II. Perhaps I,43 (complements) was in mind (see Euclid-Frajese), but see the note on II,4. (2) The origin of this result is suggested by the proof of XIII,17[a]; see the note on that result.

THEOREM XIII,6. If a rational line (X,def.3) is cut in EMR then each of the segments is the irrational line called (in X,73) apotome.
Uses: VI,17; X; XIII,2.
Used in: XIII,17[b].
Notes: (1) The proof of this result involves a torturous journey into the Euclidean world of the classification of irrationals. For a discussion, see Taisbak [1982, 49] and Mueller [1981, 297]. (2) Heiberg [Euclid-Heiberg, 262] included this proposition in his critical edition, but for Heath [Euclid-Heath, III, 451] "it seems certain that this proposition is an interpolation." Frajese and Maccioni [Euclid, 996] do not enter into the debate but consider this as "one of the most beautiful results of the *Elements*," with its "classic simplicity." (3) Cf. XIII,11.

THEOREM XIII,8. In a regular pentagon the diagonals cut one another in EMR and the larger segments are equal to the side of the pentagon.
Proof: Circumscribe a circle about the pentagon. By the equality of the sides and angles of the pentagon we have $\triangle ABE \cong \triangle ABC$. Thus ∢$BAC$ = ∢ABE and so the exterior angle AHE = 2 · ∢BAH (I,32).

Since the sides of the pentagon are equal, III,28 says that the arcs are also equal. Thus arc EC

= arc ED + arc $DC = 2 \cdot$ arc BC. Consequently, from VI,33 we have that $\measuredangle EAC = 2 \cdot \measuredangle BAC$ and thus $\measuredangle HAE \equiv \measuredangle EAC = 2 \cdot \measuredangle BAC \equiv 2 \cdot \measuredangle BAH = \measuredangle AHE$. This shows that $\triangle AHE$ is isosceles and gives $HE = EA = AB$; that is, the segment HE is indeed equal to the side of the pentagon.

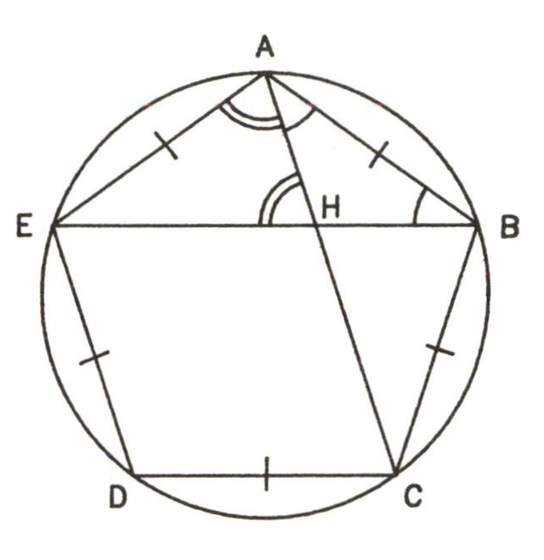

FIGURE I-29. XIII,8

Next we claim that $\triangle ABE \sim \triangle ABH$. This is so because $\measuredangle ABE = \measuredangle BEA = \measuredangle BAH$. From this we have $EB{:}BA = AB{:}BH$ or, since by above $BA = EH$, $BE{:}EH = EH{:}HB$ which is the definition of DEMR.

Since BE is bigger than EH, V,14 tells us that HE is larger than HB—that is, HE, which was shown to be equal to the side of the pentagon, is the larger segment.

Uses: I,32; III,28; IV,14; VI,33 . . .

Used in: XIII,11; XIII,17[a],corollary.

Discussed in: Sections 2,C; 22,A.

THEOREM XIII,9. Take any circle and let $CD = a_6$ and $BC = a_{10}$. Then the line BD formed by adding segments BC and CD together is divided in EMR at C and $CD = a_6$ is the larger segment.

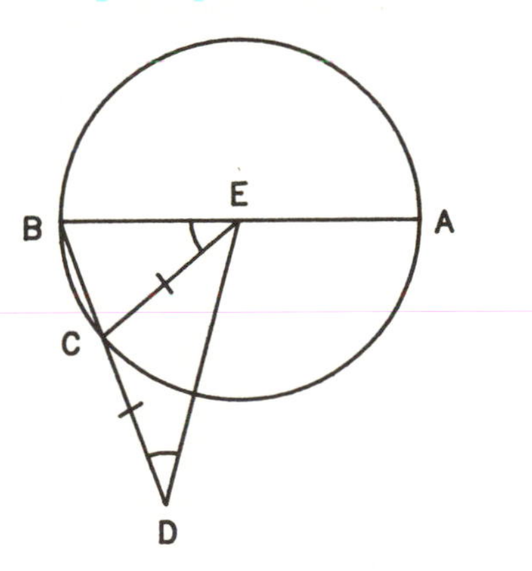

FIGURE I-30. XIII,9

Proof: Draw the diagram as shown. Since BC is the side of the decagon arc $(ACB) = 5 \cdot \text{arc}(BC)$ so that arc$(AC) = 4 \cdot \text{arc}(BC)$ and by VI,33, $4 \cdot \measuredangle CEB = \measuredangle AEC$.

Since the base angles of the isosceles triangle BEC are equal, the exterior angle $AEC = 2 \cdot \measuredangle ECB$ (I,32). Also, since the side of the inscribed hexagon is equal to the radius, we have $EC = CD$ which implies $\measuredangle CED = \measuredangle EDC$ and thus exterior angle $ECB = 2 \cdot \measuredangle EDC$.

Putting all the above information together we obtain $4 \cdot \measuredangle CEB = \measuredangle AEC = 2 \cdot \measuredangle ECB = 2[2 \cdot \measuredangle EDC] = 4 \cdot \measuredangle EDC$ and hence $\measuredangle BEC \equiv \measuredangle CEB = \measuredangle EDC$.

If we now look at $\triangle BEC$ and $\triangle BED$ we see that $\measuredangle EBD$ is common to both. Thus since we have just shown that $\measuredangle BEC$ of the first triangle is equal to $\measuredangle EDC$ of the second, these two triangles are similar. This, in turn, gives us $BD{:}EB = EB{:}CB$ or, since $EB = DC$, $BD{:}DC = DC{:}CB$. Since BD is greater than DC, V,14 implies that DC is greater than CB.

Uses: I,32; V,14; VI,33 . . .

Used in: XIII,16,18.

Discussed in: Section 2.

Note: A converse to XIII,9—which I designate as XIV,**—is discussed in Section 24,A.

THEOREM XIII,10. Take any circle and in it inscribe the regular pentagon, hexagon, and decagon. Then $S(a_5) = S(a_6) + S(a_{10})$.

Proof: Let $BA = a_5$, $BF = a_6$ and $AK = a_{10}$. Draw and extend the various lines as shown and as follows: FH is perpendicular to AB, K is obtained by extension; FL is perpendicular to AK, M is obtained by extension; N is the point where FL meets AB. By construction arc CG and chord AK will correspond to the decagon.

Again by construction we obtain the following equalities: arc $AK = 2 \cdot$ arc KM; arc $CB = 2 \cdot$ arc BK; arc $AB = 2 \cdot$ arc BK; $2 \cdot$ arc CG = arc CD = arc $AB = 2 \cdot$ arc BK. Thus arc CG = arc BK = arc $AK = 2 \cdot$ arc KM. We now have arc $GB = 2 \cdot$ arc BM so that by VI,33 $\measuredangle GFB = 2\ \measuredangle BFM$.

Next consider $\triangle AFB$. This is isosceles which means that (I,5) $\measuredangle FAB = \measuredangle ABF$. Thus (I,32) exterior angle $BFG = 2 \cdot \measuredangle FAB$. Hence $2 \cdot \measuredangle BFM = \measuredangle BFG = 2 \cdot \measuredangle FAB$ and $\measuredangle BFM = \measuredangle FAB$.

If we now look at $\triangle ABF$ and $\triangle BFN$, we see that they are similar because two angles are equal and $\measuredangle ABF$ is common. By proportionality and VI,17 we have

(1) $$S(BF) = R(AB,BN).$$

Now we turn our attention to $\triangle KNA$ and $\triangle KBA$ and show that these are similar. This is so because by construction (III,29) both are isosceles, which means that $\measuredangle KBN = \measuredangle LAN = \measuredangle LKN$, and further $\measuredangle LAN$ is common to both triangles. By proportionality and VI,17 we have

(2) $$S(AK) = R(AB,AN).$$

If we consider equations (1) and (2), we see that the quantity AB is common to both $R(AB,BN)$ and $R(AB,AN)$, whereas AN and BN together equal AB. Thus by II,2 $R(AB,BN) + R(AB,AN) = S(AB)$. Hence (1) and (2) together give $S(AK) + S(BF) = S(AB)$.
Uses: I,32; II,2; III,26,29; VI,17,33 . . .
Used in: XIII,16,18.
Note: See the discussions on XIII,16,18 for possible origins of XIII.

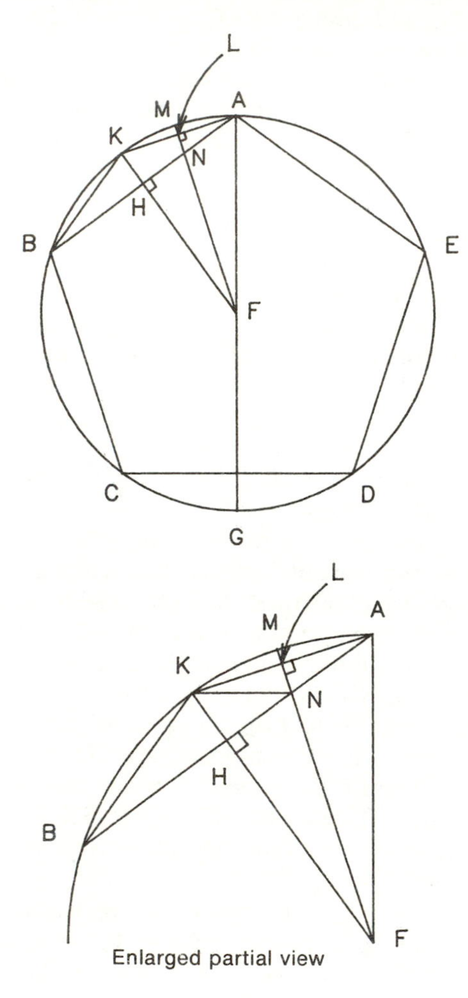

FIGURE I-31. XIII,10

THEOREM XIII,11. Take any circle which has a rational diameter (X,def.3) and in it inscribe a regular pentagon. The side of the pentagon is the irrational line called (in X,76) minor.
Uses: XIII,1,8 . . .
Used in: XIII,16[b].
Discussed in: Section 20,A.
Note: The proof of this result involves an even more torturous journey into the Euclidean world of the classification of irrationals than does XIII,6. For a discussion, see Taisbak [1982, 63] and Mueller [1981, 262].

THEOREM XIII,12. If an equilateral triangle is inscribed in a circle then $S(a_3) = 3 \cdot S(r)$.

THEOREM XIII,13. Inscription of the tetrahedron.

THEOREM XIII,13′. Let ABD be a right triangle with hypotenuse AB, and let DC be the altitude, then the following relationships hold:
(i) $S(AB)$: $S(BD) = AB$: BC (that is, S(hypotenuse): S(leg) = hypotenuse: segment [cf. VI,8]).
(ii) (lemma to XIII,13) $S(AD)$: $S(DC) = AB$: AC.

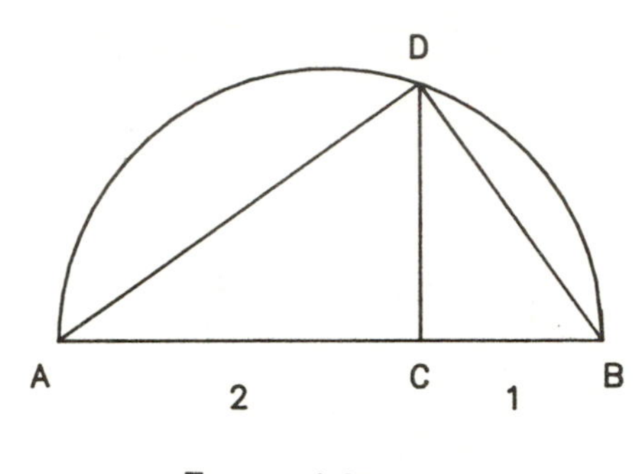

FIGURE I-32. XIII,13′

Note: As indicated (ii) is explicitly stated and proved as a lemma to the construction of the tetrahedron in XIII,13. Heath [Euclid–Heath, III, 471] considers this as a later addition. On the other hand, (i) is never stated as a lemma, but rather it simply appears without proof in connection with the cube (in XIII,15 [p. 479] and XIII,18 [p. 504]); the icosahedron (in XIII, 15 [p. 485] but not in XIII,18); and in the comparison of the edges of the icosahedron and dodecahedron in XIII,18 [p. 507]. In this last usage we are told that if three straight lines are proportional then first: third = S(first): S(second).

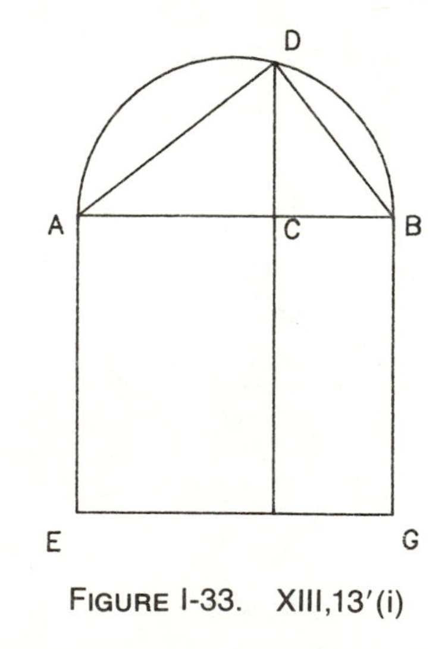

FIGURE I-33. XIII,13′(i)

Following is a proof of (i) which emulates that proof of (ii) found in the *Elements*. On AB construct square

ABGE. Then, since rectangles which have the same altitudes have their areas in the same ratio as the bases (this is VI,1), we have AB: $BC = R(AB,AE)$: $R(BC,AE) = R(AB,AB)$: $R(BC,AB) = S(AB)$: $R(BC,AB)$. However, from VI,8 AB: $BD = BD$: BC and VI,17 tells us $R(BC,AB) = S(BD)$.

THEOREM XIII,14. Inscription of the octahedron.

THEOREM XIII,15. Inscription of the cube. $S(D) = 3 \cdot S(e_6)$.

THEOREM XIII,16[a]. To inscribe an icosahedron in a sphere.
Uses: IV,11; XIII,3,9,10 . . .
Used in: XIII,18.

THEOREM XIII,16[a],corollary. Let AB be the diameter of the sphere. Let BD be the radius of the circle [in which the defining pentagon of the icosahedron is inscribed; see the discussion]. Then
(i) in Figure I-34, $AC = 4 \cdot CB$ and $S(AB) = 5 \cdot S(BD)$;
(ii) with respect to the circle of the defining pentagon, $AB = a_6 + 2a_{10}$.

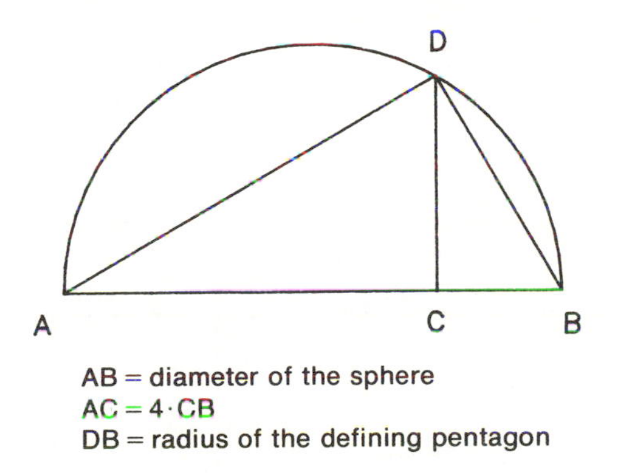

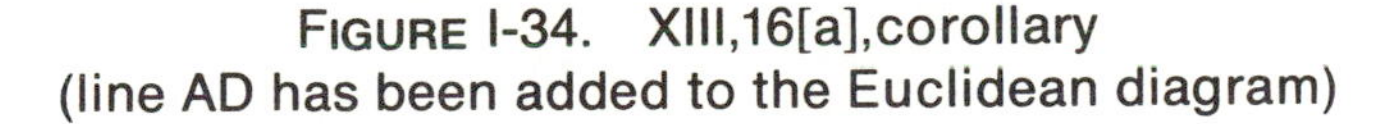
FIGURE I-34. XIII,16[a],corollary
(line AD has been added to the Euclidean diagram)

Discussion: Because pop-up models in the style of Euclid–Billingsley would quickly send the publisher into financial ruin and because the diagram in the text of the *Elements* is not terribly helpful, I have given instead (Fig. I-35) various views and sections of the icosahedron. An actual model might also prove helpful to the reader.

The important thing in considering Euclid's construction is to orient the icosahedron so that it is standing vertically, balanced on a vertex X (as opposed to resting on a face; see the construction of Pappus, Section 27). In this position (Fig. I-35a), if we look down from the top we will see a cap of five equilateral triangles with summit Z—the vertex opposite X—built upon a pentagonal base $QRSTU$ with centre W. Thus the edges of the icosahedron—that is, the sides of the equilateral triangles—are one and the same as the sides of the pentagon. I refer to this pentagon as the defining pentagon of the icosahedron. This will avoid confusion when discussing the icosahedron and dodecahedron simultaneously, as will be the case in Section 24. The bottom vertex X will correspond to the cap built upon the pentagon $LMNOP$ (Fig. I-35b) with centre V. The two pentagons are offset from one another by 36°, thus if we drop a perpendicular from Q to a point E on the circle which circumscribes pentagon $LMNOP$, the line PE (Fig. I-35b) will be the side of the decagon inscribed in this circle.

Thus our attention is directed towards the side (a_5) of the defining pentagon and the sides of the decagon (a_{10}) and hexagon (a_6) inscribed in the same circle.

There are three basic distances involved in the construction of the icosahedron. I discuss them in the order in which Euclid treats them.

(1) The distance WV between the planes of the defining pentagons. It turns out that $WV = a_6$ (Fig. I-35b,f).
(2) The height ZW of the vertex of the cap above the base. It turns out that $ZW = a_{10}$ (Fig. I-35b,f).

The statement of part (ii) of the corollary, to the effect that $AB = ZX = a_6 + 2 \cdot a_{10}$, is simply a combining of statements (1) and (2) (see Fig. I-35c).

(3) The radius a_6 of the circle of the defining pentagon and its relationship to the diameter AB of the sphere. This relationship is stated in part (i) of the corollary (or more precisely, a corollary to the theorem whose proof would essentially be the reverse of this proof).

This theorem is a construction in which we are not shown why the above three relationships are true, much less how they were arrived at in the first place. Rather, Euclid assumes that these relationships hold and then proceeds to show that he has indeed constructed the icosahedron.

Thus, with the various points located as indicated, in (1) and (2), Euclid wishes to show that the triangles, such as UQP, that lie between the two defining pentagons are indeed equilateral. If we consider the right triangle PEQ (Fig. I-35d) then, by the Pythagorean Theorem, $S(QE) + S(PE) = S(QP)$. But since $QE = a_6$ and $PE = a_{10}$, XIII,10 tells us that $S(QE) + S(PE) = S(a_5)$. Thus $QP = a_5$ so that QP does indeed equal QU.

Turning our attention to the cap we have to show that the upper triangles such as QZU (Fig. I-35a) are equilateral. If we look at the right triangle QWZ (Fig. I-35e) then, since $QW = a_6$ and $ZW = a_{10}$, the exact same argument that we just saw for QP shows that $QZ = a_5 = QU$. Since XIII,10 is only used in XIII,16 and 18 it is possible that the statement of XIII,10 was suggested by the Pythagorean relationship displayed in Figures I-35d,e.

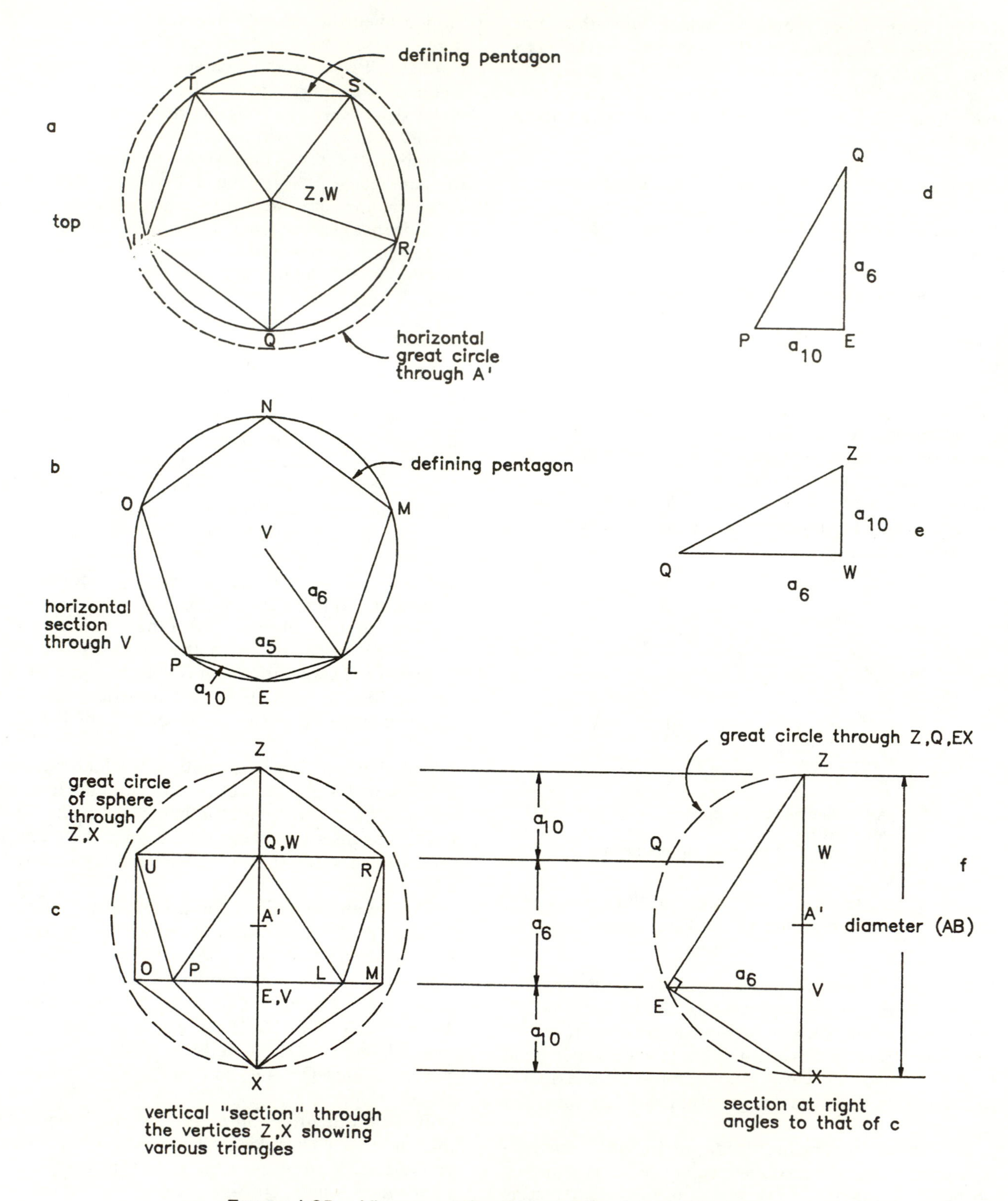

FIGURE I-35. Views and Sections of the Icosahedron (XIII,16[a])

Now Euclid wishes to show that the various vertices such as E are on the sphere with diameter ZX. To do this he considers the triangles ZVE and XEV (Fig. I-35f). Now since $VW = a_6$ and $WZ = a_{10}$ we have by XIII,9 that VZ is cut in EMR at W and since $VW = VE$ and $WZ = VX$ we have $ZV:VE = VE:VX$; that is, VE is the mean proportional between the two parts of the line ZX. This implies that $\measuredangle XEZ$ is a right angle. Thus a semicircle on the diameter XZ will pass through E which means that E is on the sphere.

Note that so far the *assumed* relationship (3) between the diameter AB of the given sphere and the radius BD of the circle of the defining pentagon has not been used. The supposed diameter AB has not even been mentioned. It is, rather, the diameter ZX that has been the basis of the construction of the icosahedron and the proof that the construction described does indeed lead to an icosahedron inscribed in the sphere of diameter ZX. This proof, as we have seen, does involve the radius a_6 of the circle of the defining pentagon via VW. Thus to finish off the proof of the theorem, or rather construction, we must use the relationship of both ZX and AB to the radius a_6 to prove that $XZ = AB$.

Euclid has picked, as stated in (3), the point C so that $AC = 4 \cdot CB$. Thus $AB = 5 \cdot BC$ and by XIII,13′(i) $S(AB) = 5 \cdot S(BD)$ which is the result of the corollary. On the other hand, if we look at the line ZX (Fig. I-35f) then Theorems XIII,9 and XIII,3 tell us that $S(ZA') = S(ZW + WA') = 5 \cdot S(WA')$ where A' is the midpoint of VW. Since $XZ = 2 \cdot ZA'$, $WV = 2 \cdot WA'$ and $WV = a_6$ by (1), we also have that $S(XZ) = 5 \cdot S(WV) = 5 \cdot S(BD)$ and thus $XZ = AB$.

As with XIII,10 it is possible that the statements of XIII,3 and 9 were suggested by the relationships displayed in this proof.

THEOREM XIII,16[b]. To prove that, [if the diameter of the sphere is rational—X,def.3—then] the edge of the icosahedron is the "irrational" line called—in X,76—"minor."
Uses: XIII,11.
Used in: XIII,18.

THEOREM XIII,17[a]. To inscribe a dodecahedron in a sphere.
Uses: XIII,4,5,15 . . .
Used in: Not used.

THEOREM XIII,17[a],corollary. If the edge of the cube inscribed in a sphere is divided in EMR then the greater segment is the edge of the dodecahedron inscribed in the sphere.
Uses: Construction of XIII,17[a]; XIII,8.
Used in: XIII,18.
Discussion: As with the discussion of the icosahedron (XIII,17) I have not used the diagram of the Euclidean text. In this case I have given an isometric drawing together with top and front views (Fig. I-36).

The important thing in considering Euclid's construction is to orient the dodecahedron so that it is standing vertically and balanced on an edge (as opposed to resting on a face, as is the case with the construction of Pappus, Section 27, or balanced on a vertex, as is the case with the construction of Bombelli, Section 31,F). In this position the opposite top edge UV will also be horizontal. In the top view we will then see the two faces I and II which share the edge UV, as well as the faces III and IV which have U and V as vertices. In addition there will be four faces, such as V, which are vertical and thus appear in edge view. In the front view the faces III and IV will appear in edge view.

We now see that the diagonals BC, EF, BE, and FC of I,II,III,IV are the sides of a horizontal square which in fact is the top of a cube whose vertical edges such as BA are diagonals of the vertical pentagonal faces. This observation, together with XIII,8, constitutes the corollary. Thus we see that one possible starting point for the construction of the dodecahedron is to first construct the cube (XIII,15), and this indeed is the way that Euclid proceeds.

Suppose that the cube were constructed, with its vertices B and E also being vertices of the dodecahedron. What we would like to do now is locate the vertex U. This vertex sits above the point R which is on the line NO which in turn connects the midpoints of the side of the top square. Since UV is the side of a pentagon and $NO = BC =$ diagonal of the pentagon, the point R can be located by dividing NP in EMR. To locate the point U we would now wish to know the distance UR. It turns out that $RU = RP$ but, this being a construction, Euclid does not prove this statement, rather he assumes it. Thus with the assumption that the cube is constructed, R obtained by dividing NP in EMR and $UR = RP$, Euclid wants to show that $BU = UV$, that the pentagonal face is equilateral.

The proof is as follows: by XIII,4 $S(PN) + S(NR) = 3 \cdot S(RP)$ or $S(BN) + S(NR) = 3 \cdot S(RU)$. By the Pythagorean Theorem applied to BN and NR in the horizontal plane we have $S(BR) = 3 \cdot S(RU)$. Now applying the Pythagorean Theorem in a vertical plane to BU, BR, and UR we obtain $S(BU) = S(BR) + S(RU) = 4 \cdot S(RU)$ so that $BU = 2 \cdot RU = 2 \cdot RP = UV$.

To show that the point U is indeed on the sphere Euclid again uses XIII,4 and the Pythagorean Theorem together with XIII,5 to obtain $S(UZ) = S(ZX) + S(XU) = S(NS) + S(SP) = 3 \cdot S(NP)$. This means, by the relationship of XIII,15, that UZ is equal to the radius of the sphere.

Note that XIII,4 is only used in XIII,17 and this together with the proof suggests that XIII,4 was developed as a lemma for XIII,17 when the author of XIII,17 was faced with this unfolding of the proof.

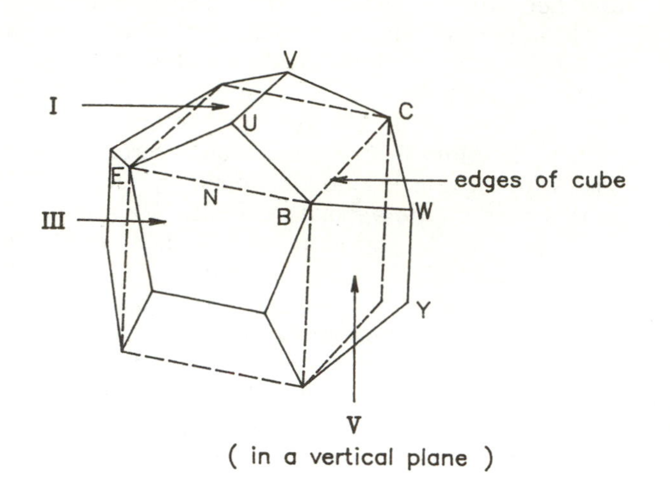

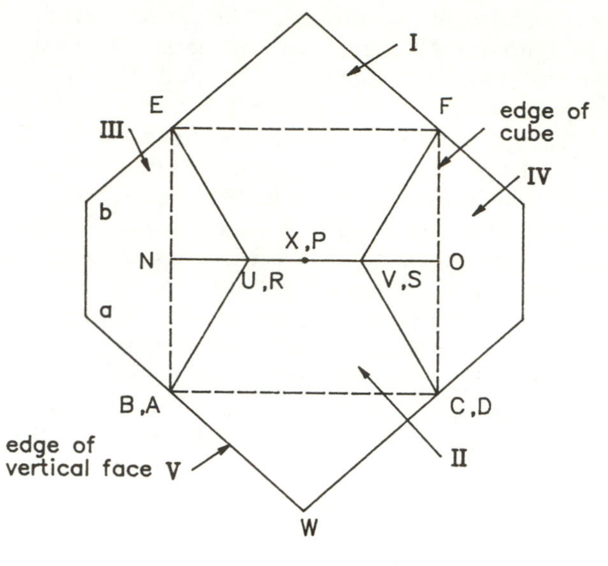

top view

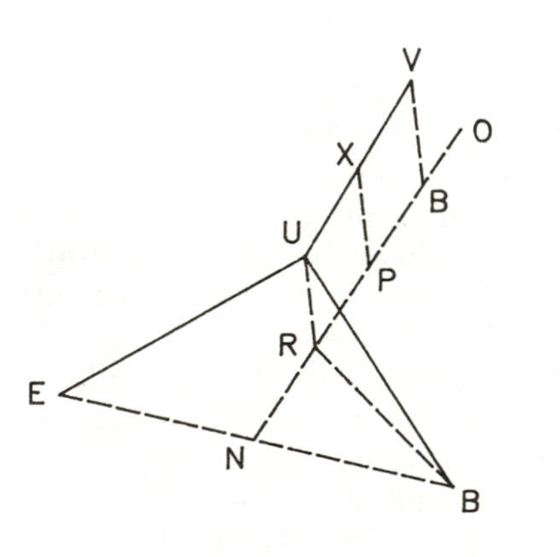

enlarged part of isometric

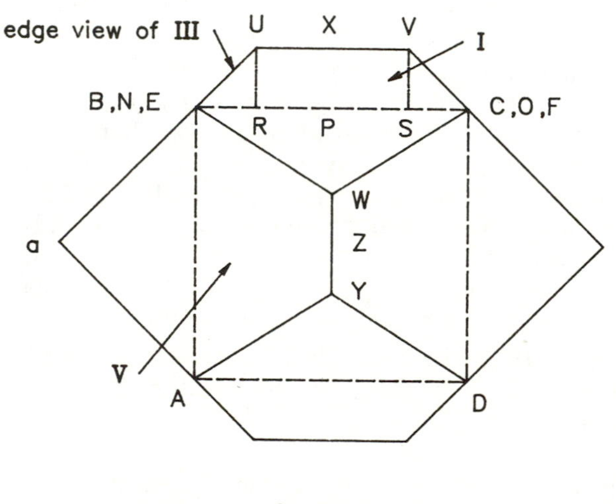

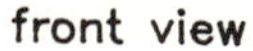

FIGURE I-36

Other parts of the proof involve such things as showing that the pentagons are planar but I have omitted them from the outline of the Euclidean proof since they do not deal with DEMR.

THEOREM XIII,17[b]. To prove that [if the diameter of the sphere in XIII,17(a) is rational (X,def.3) then] the edge of the dodecahedron is the irrational line called [in X,73] apotome.
Uses: XIII,6.

THEOREM XIII,18. "Let *AB* be the diameter of a sphere. We are to construct the edges of the five regular polyhedra which can be inscribed in the sphere. We are to compare the various lengths.

[The following statements are also true:
(1) The edges of the first three polyhedra are in rational ratio (X,def.3) with respect to one another but the edges of the icosahedron (minor) and dodecahedron (apotome) are not in rational ratio with any other edge.
(2) The edge of the icosahedron is by far greater than the edge of the dodecahedron.
(3) No other regular polyhedra exist.]

Discussion: The diagram of the Euclidean text together with the instructions for constructing it are given in Figure I-37.

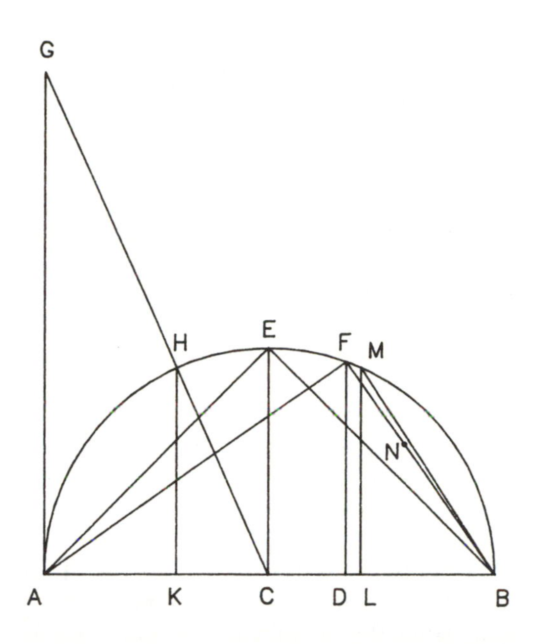

Points: C = midpoint; AD = 2·DB; GA = AB; line GC determines H,K; CK = CL; E,F,M are above C,D,L; N divides FB in EMR with NB being the larger segment.
Edges: tetrahedron, AF; octrahedron, BE; cube, BF; icosahedron, MB; dodecahedron, BN.

FIGURE I-37. Edges of the Five Regular Polyhedra (XIII,18)

Edge of the Icosahedron: As will be shown, the lengths LB and ML are, with respect to the defining pentagon (see the discussion on XIII,16[a]) a_{10} and a_6 respectively. Thus, by the Pythagorean Theorem and XIII,10, $MB = a_5$. But the side of the defining pentagon is also the edge of the icosahedron and so, as claimed, MB is the edge of the icosahedron.

To show that $LB = a_{10}$ and $ML = a_6$ one proceeds as follows: since $GA = 2 \cdot AC$ we have by similar triangles that $HK = 2 \cdot KC$ and consequently $S(AK) = 4 \cdot S(KC)$. Since both HC and CB are radii, the Pythagorean Theorem gives $S(CB) = S(CH) = 5 \cdot S(CK) = 5 \cdot S(CL)$. Thus since $AB = 2 \cdot CB$ and $KL = 2 \cdot CL$ we also have $S(AB) = 5 \cdot S(KL)$. But this is the same as relationship (i) of XIII,16[a],corollary which means that $KL = a_6$. We also learned in (ii) of XIII,16[a],corollary that $AB = a_6 + 2 \cdot a_{10}$ so that $AK = BL = a_{10}$ (thus the points K and L correspond to the points W and V of Fig. I-35f). Furthermore, we saw above that $HK = 2 \cdot KC = KL = a_6$ and, since L and K are equidistant from C, $ML = HK = a_6$.

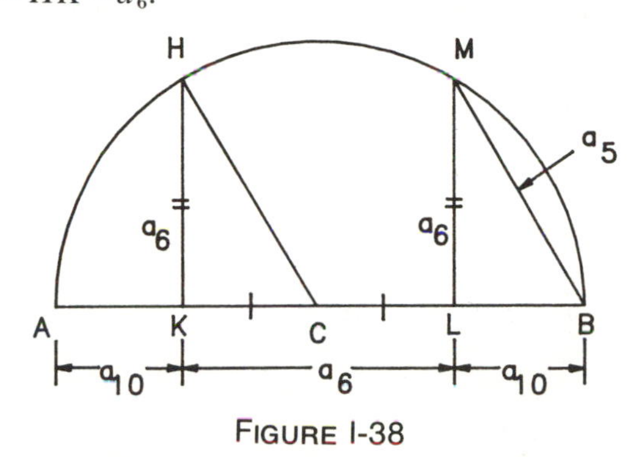

FIGURE I-38

Edge of the Cube: Just as in XIII,15 (the diagrams are identical except for the lettering), Euclid uses XIII,13′ (i) to show that $S(AB) = 3 \cdot S(BF)$ and then invokes XIII,15 to claim that BF is the side of the cube.

Edge of the Dodecahedron: From XIII,17,corollary, we know that when we divide the edge of the cube in EMR we obtain the edge of the dodecahedron. Thus by construction, BN is the edge of the dodecahedron.

The statement (1) that the edges of the icosahedron and the dodecahedron are not in rational ratio with any of the other edges follows directly from XIII,16[b] and XIII,17[b].

The proof of (2) above, which compares the sides of the icosahedron and the dodecahedron, uses XIII,13′ (i) and XIII,9 and involves a series of estimates.

Section 2. An Examination of the Euclidean Text

After having presented the development of DEMR as it appears in the *Elements*, I shall now examine the text in order to obtain a better understanding of the demonstrations that we are presented with and the relationship between them. Of particular interest is the chronological ordering of the various results. Part of my analysis will involve a consideration of certain flaws and lapses that exist in what, at first glance at least, seems to be a very smooth and polished demonstration. I will then draw some conclusions concerning the relationship between various results and the order in which they were discovered and proved.

I wish to emphasize that in this section I remain entirely within the confines of the Euclidean text. There is no drawing upon any other material such as an ancient text which indicates the state of the art at a certain time or an historical comment on some mathematician. In later sections I will present all this material and inferences obtained from them by other authors as well as myself.

It should be noted that in what follows a statement such as "Euclid showed . . ." is to be interpreted as referring either to Euclid or to the text from which he took his material. Although in some cases a proof may give indications of being original or only slightly reworked, we can never be quite sure how much editorial revision has taken place.

A. Preliminary Observations

(i) The first point we notice is that, having constructed the 36°−72°−72° triangle in IV,10, Euclid proceeds to use this triangle as a frame in the construction of the pentagon (IV,11), instead of using the 36° angle as the central angle of the decagon and then connecting alternate vertices as in Figure I-39.

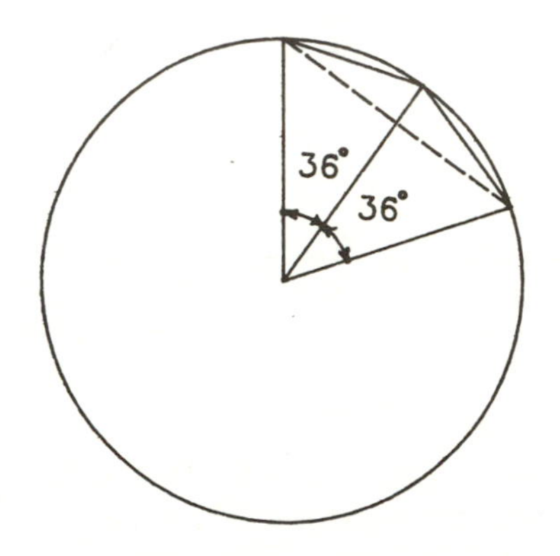

FIGURE I-39. Construction of the Pentagon from the Decagon

Indeed, if we look at the diagram accompanying IV,10 and simply consider the larger circle of radius AB as the given circle, then we see that the first step has actually been done, for BD is indeed the side of the decagon and there is no need to even transfer the triangle. Furthermore, the idea of using one regular figure to construct another appears in IV,16, although there the angle is bisected whereas here a doubling is required. This oversight has already been noted by several authors, such as Sachs [1917, 99].

One should not, however, be too hasty in drawing conclusions from this oversight. Note that in the proof of IV,10,11 there is no statement that the angle involved is one-tenth of a circle; everything depends on double angles and equal angles. Perhaps this is one of those "paradoxal failures" in the history of mathematical invention discussed by Hadamard [1945, 49], where the discoverer of a certain fact is completely unaware of a seemingly obvious immediate consequence of the discovery. Hadamard feels that most of these "failures" are due to "attention too narrowly directed."

I also note that this same sort of lack of recognition of the nature of what has been proved in IV,10 appears to be present in XIII,9 where the decagon appears once again. No formal construction is given in the *Elements*, but the idea of starting with one polygon and then using bisection to obtain another polygon appears in IV,16 so that once again the idea was available. Presumably the construction of the decagon was something to be done "in a like manner" (IV,16). In both of the diagrams accompanying IV,10 and XIII,9, the 36°−72°−72° triangle together with the bisection of one of the 72° angles appears. However, in IV,10 the principal 36° angle ($\sphericalangle BAD$) has its vertex at the centre of the circle, whereas in XIII,9 it is the 36° angle ($\sphericalangle BEC$) resulting from the bisection which appears at the centre.

We might expect somebody who was really aware of what was happening in IV,10 to prove XIII,9 using one of the following three approaches.

(1) Simply reverse the proof of IV,10 and use III,36 to show that the radius and the base of the triangle are in the relationship corresponding to DEMR and then use XIII,5 to add them together.

(2) Notice the bisection of the base angle in IV,10 and use a similar triangle argument based on this—without using the smaller circle—plus III,32 converse and III,36 to again prove that the side and base are in the relationship corresponding to DEMR and then use XIII,5 to add them together.

(3) Notice the bisection of the base angle in IV,10 and use this idea in a direct proof—rather than the round-about way of XIII,9—corresponding to the diagram of XIII,9 but which does not use XIII,9.

Again, what shall we make of this apparent lack of awareness of the possibilities of developing a proof based on the ideas inherent in IV,10?

(ii) In (i) I wondered why Euclid, once having constructed the 36°−72°−72° triangle, did not use it at the centre to construct the decagon. A related question is why he was led to the frame decomposition and not a decomposition involving ten 36°−54°−90° triangles about the centre, as is done in IV,13. The fact that this approach is used in IV,13, even though in this case it is the circle which is inscribed in the pentagon, shows that the idea was certainly available. I would suggest that the reason why this central angle construction was not used is that it would have required the construction of the 36°−54°−90° triangle. Of course this triangle can be constructed once the 36°−72°−72° triangle has been constructed, but it does not seem to be easily constructable from first principles. I note though that approximations to this triangle do appear in Babylonian and later Greek mathematics (cf. Sections 9 and 27).

(iii) There is another possibility for the construction of the pentagon and that is to use the 72° angle as a central angle. Once again this could be obtained from the 36°−72°−72° triangle. But notice the following about the proofs of IV,10 and IV,11; the actual magnitudes of the angles involved (one-fifth and two-fifths of the circle) are never mentioned. Everything depends on exactly what IV,10 states; we have one angle which is the double of another angle. Yet in IV,15 the central angle of the pentagon—one-fifth of the circle—is explicitly stated, and in IV,15 the central angle of the hexagon—one-third of two right angles—is also given.

Again we are led to the suggestion that the original development of the construction of the pentagon was done in the manner shown; the mathematician involved simply had all his efforts concentrated on the frame defined by the isosceles triangle of IV,10, and he was led to a construction in which the bisection of the doubled angles gave equal angles, with the actual magnitudes of the arcs or central angles simply playing no role.

B. *A Proposal Concerning the Origin of DEMR*

(i) While the observation of the double angle may have been the starting point for the construction of the pentagon, the proof of IV,10 is certainly non-transparent.

If we examine IV,10 we see that the principal results used in the proof are:

(1) Theorem III,32 which gives a tangent angle ($\sphericalangle BDC$) equal to an inscribed angle ($\sphericalangle DAC$).

(2) Theorem III,37 which shows that a line (BD) is tangent to a circle.

As far as III,32 is concerned, it appears in the Euclidean text as simply one of several results in Book III that have to do with the inscription of angles in circles. We note in passing that III,32 is also used for the inscription of a triangle in a circle (IV,2). Perhaps the origins, either at the informal or formal stage, were connected with this inscription problem, or perhaps its origin lies in some general study—again formal or informal—of the relationship between triangles and circles. In any case, there is no indication that III,32 was somehow originally connected with the origins of DEMR.

On the other hand, the result III,37 and its converse III,36 are of the greatest interest to us. III,36 and 37 and their relationship to IV,10 will now be considered in detail. In particular I will develop my thesis that III,36 and 37 arose out of the successful attempt (preserved in Euclid IV,10,11) to construct the pentagon and, furthermore, that the concept of DEMR, as presented in II,11, is merely a side product of III,36 and 37 and that II,6 is nothing other than the lemma that was proved as a preliminary to the proofs of III,36 and 37 and II,11. These assertions go against various statements in the literature, which will be discussed at several points in this book.

The first thing that we must look at is the use made of various results starting from II,6 and also starting from II,5, whose format and proof is very similar to that of II,6. This is shown schematically in Figure I-40. The lack of an arrow emanating from a result indicates that the result is not used in the *Elements*. Of course this does not, a priori, exclude the result having some source or use exterior to the *Elements*. Indeed, it is widely argued that II,5 and 6 were developed in connection with the solutions of equations (I shall discuss this in Section 5).

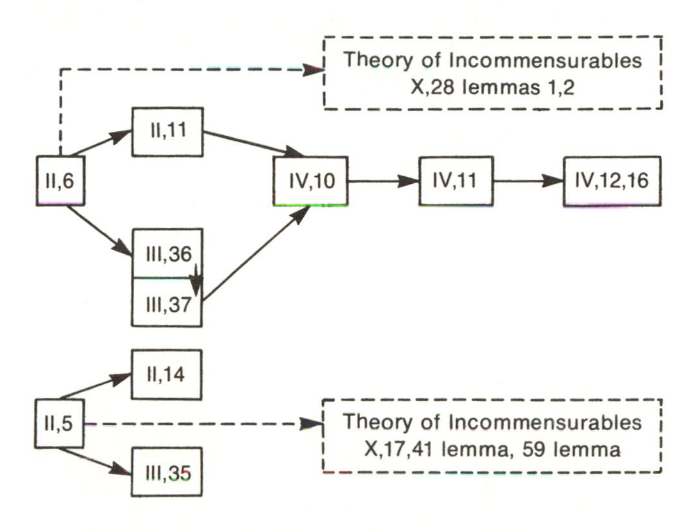

FIGURE I-40. Theorem Paths from II,5,6

I shall leave out of the discussions consideration of the results of Book X involving the theory of incommensurables. Book X uses the theory of proportions of Book V, whereas IV,10 uses the area definition of DEMR in II,11. This suggests, at least at first glance, that the development of Book X was later than that of the other results in the figure. This first glance may be deceiving and indeed it has been suggested that the results of Book II came out of the development of the theory of incommensurables (see Section 13,B). I shall, however, proceed with my discussion on the assumption that II,5 and 6 did not arise out of the development of the theory of incommensurables.

If we leave out Book X, we see that II,6 splits off into two paths—II,11 and the pair III,36 and 37—which in turn converge to a unique point, which is the construction of the pentagon in IV,10. Theorem II,5 leads only to the transformation of areas theorem II,14 and to III,35, which is very close in spirit to III,36 and 37 and in fact can be thought of as an "internal version" of III,36.

(ii) We thus turn our attention to the proof in IV,10. Here we are immediately faced with a contextual problem and that is whether we should look at the construction of the pentagon as given in IV,10 from the viewpoint of somebody who is familiar with the theory of similar triangles. The formal theory of similar triangles is only presented in Book VI because it uses the theory of proportions as developed in Book V. However, just as with the case of Book X mentioned in (i), this may be deceiving because it is possible that a theory of similar triangles existed in some cruder form at an earlier date. If an original construction of the pentagon was based on this "theory" of similar triangles, then what we may be presented with in the Euclidean text is the translation of that earlier construction into an area formulation. In this way the construction could be placed ahead of the theory of proportions and similar triangles.

For the purpose of discussion I have given the proofs of both III,36 and IV,10 converse in the language of similar triangles. My reason for presenting the converse of IV,10 rather than IV,10 itself is that the latter is a construction rather than an analysis and this obscures the matter.

From a similar triangle viewpoint, as the proofs indicate, IV,10 appears to be nothing more than the special case of III,36 and 37 for which the size of the chord AC is fixed by an additional condition, namely, the one imposed on the angles.

(iii) Despite the possibility of a similar triangles origin, I believe that the area formulation proof given by Euclid is essentially the original. There are several reasons for believing this, even at this point in the discussion—namely, the inelegant use of the triangle of IV,10; the fact that the actual angle sizes are not mentioned; and the double path from II,6 to IV,10.

Indeed the overall impression obtained from reading in full detail the development as presented is that we are reading an original work by a powerful mathematician who has invented the tools (III,36 and 37) needed to push through the original idea (one angle twice another) but who has not been able to stand back far enough to give it some final refinements.

(iv) Consider then the following possible scheme for the development of DEMR:

(1) Desire to construct the regular pentagon, perhaps as part of a program to rigorously construct as many of the regular polygons as possible.

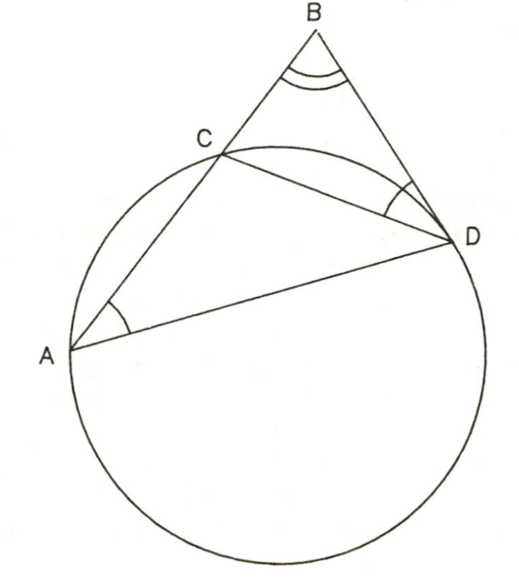

Given: BD tangent
To show: R(AB,BC) = S(BD)
1) ∡BAD = ∡BDC (III,32)
2) ∡ABD = ∡DBC
3) ΔABD ≈ DBC (1,2)
4) R(AB,BC) = S(BD) (area formulation of the proportionality of the sides)

III,36
(oriented and labelled as in IV,10)

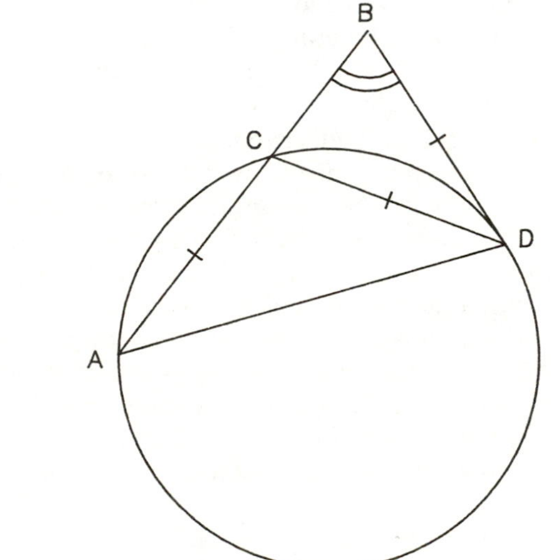

Given: BD tangent, ∡ABD = ∡BDA = 2 · ∡BAD
To show: R(AB,BC) = S(AC)
1), 2), 3), 4) as per III,36 (opposite)
5) ∡BDC = ∡CAD = ∡CDA (1,hyp.)
6) ∡CBD = ∡BCD (4,hyp.)
7) BD = CD (6)
8) BD = CD = AC (7,5)
9) R(AB,BC) = S(AC) (5,8)

IV,10, Converse
(labelled as in Euclid IV,10)

FIGURE I-41

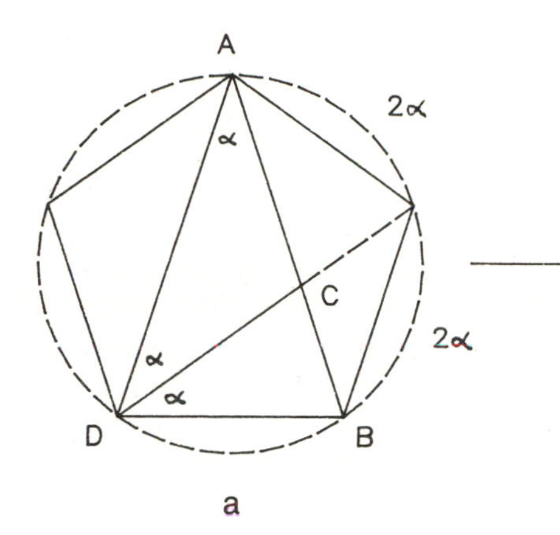

"double angle"

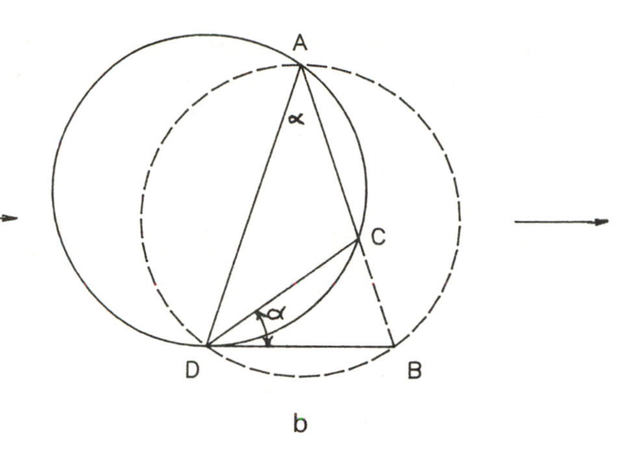

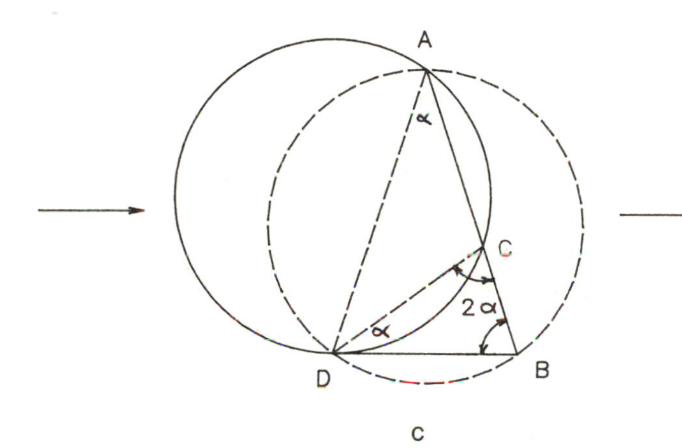

AC=DC=DB; what is the relationship between AC, CB, DB ?

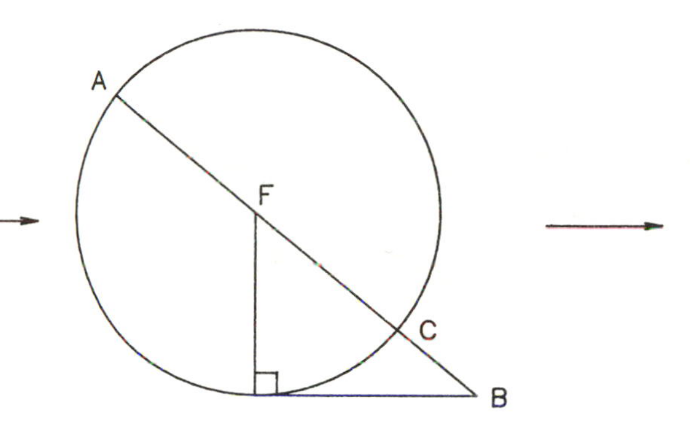

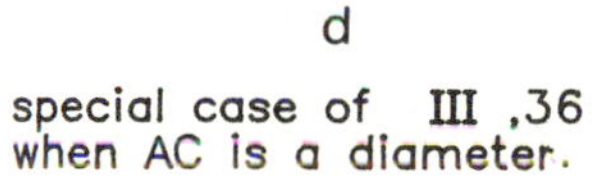

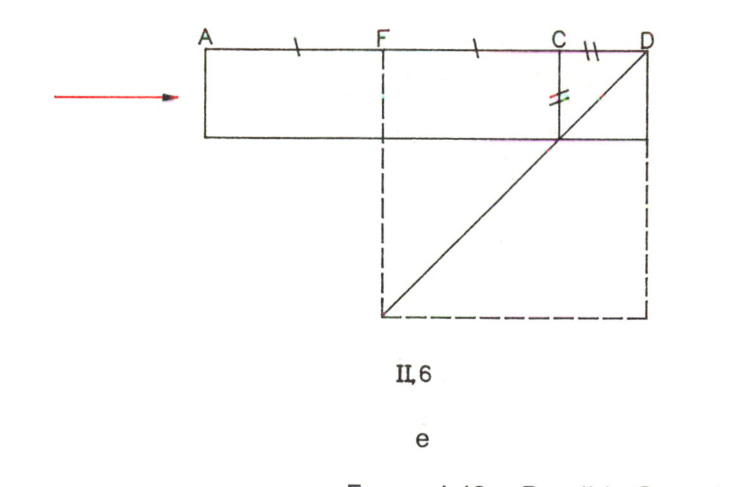

FIGURE I-42. Possible Genesis of III,36 and II,6

(2) (Fig. I-42a.) Attention fixed on the repetition of angle α (equal to one-tenth of a circumference). Observation that angle ADB involves a doubling of α. The mathematician(s) involved had a very strong feel for the "angles in a circle" theorems now preserved as Euclid's III,20, 21, 26, 27, and 32. There may have been an initial attempt to work with central triangles as in IV,13.

(3) (Fig. I-42b.) The equality of angles DAB and CDB suggests the set-up of III,32 and leads to the circumscription of a circle about $\triangle ACD$. Note that, because of the repetition of angles in the frame of Figure I-42a, the circumscribed circle will have the same diameter as the original circle.
(4) (Fig. I-42c.) The equality of the angles determines isosceles triangles and shows that $AC = DC = DB$.
(5) (Fig. I-42c.) Attempt to locate the point C using the relationship of (4) and BD being tangent to the circle.
(6) (Fig. I-42d,e.) Phase (5) leads to III,36. Case 1 of III,36, where AC is a diameter, suggests the format of II,6 which is needed to prove III,36.
(7) Substitution of the relations of (4) in III,36—$R(AB,CB) = S(DB)$—leads to the desired location of C, that is, $R(AB,CB) = S(AC)$.
(8) Equality of AD and AB in Figures I-42b,c suggests the use of a big circle with centre at A in the formal proof (i.e., in the diagram of IV,11).

(v) There remains the question of II,11. Once having arrived at the requirement $R(AB,BC) = S(AC)$ in the development of IV,10, there still remained the problem of actually constructing the required division point. Referring to the proof and accompanying diagram of II,11, we see that the key step is the drawing of the line EB where E is the midpoint of AC. It is this step which is, to say the least, not transparent. How did the creator of this construction hit upon the idea? I would like to suggest the following sequence as possible steps in the development of the proof:
(1) (Fig. I-43a.) The problem is to find point H or point F so that $\mathrm{I} \equiv S(AH) = R(AB,BH) \equiv \mathrm{II}$.
(2) (Fig. I-43b.) To simplify the problem, by avoiding

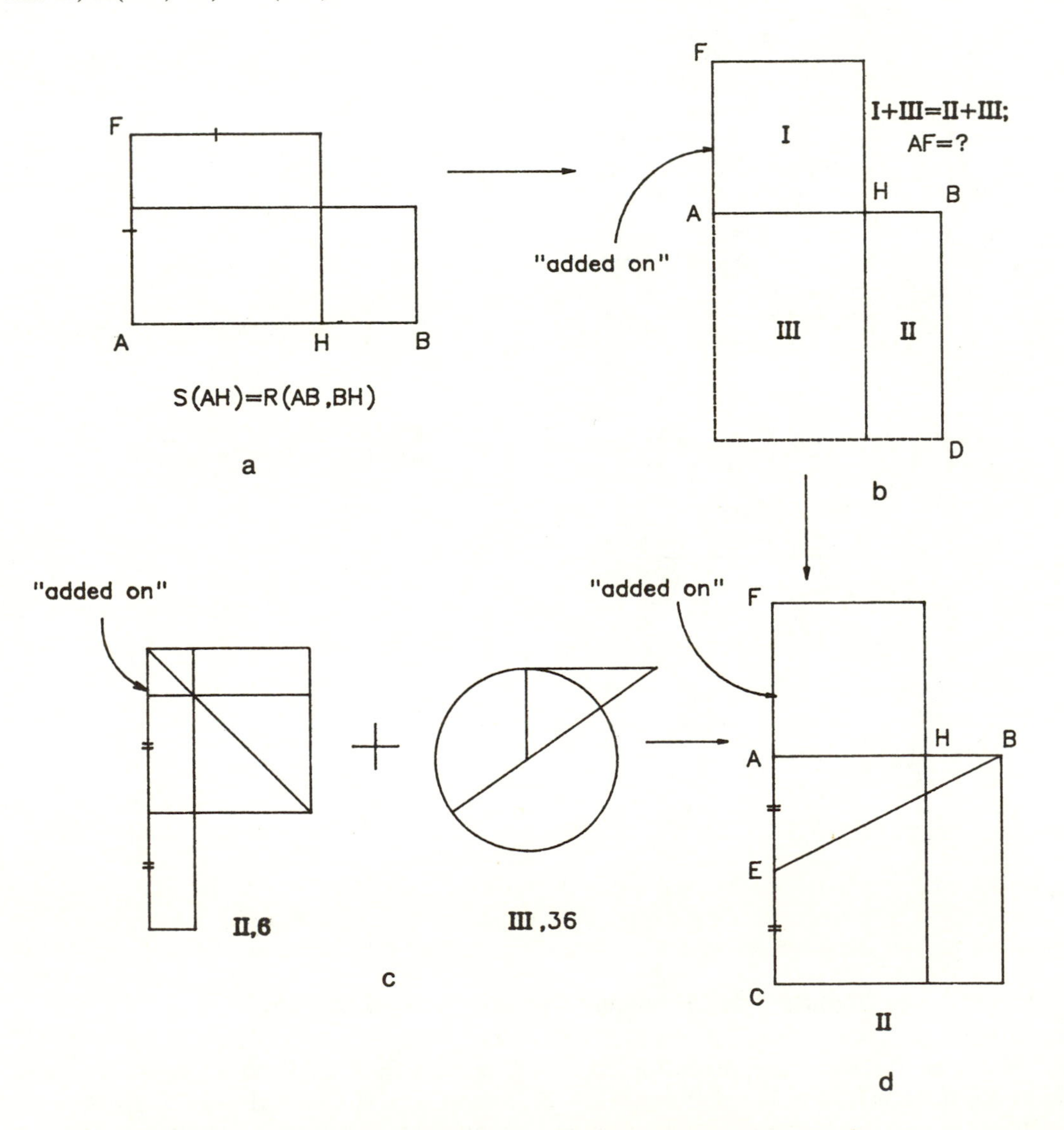

FIGURE I-43. Possible Genesis of II,11

the overlap of square I and rectangle II, the latter is rotated and moved down so that its base is *HB*. This in turn suggests "completing" the figure by the addition of rectangle III.

(3) (Fig. I-43c,d.) Now let us look at the construction of II,11 once again. Notice the division of line *AC* into two parts and the addition to *AC* of the segment *AF*. This is in fact the set-up of II,6. Notice also the right triangle *BAE* in Figure I-43d and recall the importance of the Pythagorean Theorem in the proof of III,36. I have already argued above that III,36 was developed for the construction of the pentagon and that this in turn led to II,6. Thus it was perhaps this development of III,36 and II,6 which was the inspiration, most probably subconscious, for the splendid move of drawing *EB*.

The conclusion that II,6 and II,11 arose out of the pentagon construction problem also leads us to conclude that Theorems II,1-10 cannot be considered to form a unity. Rather, at least some of the results are to be considered as lemmas originally proved in connection with other results. In view of the formal structure of the *Elements*, the lumping together of different theorems dealing with the "geometry of areas" in a single book has nothing surprising about it.

In making this suggestion about the origin of II,11, I have made the implicit assumption that the geometric version of DEMR, as given in II,11, came first and that the proportion definition and construction of VI, def.3 and VI,30 were added when the formal theory of proportion was developed. It is possible that VI,def.3 and VI,30 came first and that II,11 was added so that the construction of the pentagon could be given in Book IV. Indeed Mueller [1981, 169] considers the construction of II,11 to simply be a reworking of the proof of VI,30. It seems to me, however, that the relatively crude format of IV,10 and the use of II,6—via III,36—as well as II,11 argue against a backward move from proportion theory to an "area" development. Furthermore, we can ask why Euclid would not have avoided much redevelopment simply by placing Books III and IV, with perhaps some changes in terminology, after Books V and VI.

C. *Theorem XIII,8*

We now turn our attention to XIII,8 which tells us not only that the diagonals of a pentagon cut each other in DEMR but also that the larger of the two segments is equal to the side of the pentagon. We shall see later on (Sections 12 (viii), (ix)) that some authors assume that this result was known at the time or even before the time that the construction of IV,10,11 took place. I shall argue that, on the contrary, XIII,8 was not even known intuitively at the time of the discovery of IV, 10 and 11. My starting point is the observation that the proof of II,11 contains, without any further remarks and without any change in the order of the proof being necessary, the following result.

THEOREM II,11′ (extension version of II,11). To construct a line *AL*, on a given line *AB*, so that $S(AB) = R(AL,BL)$.

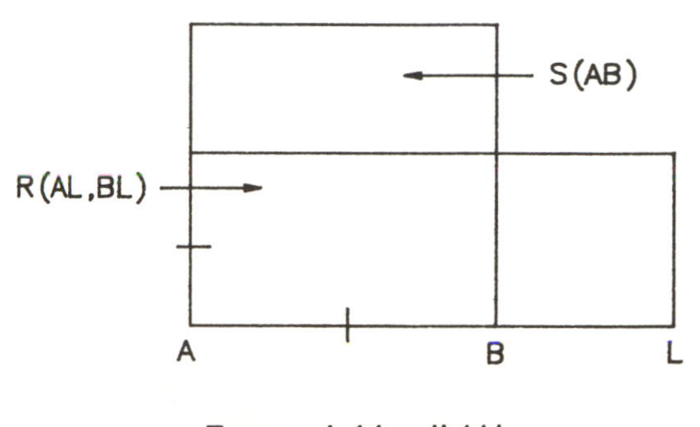

FIGURE I-44. II,11′

Indeed, the desired construction is completed at the line marked (*), in the proof of II,11 in which we can either think of *CA* as the line and *FA* as the extension or simply transfer distance *CF* to form line *AB* extended.

This result ties in with XIII,8 where we could think of the side of the pentagon as the given segment in II,11′. The extension of the side given by II,11′ is then the diagonal of the pentagon. This relationship between XIII,8 and II,11′ suggests a different proof of IV,11. Not only is this new proof shorter and more direct than the extant proofs of IV,10,11; as well, it only uses results already used in IV,10 and 11.

THEOREM IV,11 (alternate proof). To inscribe a regular pentagon in a given circle.

Proof:

(1) (Fig. I-45a,b.) Draw the isosceles triangle $(36°-36°-108°)$ whose sides are in the ratio s,s,d where s is arbitrary and d is obtained from II,11′. In the given circle draw the triangle *ABE* similar to the s,s,d triangle.

(2) (Fig. I-45b.) Next obtain *H* on *EB* with $HE = AB$. Point *C* is obtained by extension of *AH*. Now circumscribe a circle about $\triangle AHE$. By the relationship between s and d (II,11′) and III,37 we have that *AB* is tangent to the circle. Then by III,32 $\angle BEA \equiv \angle HEA = \angle HAB \equiv \angle CAB$. This in turn implies, via III,26, that arcs *AB* and *BC* are equal. Thus we have from III,28 the equality of arcs *AE*, *AB*, and *BC*.

(3) (Fig. I-45c.) Since $HE = AE$ we have from I,32 and III,27 that $\angle EAC \equiv \angle EAH = \angle AHE = \angle HAB + \angle ABH = 2\angle ABH$. If we now obtain *D* by bisection of arc *EC*, all the sides and angles of the pentagon will be equal.

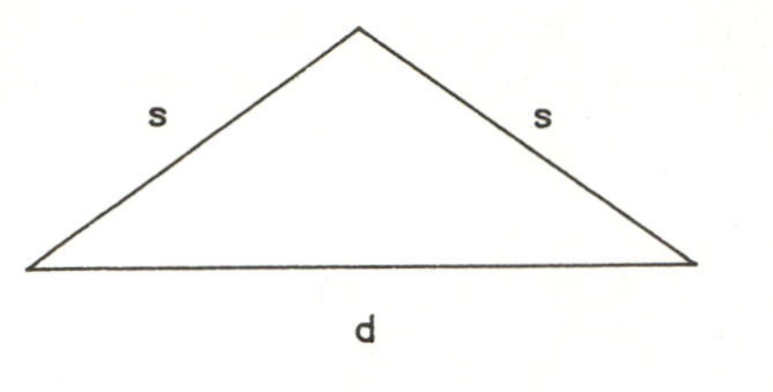

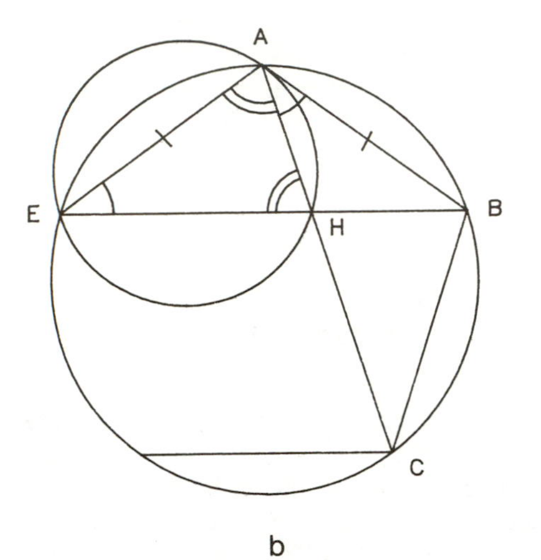

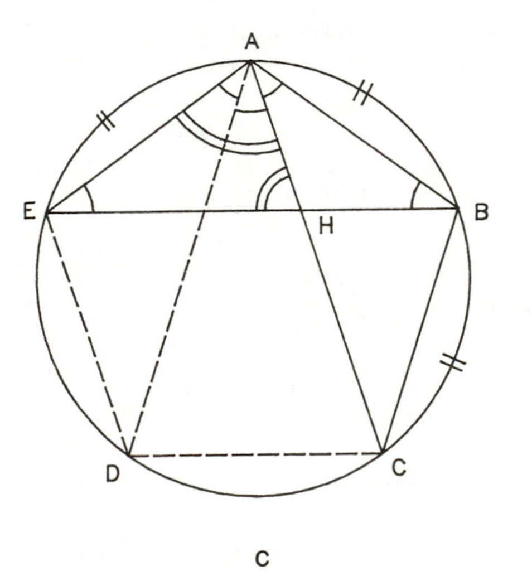

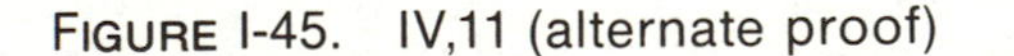

FIGURE I-45. IV,11 (alternate proof)

If XIII,8 were known when the pentagon was constructed, we could reasonably expect to see a simplified construction of the pentagon, which displayed evidence of the result of XIII,8. Of course, we might tend to counter this suggestion with the statement that perhaps the result of XIII,8 was known only intuitively.

This counter-argument is not valid. As I will now show, it is possible to give a proof—indeed based on the above proof of IV,11 in reverse order—of the area version of XIII,8 which once again only uses results already used in IV,10 and 11.

THEOREM XIII,8′ (area version of XIII,8). In a regular pentagon the diagonals cut one another so that the square on the larger segment is equal to the rectangle contained by the diagonal and the smaller segment. Further, the larger segments are equal to the sides of the pentagon.

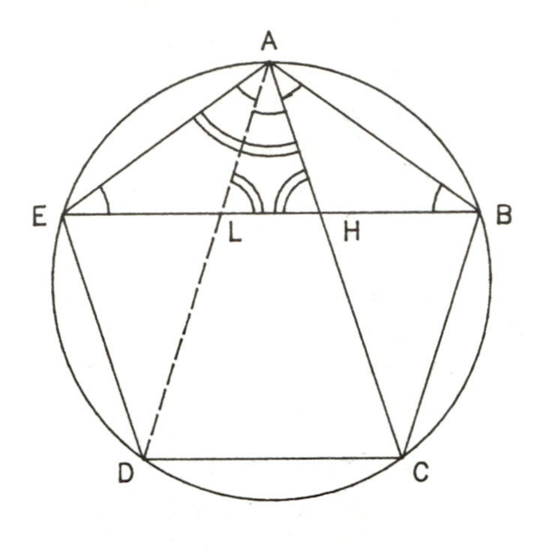

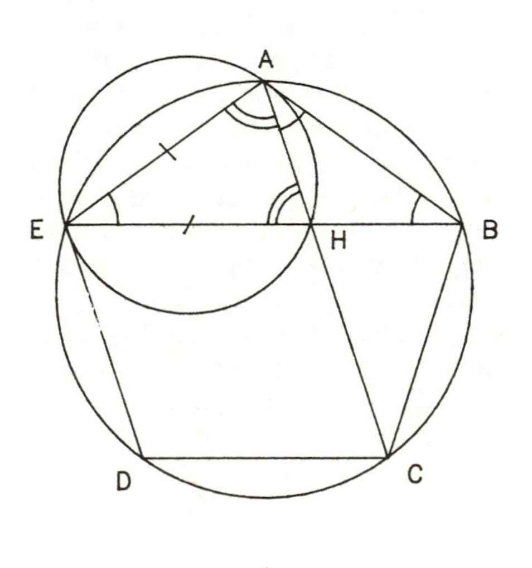

FIGURE I-46. XIII,8′

Proof:

(1) (Fig. I-46a.) Draw diagonals AD and AC. Since the sides of the pentagon are equal, III,28 tells us that the arcs are equal. Thus by III,27 $\measuredangle ABH \equiv \measuredangle ABE = \measuredangle DAC = \measuredangle EAD = \measuredangle BAC \equiv \measuredangle BAH$. Therefore

(i) $$\measuredangle EAH \equiv \measuredangle EAC = \measuredangle EAD + \measuredangle DAC = 2 \cdot \measuredangle BAC \equiv 2 \cdot \measuredangle BAH.$$

(Notice how we are able to use III,27, which talks about equality of arcs, instead of VI,33 as in the Euclidean proof of XIII,8, which talks about ratios of arcs, because we are dealing with integer multiples. The only other uses of VI,33 are in XIII,9 where we have four times an arc, and XIII,10 where we have two times an arc. Thus once again III,27 could be used. In addition, the first part of the proof should be compared with Euclid's proof of XIII,8. Instead of simply using III,28 and 27 as I have done, Euclid had recourse to congruent triangles. These details were overlooked in Fischler [1979a].)

But also, by I,32

(ii) $\measuredangle AHE = \measuredangle BAH + \measuredangle ABH = 2\cdot\measuredangle BAH$.

From (i) and (ii) $\measuredangle AHE = 2\cdot\measuredangle BAH = \measuredangle EAH$. This shows that $\triangle EAH$ is isosceles which in turn implies that $HE = EA = AB$, the side of the pentagon.

(2) (Fig. I-46b.) Now circumscribe a circle about $\triangle EAH$. Since $\measuredangle HAB = \measuredangle HEA$, the converse of III,32 (proved from III,32 by reductio ad absurdum) tells us that AB is tangent to this smaller circle. Then by III,36 $R(BE,HB) = S(AB) = S(EH)$.

(3) It remains to be shown that EH is indeed larger than HB. To see this, note that EH is greater than EL which in turn is equal to AL because $\measuredangle EAL = \measuredangle AEL$. Again, since exterior angle $ALH = 2\cdot\measuredangle EAL = 2\cdot\measuredangle EAD = 2\cdot\measuredangle BAH = \measuredangle AHE = \measuredangle AHL$, we have that $\triangle LAH$ is isosceles so that $AL = AH$. But $AH = HB$ since, in $\triangle HAB$, $\measuredangle BAH = \measuredangle HBA$. In other words, EH is greater than $EL = AL = AH = HB$. (Note that in VI,33 Euclid seems to have taken for granted that III,27 can be extended to the greater than case. If we allowed this and used III,29, then by working with $\triangle EAH$ inscribed in the circle, we would have EH greater than AH because $\measuredangle EAH > \measuredangle AEH$.)

To summarize: Euclid has preserved the inefficient path, II,11 → IV,10 → IV,11 for the construction of the regular pentagon instead of the path II,11′ → IV,11 which could have been developed in exactly the same mathematical context. Furthermore, II,11′ is contained in II,11 and its significance was available to anyone familiar with XIII,8, which is itself provable using only the tools that are used in the *Elements* to prove IV,10 and 11. Therefore there is reason to conclude that XIII,8, even in intuitive form, is of a later date than the construction of the pentagon IV,10,11.

In Section 20,C(iii) we will see that our conclusion of a later discovery of XIII,8, contrary to what one might expect, is not at all surprising when viewed within the total context of the development of DEMR.

D. Theorems XIII,1-5

If we examine the proofs of XIII,1-5 we see that none of the results from Book II is ever used (a possible exception is a lemma to XIII,2 which Heiberg [Euclid-Heiberg, IV, 223] does not consider genuine). Thus, whenever the area version of DEMR (the statement of II,11) is needed, VI,17 is used to make the conversion from the proportion statement of VI,def.3. In XIII,5 this operation is performed in both directions whereas, as an alternate proof in a manuscript shows [Euclid-Heath, III,449], this can be done within the proportion framework of Book VI. Alternate proofs of XIII,1-5 based on results of Book II can also be given (see, for example, Euclid-Heath).

What are we to make of this? Looking at the proofs, we see that they are very straightforward and based on the addition of the rectangles and squares. All five results use two basic techniques: translation of VI,def.3 into essentially the statement of II,11 by means of VI,17, as noted above; and the use of I,43 which implies that the complements of certain squares are equal. Then, aside from the use of V,14 in XIII,2 and 5 to obtain an inequality from a proportion, nothing else is used except VI,1 in XIII,1 and 2 and I,10 (bisection of a line) in XIII,1 (however, see above on XIII,2 lemma which uses II,4).

Thus, given the very elementary nature of the proofs, I suggest that the mathematician(s) who developed these theorems simply sat down to prove these results as lemmas when they were needed later on in XIII (XIII,2 is not used but is essentially the converse of XIII,1). Indeed it may be that, when XIII,1-5 were written down, Book II as we know it may not have existed and some of the theorems may not have been known or written down. With respect to this we should not forget that the proofs of II,4, 5, 6, 7, and 8 are themselves based on the method of complements. Furthermore, as I shall point out in more detail in Section 5, the method of application of areas as seen in VI,27, 28, and 29 is also basically a complements theory.

In Section 14 I will summarize the views of various authors on XIII,1-5. In Sections 20,C(iii) and 22, I will return to the question of XIII,1-5.

E. Stages in the Development of DEMR in Book XIII

In view of my conclusions that XIII,8 is of a later date than the construction of the pentagon and that XIII,1-5 simply represent lemmas which were proved specifically for the results of Book XIII, I believe that Book XIII developed in three stages as far as DEMR is concerned (Fig. I-47): (1) the constructive parts 16[*a*], 17[*a*] of XIII,16,17 (icosahedron and dodecahedron) and their lemmas XIII,3, 9 and 10 and XIII,4 and 5, respectively; (2) the classification part 17[*b*] of XIII,17 and its lemmas XIII,1 and 6 (apotome—one of the original irrationals); (3) the classification part 16[*b*] of XIII,16 and its lemmas XIII,8 and 11 (minor, one of the new irrationals)—XIII,11 in turn uses XIII,1 as did XIII,6 in (2).

In Sections 20,C(iii) and 22, I will combine the conclusions obtained in this section with various historical conclusions to present a more complete view of the historical development of DEMR.

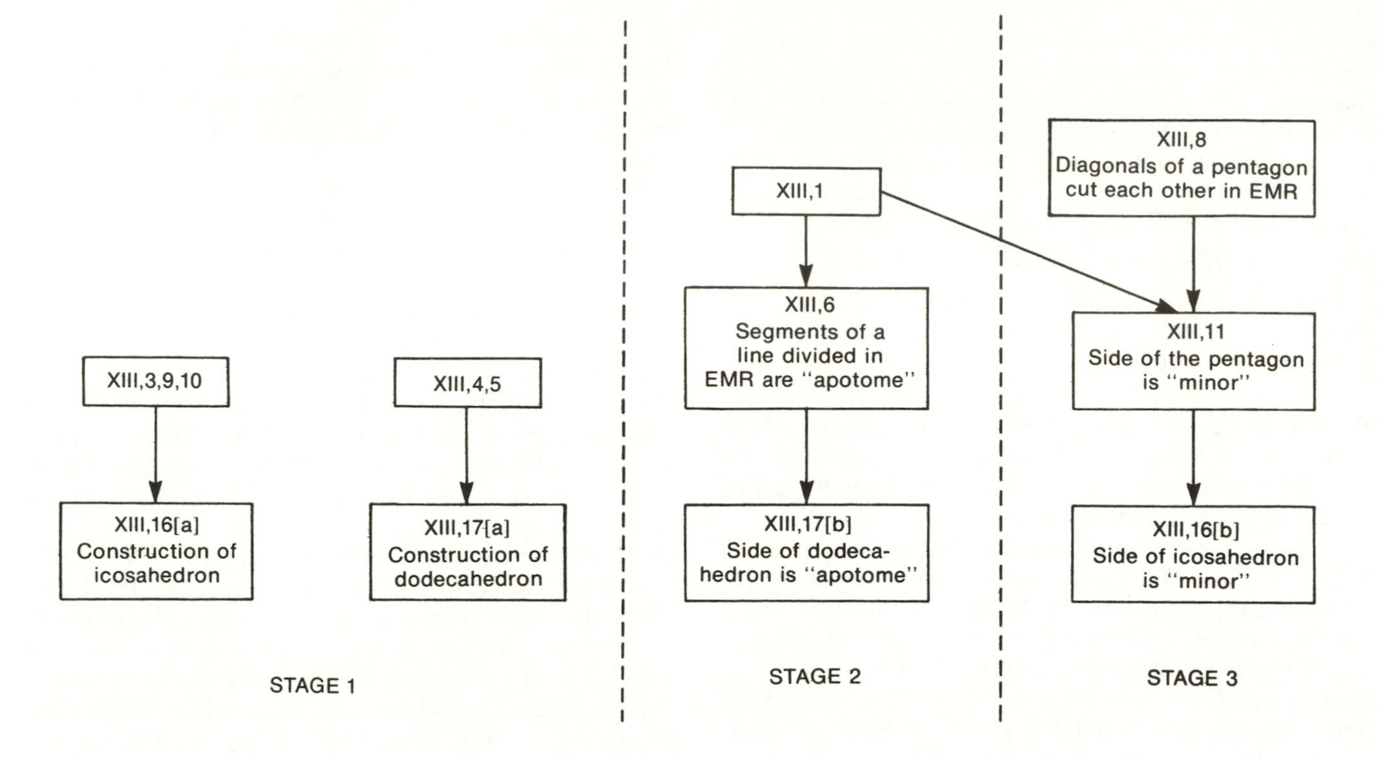

FIGURE I-47. The Three Stages in the Development of DEMR in Book XIII

CHAPTER II

MATHEMATICAL TOPICS

In this chapter I discuss several mathematical topics which either are directly related to DEMR or have been associated with DEMR in the literature. For the discussions of complements, gnomon, and application of areas in Sections 3 and 4, I thought it best to give the proofs of the key results here, rather than in Section 1, so that the true nature and meaning of these results and their relationship to DEMR in the *Elements* could best be brought out. Section 5, which investigates the topic of "geometrical algebra," is a multifaceted discussion of what in reality is a group of related topics. In the twelve subsections I not only summarize the debate and various theories but also bring in new evidence, particularly in Subsections F through I where I discuss problems in interpretation, division of figures, and Euclid's *Data*. For the last three topics—side and diagonal numbers, incommensurability, and anthyphairesis—I could not possibly expect to do complete justice to such vast subjects, and so I have simply tried to bring out the salient features and give the key quotations while at the same time giving references to the literature.

Section 3. Complements and the Gnomon

We start with a parallelogram *ABCD* and draw the diagonal *AC* as well as two lines *EF* and *GH* parallel to the sides of the parallelogram and intersecting at point *K* on the diagonal. The original parallelogram is now split into the two parallelograms II and IV which are similar to the original plus the parallelograms I and III. These latter regions are called the complements, as we learn in I,43, which also states that the complements have equal area.

This result, despite the fact that it is used to prove some very important and, at first glance, deep theorems, is extremely easy to prove.

THEOREM I,43. The complements I and III have equal area.

Proof: The diagonal *AC* divides the big parallelogram *ABCD* and the two subparallelograms *AFKG* and *KHCE* into two pairs of triangles having equal areas. If we subtract triangles on each side of the diagonal, the remaining regions I and III must also have equal area.

If we take the two complements together with one of the parallelograms, say II, about the diagonal we obtain the L-shaped region I + II + III called the gnomon (II,def.2). In other words, the gnomon is what is left of the original parallelogram when we take away the similar parallelogram IV.

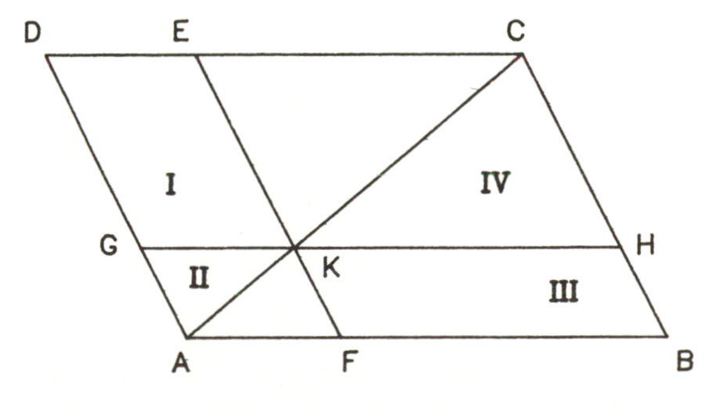

FIGURE II-1. Complements and Gnomon

Typically the gnomon acts as a sort of intermediary region in proofs in that rectangles or parallelograms of interest are shown to have their areas equal to that of the gnomon defined by the construction. In a sense we could call DEMR, as presented to us in Euclid, a complements-gnomon oriented theory, for the use of complements and the gnomon is the main tool in all the results—II,6, VI,29, XIII,1,2,3,4—which in turn play a key role in the DEMR theorems—II,11, VI,30, and XIII,16,17.

Section 4. Transformation of Areas

The term "transformation of areas" is used in this book as a general term for a theorem or problem in which one "transforms" one figure into another which has the same area, but which in addition meets certain conditions pertaining to its shape.

An example of this transformation is given in I,42 where we are to construct a parallelogram, one of whose angles is given, with an area equal to that of a given triangle. In I,45 the triangle is replaced by a rectilineal figure. Then in II,14 a given rectilineal figure must be transformed into a square. In VI,25 a rectilineal figure is transformed into another rectilineal figure which is similar to a given rectilineal figure.

There is a group of problems that from a strictly logical viewpoint can be considered as special cases of the general transformation of areas problem. These are the so-called application of areas problems where there is the added proviso that the resulting figure is a parallelogram which must have one of its sides sitting on—but not necessarily coincident with—a given straight line. Thus I,44 is the application of areas version of I,42 in which the side of the parallelogram actually coincides with the given line. In VI,28 and VI,29 we find application of areas problems involving "defects" and "excess," that is, the side of the parallelogram does not coincide with the given line.

One reason why I wish to distinguish the application of areas problems from the general transformation of areas problem is that it is far from clear that the former developed historically just as a special case of the latter. The whole question of applications of areas will be discussed in detail, and not just from the transformation of areas viewpoint, in the next section.

To fix ideas and to show the spirit of the approach taken, I will indicate here the steps, including those of I,42 and I,44, which constitute the proof of I,45.

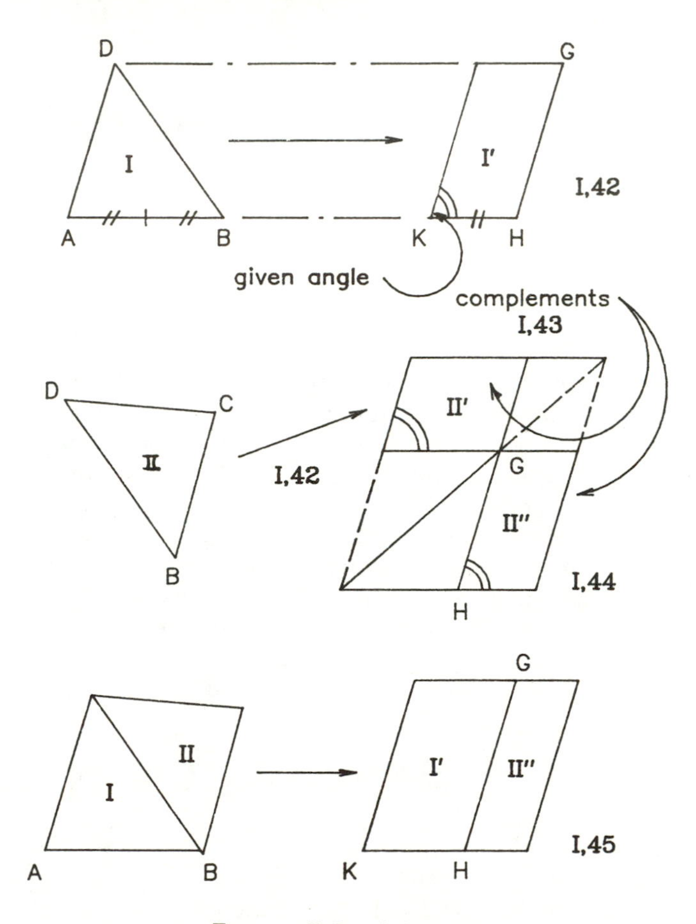

FIGURE II-2. I,42,44,45

THEOREM I,45. To construct a parallelogram, one of whose angles is given with an area equal to that of a given rectilineal figure.

Sketch of Proof: Suppose that the rectilineal region is made up of the two triangles I and II. If we construct a parallelogram I′ whose altitude is the same as that of I, whose base is half that of the base of I and such that one angle is the same as the given angle, then I′ will have the same area as I. This is the essence of the proof of I,42.

Next use I,42 to construct parallelogram II′ equal to triangle II, and place II′ on the extension of *GH*. Now comes one of those super-slick moves that brings joy to the reader of the *Elements* and makes the reader forget all of the drudgery of everyday life (and of reading a lot of the rest of Euclid). One simply draws the parallelogram indicated and invokes humble Theorem I,43 to say that parallelogram II″, which sits on *GH*, is the complement of II′ and therefore has the same area. This is the essence of the proof of I,44 where we are required to apply a triangle (II) to a given line (*GH*). By sticking I′ and II′ back to back on *GH*, the construction is completed.

Did the original idea for the application of areas problem (I,44) arise out of the transformation of areas problem in the way indicated in the *Elements*? We simply do not know, although the impressive proof of I,44 and deft mixing of I,42 and I,44 to obtain I,45 certainly suggest this as a possibility. We shall see other suggestions for the origin of application of areas in the next section.

There are other examples of application of areas theorems from early Greek mathematics. Hippocrates of Chios (late vth century) is said to have obtained certain results involving the quadrature of lunes (the moon-shaped region between semi-circles on different line segments). One theorem ascribed to him asserts that, if one inscribes a right isosceles triangle in a circle, then the lunes formed by the semi-circles on the side and the original semi-circle have an area equal to that of the triangle. The triangle in turn can be converted into a square (II,14). For texts, history, and commentary related to the work of Hippocrates on these quadrature problems see Van der Waerden [1954, 131], Heath [1921, 183], Freeman [1946, 218], and Diels [1934, I, 395].

The work of Hippocrates related to the lunes was part of ancient attempts at the "squaring of the circle"—the finding of a square whose area was that of a given circle (see the above references). Proclus writing in the 5th century says in his commentary on I,45: "It is my opinion that this problem is what led the ancients to attempt the squaring of the circle" [Proclus-Morrow, 335]. Of course, as Proclus himself says, the source of the quadrature of the circle problem is uncertain just as is the origin of the other transformation of areas problems I,42,45. With respect to this I note the articles by Seidenberg [1960; 1977], where it is suggested that transformation of areas problems possibly originated from old Vedic mathematics problems associated with the shapes of altars.

Section 5. Geometrical Algebra, Application of Areas, and Solutions of Equations

To introduce the concepts of this section I start with the following brief description of the first two topics mentioned in the title; further refinement will be necessary because of the loose ways in which these terms are often applied.

"Geometric algebra" is the term sometimes given to those results, in particular Euclid II,1-10, which can be interpreted as "algebra in geometric formulation." "Application of areas" refers to the problem of constructing, subject to certain constraints, on a given straight line a parallelogram whose area is the same as that of a given figure.

Our particular interest in these concepts will be the association that has been made between them and II,11 (the area formulation of DEMR) and with the dual pair II,6 (used in II,11, III,36) and II,5.

The association of certain geometrical results found in Euclid with what are generally considered to be algebraic results was made by Zeuthen and Tannery at the end of the 19th century. Biographical details concerning these two authors as well as their predecessors, supporters, and critics are given by Unguru [1975]. Indeed Unguru [1975; 1979] has strongly criticized the geometrical algebra theory. In turn he has been responded to by Van der Waerden [1975], Freudenthal [1976], and Weil [1978]. Additional comments on Unguru's paper are made by Seidenberg [1977]; see also the discussion of Mueller [1981, 41].

Much of the recent polemic has been rather strongly worded and of a type not usually seen in such august circles. It seems to me that much of the discussion involves questions of semantics and the definition of algebra. I shall try to avoid such problems by distinguishing different levels of geometric algebra and by concentrating on one aspect about which the lines of dispute are very clearly drawn and which is of central importance to our discussion: Did certain results in the *Elements* develop out of procedures for solving equations?

A. *Geometrical Algebra—Level 1*

At the simplest level the geometrical algebra point of view considers II,1-10 as representing what present-day mathematicians would call algebraic identities. This may be illustrated by II,1, where the line segments have been assigned letters which represent their lengths.

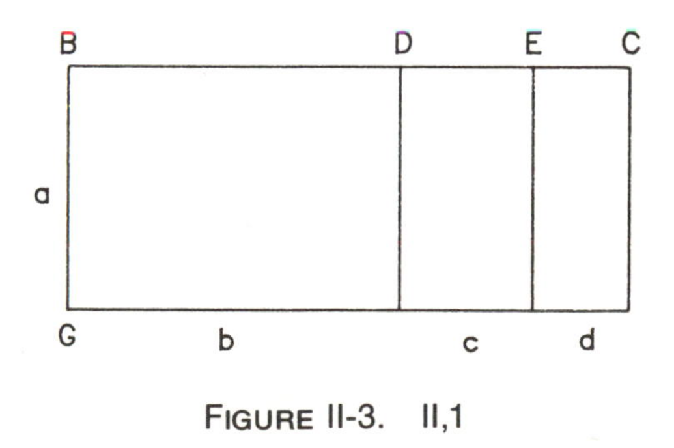

FIGURE II-3. II,1

The statement $R(BG,BC) = R(BG,BD) + R(BG,DE) + R(BG,EC)$ of II,1 then becomes in modern symbology $a(b+c+d) = a.b + a.c + a.d$. This is the so-called distributive law [Jacobson, 1951]. The current opinion held by those who support the view that these geometrical statements are really algebraic or arithmetic identities in geometric formulation is that these results were taken over by the Greeks from the older Babylonian arithmetic and were given this format by the Greeks because of logical necessities and in particular because of difficulties associated with the concept of irrational quantities (see Van der Waerden [1954, 118] and Szabó [1974b, 209] for a discussion of this view as well as the algebraic form of the other identities). Those who oppose this view argue that the Greek mathematicians did not obtain these results from arithmetic operations but conceived and developed them from a strictly geometrical viewpoint.

To a certain extent we may compare the treatment of results, such as II,1, from a completely geometrical viewpoint as being similar to the modern-day situation wherein the abstract algebraist, for example, Jacobson [1951], applies rules, such as the distributive law, to abstract systems—without even thinking of the real numbers—and proceeds to prove theorems. However, we know that modern abstract algebra in fact started by abstracting the model provided by the real numbers, whereas the whole problem in the Greek case is to know whether the Greeks too used the real numbers as a model or if it was the geometrical entities themselves that provided their own model.

B. Geometrical Algebra—Level 2

At the next level, the geometric algebra point of view would consider certain results, not as simple arithmetical identities, but rather as identities which can be used to solve equations. This may be illustrated by II,4.

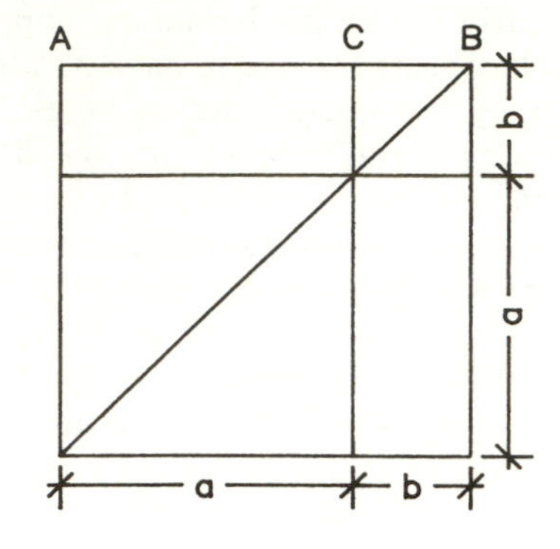

FIGURE II-4. II,4

The statement $S(AB) = S(AC) + S(CB) + 2 \cdot R(AC,CB)$ becomes in modern symbols: $(a+b)^2 = a^2 + b^2 + 2ab$. To show how this may be used to solve a quadratic equation I use an example and follow the method contained in one of the earliest, if not the earliest, extant

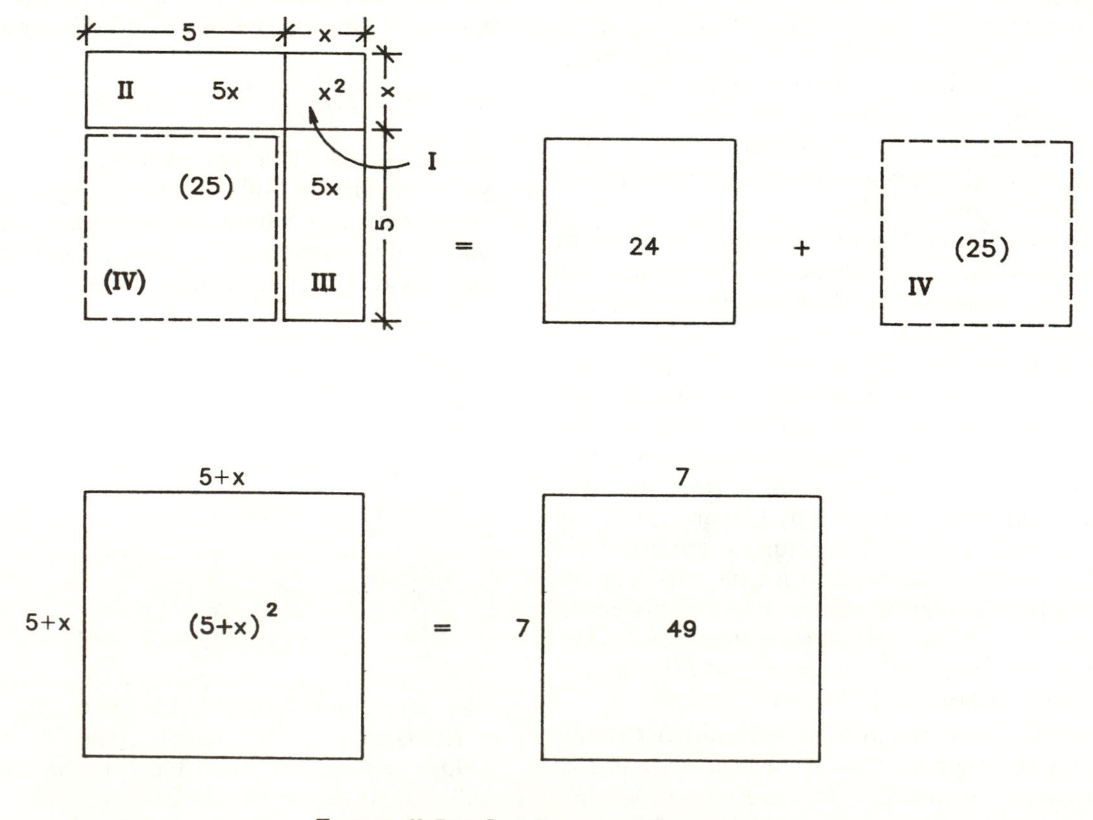

FIGURE II-5. Solving $x^2 + 10x = 24$

Arab algebra texts (probably from the early 9th century), *Logical Necessities in Mixed Equations of Ibn Turk* [Ibn Turk–Sayili, 163]. The equation in modern notation—the text uses the words "roots" and "square quantity"—is $x^2 + 10x = 24$. In Figure II-4 we represent x^2 by the square I whose sides are equal to x. Then the quantity $10x$ is represented by the rectangles II and III whose dimensions are 5 and x. To "complete the square" we add square IV whose sides are 5. This area is also added to the constant quantity 24. We now have a square of sides $5 + x$ and its area is equal to the square of area 49 whose sides are thus 7. Thus $5 + x = 7$ and so $x = 4$.

Now all this is done with words and a picture in the text; these constitute at the same time the method and the "proof." A formal proof of the method would of course use II,4. The same picture constitutes the second method of solution for al-Khwarizmi [al-Khwarizmi–Rosen, 16] again without proof.

We now turn our attention to the dual pair II,5,6. First of all, these can be translated into simple modern algebraic expressions in the same way as was done above for II,1.

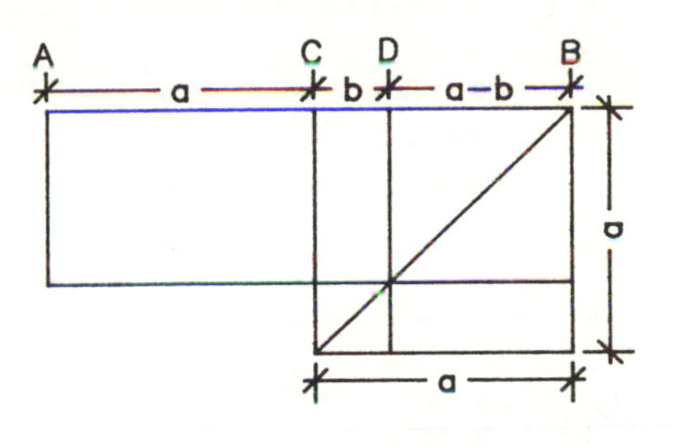

FIGURE II-6. II,5

The statement of II,5—that is, $R(AD,DB) + S(CD) = S(CB)$—becomes simply $(a+b)(a-b) + b^2 = a^2$.

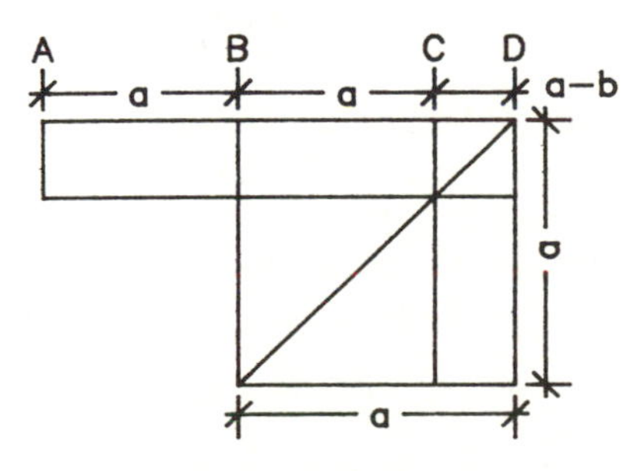

FIGURE II-7. II,6

The statement of II,6—$R(AD,BD) + S(CB) = S(CD)$—gives once again $(a+b)(a-b) + b^2 = a^2$.

Now, however, let us consider the lengths $AD = y$ and $DB = x$ as unknowns.

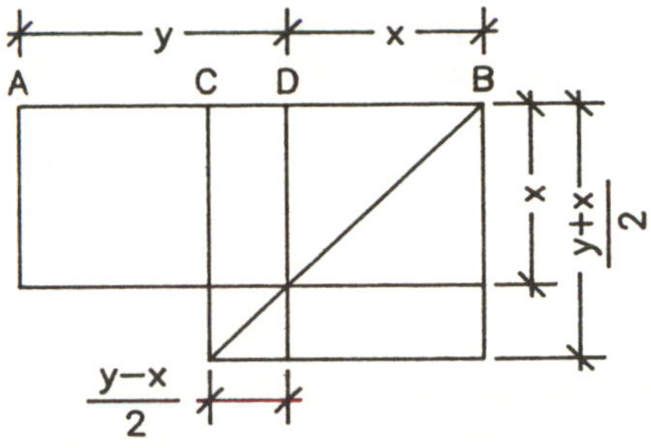

FIGURE II-8. II,5

For the case of II,5 we have that $2 \cdot AC = CB = y+x$, $CD = y - AC = (y-x)/2$ and so the statement $R(AD,DB) + S(CD) = S(CB)$ becomes

$$(1) \qquad xy + ((y-x)/2))^2 = ((y+x)/2)^2.$$

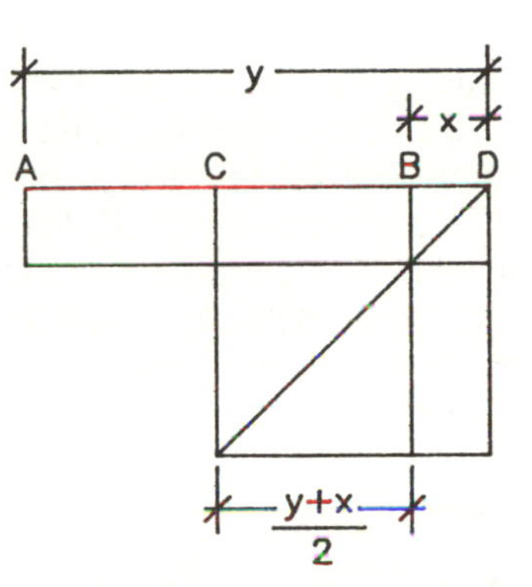

FIGURE II-9. II,6

For the case of II,6 $2 \cdot AC = 2 \cdot CB = y - x$; $CD = CB + x = (y+x)/2$ and so the statement $R(AD,BD) + S(CB) = S(CD)$ becomes once again

$$(2) \qquad xy + ((y-x)/2)^2 = ((y+x)/2)^2.$$

I note that there are variations in the forms of the identity (e.g., Gandz [1938, sect. 6], Unguru [1975, 94]) and a certain "ambiguity" (Seidenberg [1978, sect. 4 bis]), but these do not affect the basic arguments.

When we look at the forms (1) and (2) we see that they are identical—the translation into modern algebraic symbolism of II,5,6 gives identical results. Of course from a geometric viewpoint, as stated in the *Elements*, the results are not the same. Furthermore, the uses made of them are not the same. If we look at III,35,36, we see that the former involves chords and thus an internal cutting point which calls for II,5, whereas the latter involves a secant and thus an extension which calls for II,6.

An explanation has been given about why both II,5 and II,6 appear even though both have the same algebraic form (see Gandz [1938, sect. 6], Van der Waerden

[1954, 121; 1975, sect. 11]): both are to be considered as "solutions" to different pairs of equations. We can see this as follows.

In II,5 let the length AB be considered as a known quantity a and the area $R(AD,DB)$ as a known quantity b. In algebraic terms this amounts to

$$y + x = a; \quad yx = b. \tag{3}$$

Geometrically we would like to find the division point D which in turn gives the lengths x and y. Algebraically this can be done by using (1) to solve for $((y-x)/2)^2$ and then $y-x$ in terms of a and b. Since $y+x = a$ is given, x and y can easily be found by addition and subtraction.

In the same way suppose in II,6 that $AB = a$ and $R(AD,BD) = b$ are known. In algebraic terms this amounts to

$$y - x = a; \quad yx = b. \tag{4}$$

Geometrically this corresponds to finding the extension point D which gives the lengths x and y. Algebraically one uses (2) to determine $(y-x)/2$ and then $y+x$. Again by addition and subtraction x and y are found.

I mentioned in connection with II,4 that at least the diagram was used by Arab mathematicians. I also note here that II,5,6 were explicitly used by other Arab mathematicians to solve equations but not in the manner indicated here. This is discussed in more detail in Section 5,G.

C. *Application of Areas—Level 3*

Note that II,5,6 do not themselves explicitly tell us how to find point D. This remark brings us to the second of the topics mentioned at the beginning of this section, for it is the type of construction indicated under the general name "application of areas" which can be used to actually provide solutions to various equations. Some authors subsume the topic of application of areas under the heading of "geometrical algebra"; for instance, Neugebauer [1957, 149] refers to it as "the central problem of geometrical algebra," and Van der Waerden [1954, 123] speaks of it as "this important part." However, it seems to me that not enough distinction has been made between the identity approach which does not actually provide a geometric solution and the application of areas method which can provide geometric solutions. I shall thus consider the actual solution of equations via application of areas methods as geometrical algebra at the third level.

Recall that in Section 4 I spoke about I,44 in which one had to construct a parallelogram whose angle was given and whose area was that of a given triangle, but with the added proviso that one side of the parallelogram coincided with a given line. Euclid uses the terminology "apply the parallelogram to the line." To see how this can be interpreted in terms of equations, let us specialize I,44 to the case where the angle is a right angle as in Figure II-10. Suppose that the length of the line is a and that the area of the triangle is b. If we represent the height of the desired rectangle by x, then the geometrical problem can be translated into the algebraic problem of finding x such that $ax = b$.

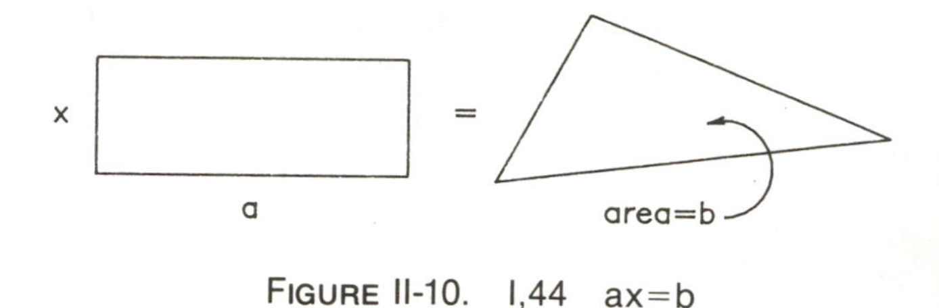

FIGURE II-10. I,44 ax=b

The other place where Euclid uses the terminology "apply a parallelogram to a line" is in VI,28,29. But in these cases we are given not just an angle but rather a parallelogram, and we are told that the applied parallelogram must be deficient (exceed) by a parallelogram similar to the given one (see the formal description of VI,28,29 in Section 1).

Again let us specialize and assume that the given parallelogram is a square. These are illustrated in Figures 11 and 12, where the applied parallelogram is shown with cross-hatching.

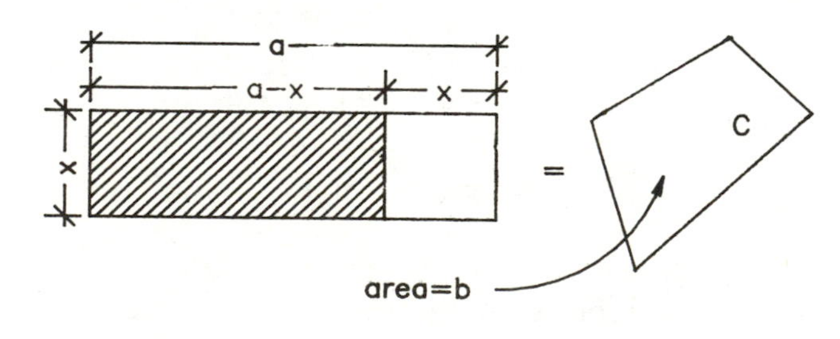

FIGURE II-11. VI,28 (defect) (a−x)(x) = b

In the case of VI,28 the applied parallelogram of dimensions $a-x$ and x "falls short" of the given line of length a by the square of side x (the defect). If the given rectilinear figure C has area b then the algebraic equivalent of VI,28, for the special case of the square, is

$$(a-x)(x) = ax - x^2 = b. \tag{5}$$

In the case of VI,29 the applied parallelogram of dimensions $a+x$ and x "exceeds" the given line of length a by the square of side x (the excess). If the given rectilineal figure C has area b then the algebraic equivalent of VI,29 for the special case of the square is

$$(a+x)(x) = ax + x^2 = b. \tag{6}$$

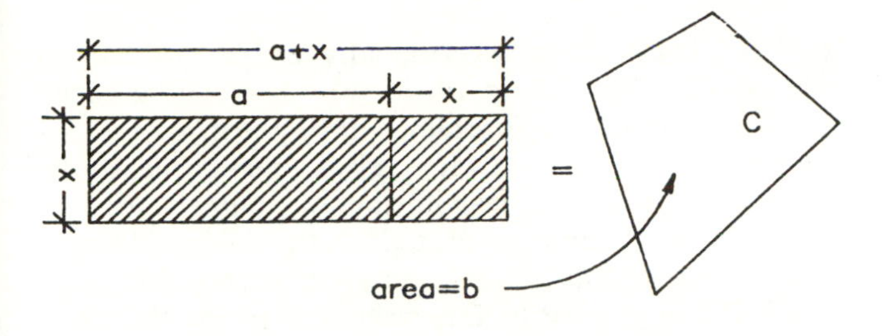

FIGURE II-12. VI,29 (excess) (a+x)(x) = b

I note that the proofs of VI,28,29 would not in any way be simplified by assuming that the applied parallelogram was a square. Other proofs for the case of a square have been suggested (see Euclid–Heath [I, 384, 387], Van der Waerden [1976, sect. 13]) but these are not attested to in any Greek text.

I have written the algebraic form of the special cases of VI,28,29 in terms of one variable. We can, however, write them in terms of two variables as follows.

If in the first problem we call one segment x and the other y, instead of $a-x$, then in place of (5) we obtain $y+x=a$ and $yx=b$. But this is precisely equation set (3). Thus the geometric solution to the application of areas problem with the defect equal to a square is the same as the solution of the algebraic problem obtained from II,5.

In the same way, if in the second problem we call the length of the extended line y instead of x, the two variable problems obtained is just $y-x=a$ and $yx=b$ which is the same as (4). Thus the geometric solution to the application of areas problem with the excess equal to a square is the same as the solution of the algebraic problem obtained from II,6.

We shall see that the format, single variable or two variable, will make a difference when considering possible Babylonian influences on Greek geometry.

D. *Historical References*

Before examining further the application of areas results from an algebraic viewpoint, I will present some historical references where the word "application" is used. The first two are from Plato.

(Q.1) [Socrates] For all their talk of squaring and applying and adding and the like whereas in fact the real object of the entire study is pure knowledge.

(Plato, *The Republic*, Book VII, 527A [Plato–Shorey, II, 171])

In the next quotation Socrates is discussing the methods of mathematicians and how they consider the effect of different hypotheses.

(Q.2) I mean by hypothesis what the geometricians often do in dealing with a question put to them; for example, whether a certain area is capable of being inscribed as a triangular space in a given circle: they reply—"I cannot yet tell whether it has that capability; but I think, if I may put it so, that I have a certain helpful hypothesis for the problem, and it is as follows: If this area is such that when you apply it to the given line of the circle you find it falls short by a space similar to that which you have just applied, then I take it you have one consequence, and if it is impossible for it to fall so, then some other."

(Plato, *Meno* 86E-87A [Plato–Lamb, 325])

This text has been subject to many different interpretations. For recent discussions of some of them, see Plato–Bluck [Appendix] and Brumbaugh [1954, 32]. I simply remark here that some, but not all, commentators take the given line to be the diameter and that some, but not all (for example, see Plato–Bluck [448, 452]), interpret the term "application" in the way that it appears in VI,28.

In the works of Plutarch (1st century) we read:

(Q.3) and when Pythagoras discovered his theorem he sacrificed an ox in honour of the occasion, as Apollodorus says:

When for the famous proof Pythagoras
Offered an ox in splendid sacrifice—

whether it was the theorem that the square on the hypotenuse is equal to the sum of the squares on the sides of the right angle or a problem about the application of a given area.

(Plutarch, "A Pleasant Life Impossible," 1094B [Plutarch–Einarson, 67])

(Q.4) Now among the most characteristic theorems, or rather problems, of geometry is this: given two figures, to construct a third equal to one and similar to the other. They say, in fact, that Pythagoras offered sacrifice when he solved this problem; for it is surely much more elegant and inspired than that famous theorem which gave the proof that the square on the hypotenuse is equal to the sum of the squares on the sides enclosing the right angle.

(Plutarch, "Table Talk," VIII,2, 720A [Plutarch–Minar, 127])

The next quotation from Proclus (5th century) is from his commentary on I,44 and brings in the idea of excess and deficiency.

(Q.5) Eudemus and his school tell us that these things—the application of areas, their exceeding and their falling short—are ancient discoveries of the Pythagorean muse.... Something like this is the method of "application" which has come down to us from the Pythagoreans.

(Proclus, *On Euclid I*, commentary on I,44 [Proclus–Morrow,332; Thomas, I, 187; Heath, 1921, I, 150])

Knorr [1975b, 208 fn. 45] suggests that Proclus' source was not in fact Eudemus and that the phrase be rendered as "sources ostensibly based on Eudemus." (On this point, see also Szabó [1974a, 291, 311 fn. 3].)

E. *Setting Out the Debate*

The situation at this point is that I,44 and VI,28,29 can be interpreted as giving solutions to certain linear and quadratic equations. Furthermore, the quadratic equations correspond to the algebraic version of II,5,6. The question of interest then is: Do II,5,6 and VI,28,29 represent the Greek mathematicians' deliberate geometrization of equations and their solutions? Since not only does II,11 depend on II,6 for its proof but, as I shall discuss in Subsection K, it can also be interpreted in terms of equations, and since VI,30 depends on VI,29, the answer to these questions would be of the utmost interest to the study of the history of DEMR.

What I shall do in the following subsections is present some of the major arguments that have been advanced for and against the interpretation of parts of Euclid as geometric algebra and in particular as representing solutions to equations. I shall also make some comments of my own and consider the case of DEMR in more detail.

F. *Other Interpretations in Terms of Equations*

Aside from I,44 and II,5,6, and VI,28,29, other results in Euclid have been interpreted in terms of equations.

If we reconsider II,14, in which we are to find a square whose area is that of a given rectilineal figure, and we call the side of the square x and the area of the rectilineal figure b, then II,14 can be interpreted as being the geometric version of the simplest quadratic equation $x^2 = b$. Note that this is what was called, in Section 4, a transformation of areas result.

In the same way that II,5 and II,6 correspond to the equations (3) and (4) in two variables, the Theorems II,9,10 also correspond respectively to the equation pairs:

(7) $\quad y + x = a, \quad x^2 + y^2 = b$

(8) $\quad y - x = a, \quad x^2 + y^2 = b.$

(See Gandz [1938, 464] and Van der Waerden [1954, 123]; compare with the interpretation of Heath [1921, I, 93].) Furthermore, these four sets of equations in two unknowns correspond to four standard types of problems and equations which appear in Babylonian mathematics [Gandz, 1938, 405, types B I–B IV; Van der Waerden, 1954, 80, types B I–B IV]. Van der Waerden [1954, 124] writes, "Apparently the Pythagoreans formulated and proved geometrically the Babylonian rules for the solutions of these systems."

An objection can be raised which shows the care that must be taken when looking for algebraic interpretations of geometric statements and vice versa; for note that $x+y = a$, $x^2+y^2 = b$ implies that $b+(2xy) = a^2$ or $xy = (a^2-b)/2$. Thus algebraically every problem of the form of (7)—$x+y = a$, $x^2+y^2 = b$—can be easily transformed into a problem of the form of (3)—$x+y = a$, $xy = b$—and vice versa. Furthermore, the result needed for the transformation, namely, $(x+y)^2 = x^2+2xy+y^2$, corresponds to II,4. In addition, as mentioned previously, the geometric formulations of Book II can at most be interpreted as giving the statement of the identities needed to solve the equations as opposed to the actual solutions of the equations. Why then, if II,5 and II,9 are meant to be geometric formulations of the identities (3) and (7) associated with equations, does Euclid give both of them when they can be transformed into one another?

In response to this objection it could be countered that the Babylonians did not in fact transform the equation set (7) into equation set (3) (see Gandz [1938, 423, example 4; 497, example 30, nos. 8,12]). Rather, by means of substitutions, they reduced both types to simpler problems. Indeed the solution of equations is a matter not simply of identities but also of transformations and artifices. But again Euclid does not give these transformations. The point is that it is necessary to dig more deeply to find and solve these quadratic equations than might be expected if indeed these various results were supposed to represent equations.

In the same vein, when Van der Waerden [1954, 124] points out that an equation of the form $x^2(x \pm 1) = a$ appears in Babylonian mathematics and in the works of the Greek mathematician Nichomachus (1st to 2nd century), we may ask, as does Fowler [1979c, 5], why no sign of higher order equations appears in Euclid: "Euclidean mathematics appears to be solely concerned with linear, quadratic and biquadratic procedures." Fowler also points out that if the *Elements* are to be considered as a rigorous reformulation of Babylonian procedures we could expect to find such things as the binomial expansion of the cube of a sum.

Again consider the following equation sets:

(9) $\quad x+y = a, \; x^2-y^2 = b; \quad x-y = a, \; x^2-y^2 = b.$

These appear in Babylonian mathematics, according to Gandz [405, types B V and B VI; 451, example 21, nos. 3,11,12—there is some disagreement over the interpretation of these problems (p. 458), and this may be the reason Van der Waerden (p. 80) does not list them]. The first type as well as a variation of the second type appears in the *Arithmetic* of Diophantus (fl. 250 ?) [Diophantus–Heath, 1910, 141; I,29, 144; II,6,7 and Gandz, 416, 447]. I also remark that both Gandz [p. 417] and Van der Waerden [p. 281] agree that Diophantus' completely arithmetical—as opposed to geometric—methods simply preserve the Babylonian methods.

Note, however, that problem set (9) also has a very direct geometrical interpretation; if we think of x as the diagonal of a rectangle and y as one of the sides, then x^2-y^2 represents the other side. Thus we can think of being given the sum or difference of the diagonal and one side as well as the other side. The point is that the set (9) has just as much of a geometrical interpretation as do (7) and (8), and yet no trace of them appears in Euclid.

While it is possible to use VI,28,29 to obtain solutions to the equation sets (5) and (6), there are simpler and more direct ways of doing so; see, for example, the method of Simpson [Euclid–Heath, II, 384, 387] and those of Van der Waerden [1976, sect. 13; also Seidenberg, 1978, 314]. An obvious question is why, if the Greeks wanted to display geometrically the solution to algebraic equations, did they not use these simpler methods which are based on results of Books I, II, and III.

Van der Waerden [1976], on the contrary, argues that the more complicated aspect of VI,28 is indeed a sign: "Their starting point was not geometry but algebra, and they [the Pythagoreans] translated the algebraic solution, step by step, into the language of geometry."

What seems to be the case, then, is that all or almost all of the Babylonian equation types can be found in later Greek mathematics and that some, but not all, can be interpreted as appearing in Euclid. Furthermore the "reflection" of those equations that do "appear" in Euclid is not consistent.

G. Problems in Interpretation

The Babylonians had problems involving both one and two variables, and this fact causes problems of interpretation of any geometrical result in terms of Babylonian algebra.

We saw that the algebraic version of the application of areas problem with "defect" can be written either in terms of one variable $ax-x^2=b$ or in terms of two variables $x+y=a$, $xy=b$. Area problems seem always to have been treated by the Babylonians as two variable problems [Gandz, 1938, sect. 4]. On the other hand, the pure quadratic equation $x^2+b=ax$ was apparently never used by the Babylonians, presumably because of the problem of the double positive solution [Gandz, 405, type B IX, 411, 480].

When we turn to the excess problem $y-x=a$, $xy=b$ or $x^2+ax=b$, we find that this pure quadratic type did exist in Babylonian mathematics. We shall see further problems with the dual possibility of two variables or one variable when we turn our attention to the case of DEMR (II,11).

When we examine the use that Arab mathematicians made of geometrical statements to solve quadratic equations, we again see what difficulty there may be in analysing results and methods. We saw how Ibn Turk and al-Khwarizmi seem, from the diagram used, to have had II,4 in mind for the solution of equations of the form $x^2+ax=b$ (Arab type I [Gandz, 406]). But when we examine a later work by Thabit ibn Qurra [Thabit ibn Qurra–Luckey, 105], we see that he explicitly mentions that II,6 is to be used to solve this type of equation. In the algebra of Abu Kamil [Abu Kamil–Levy, 30], we are shown how to solve the equation by the method of al-Khwarizmi (Euclid is not mentioned) and also by using II,6 (Euclid is mentioned).

Again consider how two different historians interpret the algebra of al-Khwarizmi and the sources of his methods. On the one hand, Gandz [524] says, "these geometric demonstrations are the strongest evidence against the theory of Greek influence. They clearly show the deep chasm between the two systems of mathematical thought, in algebra as well as in geometry." On the other hand, Sayili [Ibn Turk–Sayili, 137] writes: "It is therefore quite clear that the geometrical figures of al-Khwarizmi's algebra, far from being totally different from the corresponding figures found in Euclid's *Elements*, are essentially the same as the latter, and the nature of the geometrical demonstrations in the two cases are, for all intents and purposes, and as geometrical demonstrations identical." The same view is taken by Trofke [1934, 97].

These examples all illustrate that a great deal of caution is required when trying to connect algebra and geometry. I conclude, then, that even the interpretation of geometrical statements in terms of equations, without even considering the more difficult question of possible influence, is a question that is not to be treated lightly.

H. Division of Figures

Neugebauer [1957, 150] writes: "Attempts have been made to motivate the problem of 'application of areas' independently of this algebraic background. There is no doubt, however, that the above assumption of a direct geometrical interpretation of the normal form of quadratic equations [i.e., equation sets (5) and (6)] is by far the most simple and direct explanation."

There seems to be here the implicit a priori assumption that both the simplest form of the application of areas problem as in I,44 and the more complicated versions with defect and excess of VI,28,29 developed in the same way and at about the same time. This is taking the statement of Proclus (Q.5) to its extreme. I have already suggested in Section 4 that it was possible that the simple application of areas concept arose out of the transformation of areas problem I,45. In view of this one should ask if the simpler and more complicated versions of application of areas did not arise at different times and were not motivated by different problems.

Indeed, it seems to me that there is another context in which the application of areas formulation of VI,28,29 fits in more naturally than the equation hypothesis, for which we have written evidence that associates it with at least the period of Euclid, namely, the problem of division of figures.

Proclus, in his summary of the history of Greek mathematics [Proclus–Morrow, 57], mentions a work by Euclid on the division of figures. Further, an Arab manuscript entitled *The Treatise of Euclid on the Division [of Plane Figures]* has been published [Euclid–Woepcke; Euclid–Archibald] and gives evidence [Euclid–Archibald, 66] of corresponding to the work mentioned by Proclus.

Of interest to us is the following result from the manuscript [Euclid–Woepcke, 236; Euclid–Archibald, 50].

PROPOSITION 18. To apply to a straight line [AB] a rectangle equal to $R(AB,AC)$ [C is a point on AB] and deficient by a square.

This is just a special case of VI,28 when the defect is a square and the figure which determines the area is a rectangle constructed on the given line itself.

Proposition 18 in turn is used in the proofs—given in the Arabic manuscript—of Propositions 19 and 20.

PROPOSITION 19(20). To divide a given triangle into two equal parts (to cut off a certain fraction) by a line which passes through a point which is situated in the interior of the triangle.

Proposition 18 dealt with application with deficiency and the obvious question is whether the case of excess appears. While no direct analogue of Proposition 18 appears in the Arabic manuscript, there is evidence to indicate that application with excess did appear in the original. To see this we consider the following.

PROPOSITIONS 26(27). To divide a given triangle into two equal parts (to cut off a certain fraction) by a line drawn from a given point which is situated outside of the triangle.

These propositions are stated without proof in the Arabic manuscript, but proofs are given—and do indeed involve application of areas with excess—in the corresponding theorems in the *Practica Geometria* of Fibonacci (Section 31,B). Furthermore, the statements of most of the Arabic text, including Propositions 26 and 27, are close to those of the more extensive Fibonacci text and many of the proofs, including those of Propositions 19 and 20 given above, are virtually the same in the two texts. It therefore seems safe to conclude that the original Greek text also used application of areas with excess in the proofs of Propositions 26 and 27.

Thus the statement of Proclus, the title and content of the Arabic text, and Fibonacci's work lead us to the conclusion that application of areas with excess and defect was associated with the problem of division of figures at the time of Euclid. This of course does not necessarily mean that application of areas with excess and defect had its origin in the problem of division of figures, but it appears that since there is ancient evidence for the association, the possibility is certainly as likely as the equation hypothesis which is unsupported by any ancient text.

It is possible to speculate on the origin of the problem of division of figures. It may have arisen as a purely mathematical question or it may have as its ultimate origin, albeit very distantly, ritual problems (see Seidenberg [1966; 1977] or land division problems. There is a tradition associating the origins of geometry with Egyptian land measurement problems, but there is no consistency among the ancient texts discussing the matter, and indeed some of the existing Egyptian records would seem to give a counter-indication (see Fowler [1983; 1985, chap. 8]). Various texts are presented in Fowler [1983; 1985; Heath, 1921, I, 121]; see, in addition, the preface to Books I and III of Hero's *Metrica* [Hero–Bruins, III, 182, 313].

I. Theorems VI,28,29 vs II,5,6

We have seen how attention has been drawn to the fact that VI,28,29, when specialized to the case when the defect and excess are squares, correspond algebraically to II,5,6. Notice the following: the arguments used at the ends of the proofs of VI,28,29 to show that the parallelograms have an area equal to that of the gnomon are in the same spirit as the proofs of II,5,6. Indeed, in the special case of the defect or excess being a square this part of the proof is precisely the statement of II,5,6. Furthermore, the proof of the parallelogram versions of II,5,6 would be precisely the same as the arguments of VI,28,29. Thus, if II,5,6 were developed out of, or related to, the application of areas solutions as given in VI,28,29, we should expect to see either a "square version" of VI,28,29 or more likely the parallelogram versions of II,5,6.

J. Euclid's Data

Various authors, such as Tannery [1882, 401; read 84,85 for 79,80], Heath [1921, I, 423], and Van der Waerden [1954, 121], have supported their association of II,5,6 with the statement of problems (1) and (2) by a reference to Propositions 84 and 85 of the *Data* [Euclid–Peyrard, III, 456; Euclid–Ito].

PROPOSITION 84. If two straight lines enclose a given area in a given angle, and one of them is greater than the other by a given line, then each of them will be given.

PROPOSITION 85. If two straight lines enclose a given area in a given angle and if their sum is given, then each of them will be given.

Let us consider the special case when the angle in these propositions is a right angle. Suppose we interpret the statement of Proposition 85 to mean that some rectangle had unknown sides of lengths x and y and that the sum of the lengths $y+x$ is known to be a and that the area yx is known to be b. Then corresponding equations would be precisely (3), which was also obtainable as the algebraic interpretation of II,5. This type of reasoning leads Heath [1921, I, 422] to say, "Euclid shows how to solve these equations in Propositions 84,85 of the *Data* . . . ." This is stretching things quite a bit, for when we read the proofs of the propositions we see that Euclid never actually finds the lengths of the sides of the parallelogram. In fact Euclid never even says "Let the given sum of the lengths be equal to the line AB" or the equivalent.

Let us then look at Proposition 85 more closely (84 is analogous) by working backwards through the chain of propositions from the *Data* and theorems from the *Elements*.

From the *Data* itself, nineteen propositions are used (I am counting those indicated by Peyrard). These in turn use the following theorems from the *Elements*: I,3,8,10,11,16,23,30,31,32,34,41,43,46; V,16,18,22; VI,1,4,6,11,13,17,18,20,22. Conspicuous by their absence are the application of areas results I,44 and VI,28,29 and the results of Book II.

However, if we look at how Proposition 85 uses earlier results from the *Data*, we see that the key result is Proposition 58 [p. 397].

PROPOSITION 58. If a given area is applied to a given line, and if this area is deficient by a figure whose shape is given, then the lengths of the deficiency are given.

The sense of the expression "shape is given" is indicated in definition 3 [p. 302], where we read: Rectilinear figures, each of whose angles is given, and the ratios of whose interlying sides are given, are said to be given in shape (cf. Mugler [1958, 161]).

The statement of Proposition 58 reminds us of VI,28, and if we go further and compare the proofs of VI,28 and Proposition 58, we find that they both use I,43 (to prove a II,5 type result; see Subsection I above) and VI,18 (to construct a similar figure). What is happening then? What is the difference between Proposition 58 and VI,28 and between Proposition 85 and the equation interpretation of Proposition 85. To help explain this I first state and prove a constructive version of 85 and show how its proof is immediate from VI,28.

PROPOSITION 85′. It is required to perform the following construction. We are given that the area of the parallelogram defined by two lines and their included given angle is equal to the area of a rectilinear region C. We are also given that the sum of the two sides is equal to a line AB. We are to construct the two sides of the region. Furthermore, we are to show that no other sides satisfy the requirements.

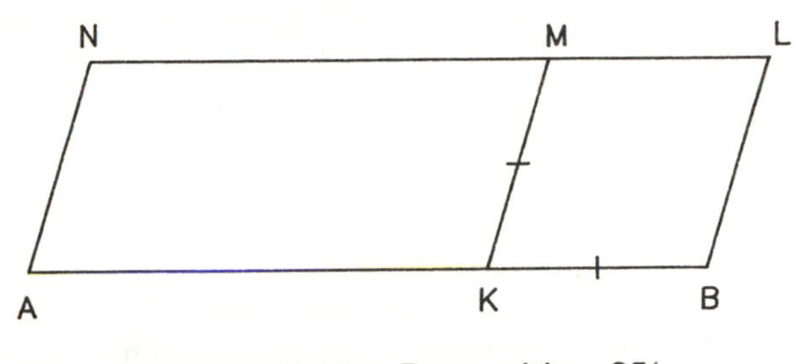

FIGURE II-13. Proposition 85′

Proof: Using VI,28, apply to line AB a parallelogram $AKMN$ whose area is the same as C and with the defect being a rhombus $KBLM$ whose angle is the same as that of the given angle between the lines. Then, since $MK = KB$, the sum of the sides AK and KM of the parallelogram $AKMN$ is equal to AB. Thus AK and KM are solutions to the problem.

To show that these solutions are unique, I borrow some ideas from the proof of VI,28. Referring to the diagram accompanying VI,28, suppose that S is a division point of AB, say to the right of K. Since the sides of the required parallelogram must add to AB, we consider the parallelogram $ASQT$ with $QS = SB$. But since in the above solution obtained from VI,28 we had $KB = KM$, we have that both M and Q lie on the diagonal GB. Since S is to the right of K, we then have that the gnomon defined by Q has a smaller area than the gnomon defined by M. But the area of each parallelogram is precisely the area of the corresponding gnomon—as was demonstrated in the proof of VI,28—so that the parallelogram $ASQT$ will have a smaller area than parallelogram $AKMN$. In the same way, if S is assumed to the left of K the area will be larger. Implicit in the above was the assumption that both points K and S were to the right of the midpoint E. Indeed the maximum area occurs when S coincides with E; this is in fact the message of VI,27. Of course a dual solution exists with AK less than KM.

It is something very close to Proposition 85′ that we would expect to find in the *Elements* if Euclid were trying to give the geometric version of the equation set (3), and it is something like the above proof that we would expect to find if he had wanted to indicate how application of areas, in particular VI,28, was related to the solution of these equations.

If Proposition 85′ represents what the solution of equation set (3) looks like in geometric form, then one may legitimately ask what Proposition 85 represents. From the viewpoint of a modern mathematician, one possible answer is that, whereas Proposition 85′ represents a constructive proof that the set of equations (3) has a (unique) solution, Proposition 85 represents a non-constructive existence proof. This situation may be compared to two proofs that two linear equations with non-zero determinant have a solution: first (existence and uniqueness), the corresponding lines are not parallel and therefore intersect in only one point; second (constructive), solve the equations—by Cramer's rule, for example.

I believe, however, that this interpretation is not valid if put in the context of Greek mathematics; it is only valid if there is an a priori assumption that we are dealing with equations and if we view it from the modern viewpoint or if we introduced a modern "existence" theory for geometrical entities (cf. a recent discussion of the meaning of the *Data* for the history of algebra and a discussion of it in terms of the theory of linear functions in Slavutin [1979]).

It seems that the *Data* must be taken at face value as a study of when certain geometrical quantities are "determined" when other geometrical quantities are "given." The study perhaps arose out of philosophical considerations. The commentary of Marinus, a student of Proclus (late 5th century) [Marinus–Michaux, chap. 2], in fact starts off: "It is first of all necessary to determine what is Datum." This is followed by a discussion of various ancient and modern (at the time of Marinus) opinions on the matter (see also Marinus–Michaux [chap. 1, sect. 2]). Marinus concludes that the best general definition of a "given" is something that can be constructed, obtained, and reproduced. Euclid in the *Data* only gives definitions of being "given" for specific things, such as area, ratios, lines, and in fact Marinus [p. 63] reproaches Euclid for not having defined the concept in general. A look at the "plan" of the *Data* [p. 52] gives further support to the idea that the *Data* is essentially an investigation of the type of hypothesis that is needed in order for there to be a "solution" to a problem and in particular it is a study of which geometrical quantities must be "given" in order that other quantities be "determined." For other historical examples of texts which seem to have an "existence proof" flavour to them, see Q.2 from Plato's *Meno*; the discussion of Leon and "diorismi" (cf. Mugler [1958, 141] in Q.3 of Section 16,B; and the discussion of al-Biruni's *Chords* in Section 28,E.

It would seem, then, that a reference to *Data* 84,85 would be misleading in any attempt to interpret II,5,6 as algebraic problems.

K. *Theorem II,11*

I will now discuss II,11—the area formulation of DEMR—in terms of equations. From an historical viewpoint we would wish to know if II,11 arose from geometrical considerations, such as in the manner suggested in Section 2,B, or whether it was treated at an early stage in terms of equations (see Tannery [1882, 399], Euclid–Heath [I, 403], Van der Waerden [1954, 101], Michel [1950, 568]). Note that, whereas Theorems II,1-10 are stated in the form of geometrical identities, II,11 is stated in the form of a construction. It thus can be considered as a third-level problem (see Subsection C) in geometrical algebra, where only pure equations without any relationship to identities are considered.

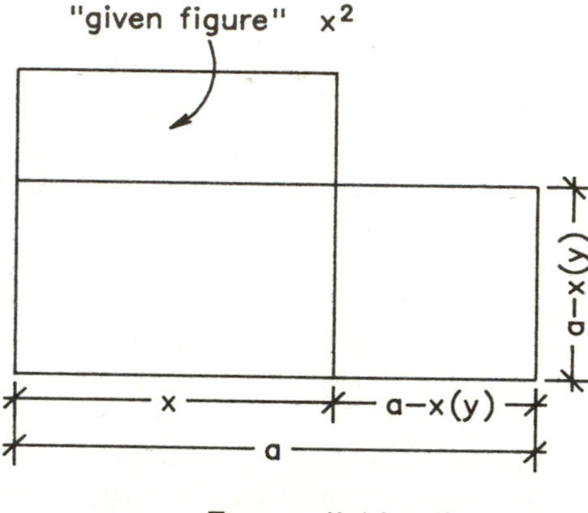

FIGURE II-14. II,11

Consider the statement of II,11 as in Figure II-14. It is required to divide the fixed line of length a into segments of length x and $a-x$ so that the area of the rectangle on the line equals the area of the square. From an application of areas viewpoint, we might at first think of this as we did in Subsection C in connection with I,44, namely, the simple application of a rectangle to a line. But this would not be valid, because here the given figure has a variable area (x^2). Again if we write the relationship in algebraic form, we obtain $a(a-x) = x^2$ or $a^2 - ax = x^2$. If we algebraically transpose the term ax, which does not have a geometric equivalent in the diagram, we obtain $x^2 + ax = a^2$. This, from the point of view of equations, is just a special case of equation (6)—$x^2 + ax = b$—which as discussed has been related, not to the simple application of areas problem, but to the application with excess result VI,29.

The difficulty with II,11 is emphasized when we write the algebraic version in terms of two variables. As pointed out in Subsection G, this appears to have been the usual Babylonian approach to rectangle problems. Indeed, the equations become $x+y = a$ and $ay = x^2$, but this equation pair does not appear in any of the lists of Babylonian standard equation sets.

Recall now the discussion in Section 2 of the result II,11′ contained in II,11 (Fig. II-15).

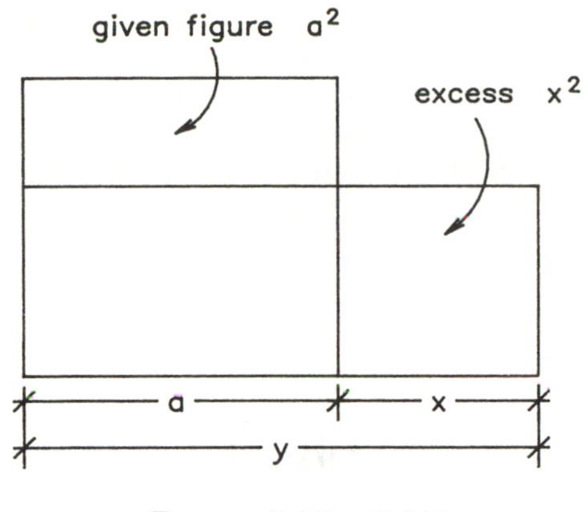

FIGURE II-15. II,11′

Theorem II,11′ is a true application with excess problem; the area of the rectangle on the extension equals the area on the fixed original line. The equation is $(a+x)x = a^2$ or $x^2+ax = a^2$ or in double variable form $y-x = a$, $xy = a^2$ which are of types B VIII and B II, respectively [Gandz, 1938, 405]. Thus, if the development of DEMR were closely linked to application of areas, we would expect to find II,11′ and not II,11 in the *Elements*.

L. *II,11—Application of Areas, Various Views*

The relationship of application of areas to DEMR has been discussed by several authors.

i. Szabó

The article by Szabó [1974a] is entirely devoted to a discussion of application of areas and the use of it by the Pythagoreans. He points out [p. 308] that as well as being used for II,11, Theorem II,6 is also used for III,36, but Szabó considers the latter as just a special case of II,11 (cf. Section 2 where I argue just the opposite). Regarding II,5, Szabó [1974a, 303, 305; also see 1968; 1969, appendix, 370] considers it as just a lemma for II,14.

In another paper [1974b] Szabó states: "So far as I know, there is absolutely no historical evidence for the conjecture [that] the Pythagorean 'application of areas' would ever have been a method for solving quadratic equations. In fact, nothing is known about any pre-Euclidean Greek algebra that could have been given a geometric formulation." Three arithmetical propositions, numbers 16, 18, and 19, from Book IX are then singled out for special attention. Theorem 18 in particular deals with the third proportional to two given numbers a and b; in other words, it deals with the question of when there is a number x such that $a:b = b:x$. From his examination of these results, Szabó concludes, "the simple application of a rectangle to a given straight line is just the general solution of a problem of proportionality, that in arithmetic could not be solved, but under certain conditions." The development of application with the defect and excess being squares is explained via a reference to II,5,6; $S(BD)$ is the square defect (excess).

There are several difficulties with these theories. The preliminary argument [p. 209] deals with simple application but involves the complements that leave a defect (the parallelogram about the diagonal). Furthermore, it is not clear how the shape of the first complement is determined; indeed that is why the construction of VI,28,29 is needed. In the square defect (excess) case it is again unclear to me how the given area is to be transformed into one of the complements or exactly how one should proceed. Further, VI,28,29 do not involve II,5,6; the proofs employ I,43 (complements) as the principal tool.

Szabó continues: "Now we understand the historic role of [II,5,6] if we realize that they are necessary lemmas for the solution of two other very important geometrical problems." Again II,5 is considered to be a lemma for II,14 and was "in all likelihood connected with the discovery of incommensurability." Theorem II,6, considered as the application of a rectangle with excess, is treated as a lemma for II,11, the area formulation of DEMR which is a "particular case of the mean proportional. . . ." Furthermore, the "application of area was in that case beyond any doubt connected with the discovery of incommensurability." In conclusion Szabó writes: "I think therefore that the Pythagorean application of areas was a powerful method of the early Greek geometers to overcome those difficulties which they encountered in discovering the existence of linearly incommensurable magnitudes that were at the same time commensurable in square."

One further counter-indication to the consideration of the extension of the integer mean proportional problem as the inspiration for the application of areas is that a much simpler solution to the problem is given in VI,2 (a line parallel to the base of a triangle cuts the sides proportionally); this solution is given in Szabó [1974a]. Of course, being in Book VI, it is there stated in the language of proportion theory.

ii. Junge

Junge [1948] not only discusses the relationship between application of areas and DEMR but believes that the first developed from the second [p. 323]: "From the two fold [i.e., II,6,11; VI,29,30] occurrence of the application of areas and DEMR in the first books of Euclid, both times together, we conclude with a certain probability that both concepts were related from the beginning, and that at first glance strange propositions dealing with the application of area were found from the goal of DEMR. As soon as the problem of the application of areas was posed and solved for the special case of division in extreme and mean ratio it was easy for the Greeks to generalize."

iii. Valabrega-Gibellato

The author's theory is that Book II and in particular Theorems II,5,6 which she associates with application of areas, are based on Pythagorean number theory, that the arrangement of dots—corresponding to integers—led to the purely geometrical statements that Euclid presents. Valabrega-Gibellato [1979, 199] finds the appearance of DEMR—in the form of II,11—already in Book II as being significant for her theory, in particular concerning the use made of II,6 in the proof of II,11.

I will summarize and discuss further the points raised in this section in Section 22.

Section 6. Side and Diagonal Numbers

The side and diagonal (or diameter) numbers are a sequence of numbers obtained from a certain recurrence relation which can be used to obtain approximations to the ratio of the diagonal of a square to its side.

The recurrence relationship can be obtained in the following way. Let s_1 and d_1 $(= \sqrt{2}_1)$ be the side and diagonal of a square. Now form a new square with side $s_2 = s_1 + d_1$. The diagonal d_2 of this new square will have length $\sqrt{2}s_2 = \sqrt{2}s_1 + 2s_1$. Thus we have

(10) $$d_2 = 2s_1 + d_1.$$

This relationship, stated in Greek geometrical language, was proved by Proclus (5th century) using II,10; extracts from the text will be given in Q.7.

Continuing in this way, we can obtain an increasing sequence of squares whose sides and diagonal are given by the recurrence relationships:

(11) $$s_n = s_{n-1} + d_{n-1}; \quad d_n = 2s_{n-1} + d_{n-1}.$$

Since the whole idea of the procedure is to obtain approximations to the ratio of the diagonal and side of a square, we start off the approximating sequences by taking $s_1 = d_1 = 1$. Geometrically this corresponds to the rhombus with side 1 and angles of 60° and 120°. We are then generating a whole sequence of rhombii whose shapes approach that of a square. Numerically the recurrence relationships (10) give us the sequence of d_n/s_n values 1/1, 3/2, 7/5, 17/12. . . . The closeness of the approximations to the desired ratio can be obtained from the relationship:

(12) $$d_n^2 - 2s_n^2 = \pm 1.$$

This relationship was stated verbally and without proof by Theon of Smyrna (fl. early 2nd century). (See Thomas [1939, I, 133] and Theon of Smyrna–Dupuis [67, 73].)

While the first explicit historical mention of side and diagonal numbers appears in the just cited text by Theon of Smyrna, there are indications that the side and diagonal numbers date from an earlier period.

The main indication of an early date for the development of side and diagonal numbers occurs in a commentary by Proclus on the following text from Plato's *Republic*.

(Q.6) But the laws of prosperous birth or infertility for your race, the men you have bred to be your rulers will not for all their wisdom ascertain by reasoning combined with sensation, but they will escape them, and there will be a time when they will beget children out of season. Now for divine begettings there is a period comprehended by a perfect number, and for mortal by the first in which augmentations dominating and dominated when they have attained to three distances and four limits of the assimilating and the dissimilating, the waxing and the waning, render all things conversable and commensurable with one another, whereof a basal four thirds wedded to the pempad yields two harmonies at the third augmentation, the one the product of equal factors taken one hundred times, the other of equal length one way but oblong—one dimension of a hundred numbers determined by the rational diameters of the pempad lacking one in each case, or of the irrational lacking two; the other dimension of a hundred cubes of the triad. And this entire geometric number is determinative of this thing, of better and inferior births. And when your guardians, missing this, bring together brides and bridegrooms unseasonably, the offspring will not be wellborn or fortunate.

(Plato, *Republic* VIII, 546B, "The Nuptial Number" [Plato–Shorey, 247].)

Brumbaugh [1954, 110, 112] says about this: "No 'neutral' translation of . . . this passage can be found," and "If Plato had anticipated the controversies and interpretations occasioned by this passage explaining the principle of political decline, he would probably also have anticipated the example of some of his recent translators and deleted it from his text." In view of this I have simply given the translation from a standard series of the classics (cf. also the translations of Thomas [I, 399] and Brumbaugh [1954, 110]).

The number referred to here is usually called the nuptial or geometric number. There is a large body of literature dealing with it and many scholars have tried to determine its value (for a discussion and reference to some of these, see Plato–Shorey [I, xliv], Brumbaugh [1954, 143], Cherniss [1959, 173, no. 863], and Brisson [1979, 282]).

Following is part of the commentary of Proclus on this passage.

(Q.7) The Pythagoreans prove numerically that the rational diagonals connected with irrational diagonals are greater or smaller than double by one. For since the unit is the begetter of everything, it is clear (they say) that it is both a side and a diagonal. So let there be 2 units, one being the side and the other the diagonal; and let one diagonal be added to the side, and two sides to the diagonal, for the diagonal is smaller than double the side by one. Thus one will be of 2 units and the other of 3 [i.e., $1+1=2$, $1+2.1=3$]. And the squares of these are 4 and 9, which is greater than double by 1. Again, let be added to the 2 a diagonal 3, and to the diagonal 3 twice the side 2. So the side will be 5 and the diagonal 7, and their squares will be 25 and 49, which is less than double by 1. Whence Plato said that the number 48 is the square of rational diagonals of 5 minus 1 [i.e., $(R(50-1))^2-1$] and of irrationals minus 2 [i.e., $(R(50))^2-2$] since the square of the diagonal is double that of the side [i.e., $(R(50))^2=5^2\times 2$]. And if we take all the squares from these [irrational] numbers, they will actually be double, while those of the rationals will be greater or smaller than double by 1, e.g. 9 and 49 in relation to 25 and 4 [i.e., $9=(4.2)+1$, $49=(25.2)-1$]. Wherefore the Pythagoreans also trusted this method. . . .

Since the rational diagonal is impossible when the side is rational (for there is no number squared equal to another number squared; by which it is also clear that magnitudes are incommensurable and that Epicurus was wrong in making the atom a measure of all bodies and Xenocrates in making the atom the line of lines), the Pythagoreans and Plato meant to say that when the side is rational, the diagonal is rational, not simply, but in squares equivalent to the double ratio which is necessary to make the diagonal either deficient by 1 or excessive by 1. For it is excessive in the ratio 4:9 [i.e., $2^2+1=3^2$] and deficient in 25:49 [i.e., $5^2=7^2+1$]. From this the Pythagoreans proposed a rather elegant theory of diagonals and sides, that the diagonal, including the side of which it is the diagonal, is a side; and the side, combined with itself and including the diagonal of itself, is a diagonal. And this is proved graphically by that man [Euclid] in the second book of elements . . . .

(Proclus, *Commentary on Plato's* Republic, chaps. 23, 27 [Proclus–Kroll, vol. 2, 24, 27])

Proclus now goes on to give a slight variation on the statement of II,10. This statement is given in translation by Heath [Euclid–Heath, I, 400]. Since the proof given by Heath contains elements not in Proclus' text and excludes others, I shall now give Proclus' proof.

(Q.8) Let there be a side AB, and BC equal to it, and CD the diagonal of AB with double its square.

By the theorem, the squares on AD,DC are double those on AB,BD. And the square of DC is double that of AB; and by subtraction therefore, the square of AD is double that of BD. For if whole is to whole as subtractor to subtractor, so also is difference to difference as whole to whole.

Therefore the diagonal CD, including side BC, is a side, and AB, including BC and its diagonal CD, is a diagonal. For the square is double the side DB.

Let it also be shown numerically for the rational diagonals, which we said were greater or smaller by 1: $1+1=2$. . . $5+7=12$; $2.5+7=17$; $17^2=289=2.12^2+1$ and so on.

(Proclus, *Commentary on Plato's* Republic, chap. 27 [Proclus–Kroll, vol. 2, 27].)

A related statement appears in Proclus' commentary on Book I of Euclid (see Proclus–Morrow [51]).

Further evidence for the early use of the side and diagonal numbers is the appearance of the approximation 7/5 by Aristarchus (c. −280) and 17/12 by Hero of Alexandra (1st century). For a discussion of these references, see Van der Waerden [1954, 126], Knorr [1975, 33], Heath [1921, I, 91], Euclid–Heath [I, 398], Schmidt [1900], and Fowler [1979a].

Side and diagonal type identities have also been suggested by Stamatis [1956] and Knorr [1975, Appendix; see Section 23,A] in connection with an approximation for R(3) given by Archimedes.

In Sections 12,ix and 23 I will discuss proposals that have been made in connection with side and diagonal type approximations for DEMR. I also note the suggestion made by Taylor [1979] that II,10 may be connected with the discovery or the proof of the incommensurability of the diagonal and side of square. For the relationship between side and diagonal numbers and anthyphairesis, see Section 8.

Section 7. Incommensurability

The concept of commensurable and incommensurable magnitudes is defined in X,def.1. From X,5,6,7,8 we have that two magnitudes will be incommensurable if and only if the ratio of their lengths (area) is what we would today call an irrational number—one which is not the ratio of two integers. The modern term "irrational number'' is not to be confused with the concept of an ''irrational'' line segment defined in X,def.3.

Unfortunately very little is known about the early history of incommensurability. This question is the subject of the book by Knorr [1975] who discusses not only the texts that have been preserved but also the various commentaries on the subject. A particularly difficult question is the dating of the discovery of the concept of incommensurability. Most writers place the date at some time between −450 and −410. The various theories are discussed by Knorr [chap. 2, sect. 3] who places it at about −430. (See also Phillip [1966, Appendix 2], Burkert [1972, 455, 465].)

What I shall do here is simply mention three texts dating from before the Euclidean era that indicate what can be ascertained from preserved texts of the period. Some other later Greek commentaries will be given in Section 18,B.

i. Aristotle writes in his *Prior Analytics*:

(Q.9) the diagonal [of a square] is incommensurable [with the side] because odd numbers are equal even if it is assumed to be commensurate.

(Aristotle, *Prior Analytics*, I,23,41a [Thomas, 1939, I, 111].)

The proof alluded to in this quotation is not given by Aristotle; however, a proof of the incommensurability of the diagonal and side of a square using an even-odd contradiction does appear in some manuscripts of the *Elements* (see Knorr [1975, 23]). Note, however, that in Euclid−Heiberg [III, 408] this proof is considered to be not in the original version.

ii. In Plato's *Theaetetus*, Theaetetus is talking about Theodorus (fl. c. −410 to −390; see introduction to Chapter III and Section 15) and says:

(Q.10) Theodorus here was drawing some figures for us in illustration of ''dȳnamis,'' showing that squares containing three square feet and five square feet are not commensurable in length with the unit of the foot, and so, selecting each one in its turn up to the square containing seventeen square feet; and at that he stopped.

(Plato, *Theaetetus*, 147D [Plato−Fowler, 148])

Aside from questions of translation, in particular of the word *dȳnamis*, there is a great deal of material concerning the way in which Theodorus might have proved the results and also whether or not he actually did the case of 17. In addition to the literature discussed by Knorr [chap. 3], see Cherniss [1959, 201, no. 961], Bulmer−Thomas [*DSB* 3,4], McCabe [1976], Giacardi [1977], Burnyeat [1978; 1979], Knorr [1979], Pappus−Thomson [Appendix A], Taisbak [1982, 25, 72], Berggren [1984, 400], Høyrup [1985]. I shall refer to the relationship of Theaetetus to the development of the irrationals in Sections 18 and 20,C.

iii. In Plato's *Laws* there is a discussion between Clinias and an Athenian.

(Q.11) *Athenian*: In addition to these there are other matters, closely related to them, in which we find many errors arising that are nearly akin to the errors mentioned.
Clinias: What are they?
Athenian: Problems concerning the essential nature of the commensurable and the incommensurable. For students who are not to be absolutely worthless it is necessary to examine these and to distinguish the two kinds, and, by proposing such problems one to another, to compete in a game that is worthy of them,—for this is a much more refined pastime than draughts for old men.

(Plato, *Laws*, 820C [Plato−Bury, 109].)

Based on this and the preceding passages of the text, some historians conclude that there was widespread ignorance of the concept of commensurability at the time in question and that therefore the discovery of incommensurability was very recent. This and such things as the unlikelihood of a long period of time between the initial discovery and Theodorus' work mentioned above lead them to a date near the end of the vth century (see Burkert [1972, 465], Heath [1921, I, 156], Knorr [1975], Sachs [1917]).

While most writers associate the discovery of incommensurability with the diagonal and side of a square, as in the quotation from Aristotle and the proof from Euclid, others associate it with DEMR. These theories will be discussed in Section 12,vii,viii,ix.

Section 8. The Euclidean Algorithm, Anthyphairesis, and Continued Fractions

In VII,1,2 we find described the so-called Euclidean Algorithm for finding the greatest common divisor of two integers b_0 and b_1. The same procedure for magnitudes—the subtracting at each stage of as many integral multiples of the smaller magnitude as is possible from the larger magnitude—is described in X,2,3,4. If the process continues indefinitely, then the two magnitudes are incommensurable (X,2). If the process stops, then the two magnitudes are commensurable and the last "remainder" (that which measures what is left) is their greatest common measure (X,3). This process of continually subtracting is referred to as anthyphairesis (see Mugler [1958, 61] for the etymology and uses of the word).

Let us write the steps in X,2,3, as follows (B_0 and B_1 are magnitudes with B_0 greater than B_1):

$$B_0 = n_0 B_1 + B_2$$
$$B_1 = n_1 B_2 + B_3$$
$$\ldots.$$

We see that the entire process is completely determined by the sequence of integers n_0, n_1, $n_2 \ldots$. Thus we can also call the sequence—as opposed to the process—the anthyphairesis of B_0 and B_1 and write ANTH (B_0, B_1). If the process stops—in the commensurable case—then we can take the sequence either to be finite or to consist of zeros after a point.

Let us look at the above situation from a modern point of view. Suppose that with the magnitudes B_0 and B_1 are associated the real numbers b_0 and b_1 and that the ratio of these numbers is the real number θ. Now we can write the following steps:

$$b_0/b_1 = \theta = n_0 + 1/\theta_1; \quad \theta_1 = n_1 + 1/\theta_2 \ldots.$$

Here θ_1, $\theta_2 \ldots$ are all real numbers greater than 1. Finally, by substituting backwards, we can write:

$$\frac{b_0}{b_1} = \theta = n_0 + \cfrac{1}{n_1 + \cfrac{1}{\ldots}}.$$

This is called the continued fraction expansion of the real number θ (see Niven and Zuckerman [1960, 137], Hardy and Wright [1938, 139]). Instead of writing θ as above we can simply write $b_0/b_1 = [n_0, n_1, n_2 \ldots]$. Thus the continued fraction expansion of the number that we have associated with ratio of the sizes of the magnitudes is identical with the anthyphairesis of the magnitudes.

If instead of continuing the fraction expansion indefinitely we stop at some point, we will obtain an approximation, called a convergent, to the number (see Niven and Zuckerman [1960, 140], Hardy and Wright [1938, 130]).

The discussion of magnitude in Book V is very vague. Further, there is a certain ambiguity with the term "magnitude" being used when it is the length or area of the magnitude that is meant. If we went one step further we could look at the ratio of the lengths of magnitudes and use this ratio to compare the magnitudes. The closest that Euclid comes to doing something of this sort is in V,def.3 where we read "A ratio is a kind of relation with respect to size between two magnitudes of the same type." This vagueness and its possible implications will be discussed in Section 13.

As an example, consider the side and diagonal of a square. By strictly geometrical means one can show that the anthyphairesis is [1,2,2,2 . . .] (see Fowler [1979, 819]). On the other hand, if we consider the real number R(2) which is the numerical value of the ratio of the lengths of the diagonal and side of a square, then by strictly numerical methods we can show that the continued fraction expansion is also [1,2,2,2 . . .] (see Fowler [1979, 842], Hardy and Wright [1938, 146]). The successive convergents for the continued fraction are 1/1, 3/2, 7/5, 17/12, 41/29 . . . .

I shall speak of the anthyphairesis of the segments of a line divided according to DEMR, as well as the anthyphairesis of the diagonal and side of a pentagon, on several occasions (see Sections 12,ix and 23). For now I merely state that the continued fraction expansion of the number $(1+\mathrm{R}(5))/2$ which is the ratio of the sizes of these magnitudes is [1,1,1 . . .] (see Hardy and Wright [1938, 44] and Fowler [1979, 842]). The convergents are 1/1, 2/1, 3/2, 5/3, 8/5, 13/8, 21/15 . . . , which in turn are ratios of successive terms of the Fibonacci sequence 1,1,2,3,5,8,13,21 . . . in which each term after the first two is the sum of the preceding two terms (see Niven–Zuckerman [1960, 92]).

Recall my discussion of the side and diagonal numbers for the square discussed in the last section. We see now that the successive ratios of diagonals and sides are precisely the same as the convergents of the continued fraction expansion of R(2). We shall see in Section 12,ix that a similar situation prevails in connection with DEMR. Some authors (Van der Waerden [1954, 127], Knorr [1975, 257], and Fowler [1979, 822]) link the development of procedures for generating approximations via side and diagonal numbers with the procedures connected with anthyphairesis in the sense that the first may have been a theoretical procedure developing out of practical experience with the latter.

CHAPTER III

EXAMPLES OF THE PENTAGON, PENTAGRAM, AND DODECAHEDRON BEFORE −400

As we shall see in Section 12, some authors associate the origin of not only DEMR but also of incommensurability with an intuitive development based on examples of the pentagon, pentagram, and dodecahedron. In particular there have been attempts to associate the Pythagoreans with the development of DEMR—among other things—because the 2nd-century writer Lucian (Section 11, Q.2) stated that the pentagram was used as a symbol of the early Pythagorean sect. Because of these various theories this chapter is devoted to an examination of early examples of the pentagon, pentagram, and dodecahedron.

The date −400 was chosen as terminal date because the ɪvth century brings us to the period when we start to have other indications, unfortunately not complete enough, of the development of mathematics. In particular, the evidence of the quotations which I will give later on indicates that by the first quarter of the ɪvth century mathematics had developed to some extent into the rigorous science of the type that we find present in the *Elements*. Furthermore, the ɪvth century brings us to the writings of Plato and Aristotle. Thus it is to these various sources, rather than to examples, that we should turn to from the ɪvth century on. I note that various authors have looked at examples of a later date and have examined these for evidence of a Pythagorean presence or influence in different places. Much of this material is discussed by Vogel [1966, chap. 3, Appendix B]. Other references may be found in Burkert [1972, 176, 460] and Cantor [1892, I, 175]. I shall mention a few of these later examples because in previous discussions of this type they were erroneously attributed to a pre-ɪvth-century date.

Before presenting the examples, something should be said about the whole idea of a study such as this. First of all, it should be always kept in mind that the pentagram was just one of many symbols, geometric and non-geometric, which appeared in various times and places in the ancient world. It is mainly because of the Lucian quotation and the widespread interest in the Pythagoreans that the pentagram has been singled out.

Indeed it is this over-enthusiasm which has led certain writers to "extend" the field of interest. Thus, for example, when we read Lindemann [1896, 732] we are led to believe that Schliemann had found examples of the pentagon at Mycenae. However, when we consult the original [Schliemann, 1878, 304], we find that the designs were essentially rosettes (Fig. III-1a). Again when Vogel [1966, Plate 10] presents us with a five-rayed star from Egyptian hieroglyphs we must ask ourselves whether this really has anything to do with the pentagon or pentagram. As we shall see later (Subsection C), the pentagram did have meanings in Sumerian and Akkadian; the principal one seems to have been "the [four] regions of the [inhabited] world or universe." Now if we look at the meaning of the five-rayed star (Fig. III-1,c,i) as an Egyptian hieroglyph [Gardiner, 1957, 487, symbols N13,14,15], we find that when the star was alone it was used as an ideogram or determinative (sometimes it was also used phonetically or as a phonetic determinative) in various words having to do with stars or constellations. When the star was enclosed in a circle the symbol means "netherworld (originally the place of the morning twilight, popularly known as the 'the Ducat')...." The number 5 here is probably accidental and may be due to nothing more than the fact that the star with five spokes is the star with the fewest spokes.

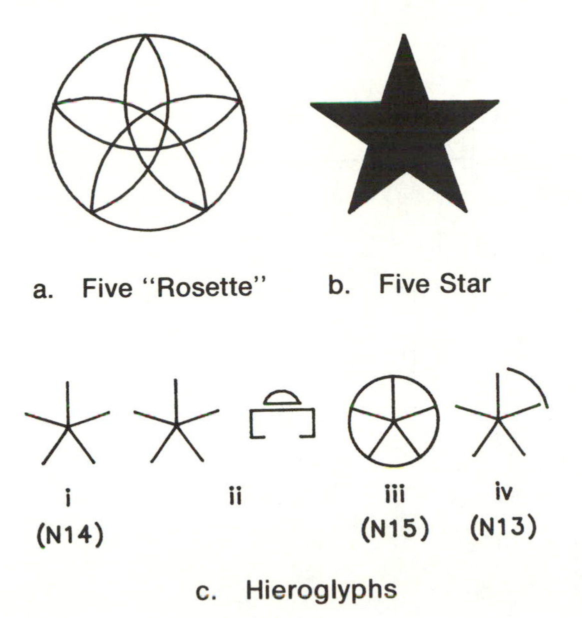

Source of C: Gardiner [1957, 487]

FIGURE III-1. Examples of Designs Not Considered

I have also excluded the true five-pointed star (Fig. III-1b)—the object that we obtain when a pentagram is completely darkened in so that the diagonals are no longer visible (see Subsection A). The literature from all periods simply abounds with stars and rosettes having five, six, eight, or ten arms (see, for example, Van Buren [1939; 1945], Goff [1963, figs. 306, 338, 461, 684], and Goodenough [1958, 179]).

The second thing that should be kept in mind when considering the following examples is that, even though this is to my knowledge the most extensive study of early examples of the pentagon, it is—like any study of this kind—influenced by chance. On the one hand, there is the chance that enters into archaeology (were there really no pentagons used in the Near East from before −3000 to c. −900?); on the other hand, many of these examples appear only in very specialized literature and it is only by chance that I stumbled upon them. For instance, a general interest in old Semitic writing unrelated to any interest in EMR and a particular interest in pentagons led me to the Tell Dan (cf. example no. 9 in Section 9); further, it was while reading the Winter 1979 issue of *Biblical Archaeologist* that I noticed that Dr. Hennessy of the University of Sydney was working at Tuleilat Ghassul (cf. example no. 8). While he had no pentagrams to offer me from there, he kindly supplied me with the early Egyptian example no. 1. All these limitations should be kept in mind.

For convenience I have grouped examples by regions. Also I have formed subgroupings of pottery, shields on vases, and coins. Again for convenience and because the dating of some of the Etruscan material is so uncertain, I have put all the vases, even those securely dated in the VIIth century, in Section 10. Each example (or group of examples) has been given a number which corresponds to the number on the map.

On the map I have indicated the locations of the various sites at which examples of the pentagram have been found. Pottery is indicated by P; coins by C; writing, whether on clay or stone, by W; mathematics by M; and dodecahedra by D. More precise descriptions are given in the sections indicated. Unknown dates are indicated by ? and are followed by very speculative guesses on my part to give some idea of the time sequence. On the other hand, dates followed by ? indicate an estimation on my part that is based on various considerations, such as the style of vase. The symbol < means before. Other dates are based on the articles cited. For more precise indications of the locations of the sites, one may consult Stillwell [1976], Grant [1971], Scullard [1967], Hammond [1981], and Avery [1962].

Section 9. Examples before Pythagoras (before c. −550)

A. *Prehistoric Egypt*

(1) *Nakada* (Upper Egypt north of Thebes; first dynasty, c. −3100)

Baumgartner [1955, 74] gives an example of a pentagram that appears as a pot mark on a jar. The sketch suggests that the incision was done in one continuous motion. I note that Baumgartner also presents a drawing of a vase with two five-pointed stars, one completely filled in and one with a circle inside of it.

For further details on the civilization and art of the first dynasty and in particular at Thebes, see Gardiner [1961, chap. 14, 30, 384] and Baumgartner.

B. *Prehistoric Mesopotamia*

The following references to pentagrams are taken from Goff's study of symbolism in prehistoric Mesopotamia [Goff, 1963]. The dates are based on Porada [1965, 176] and are to be considered as approximate. It is important

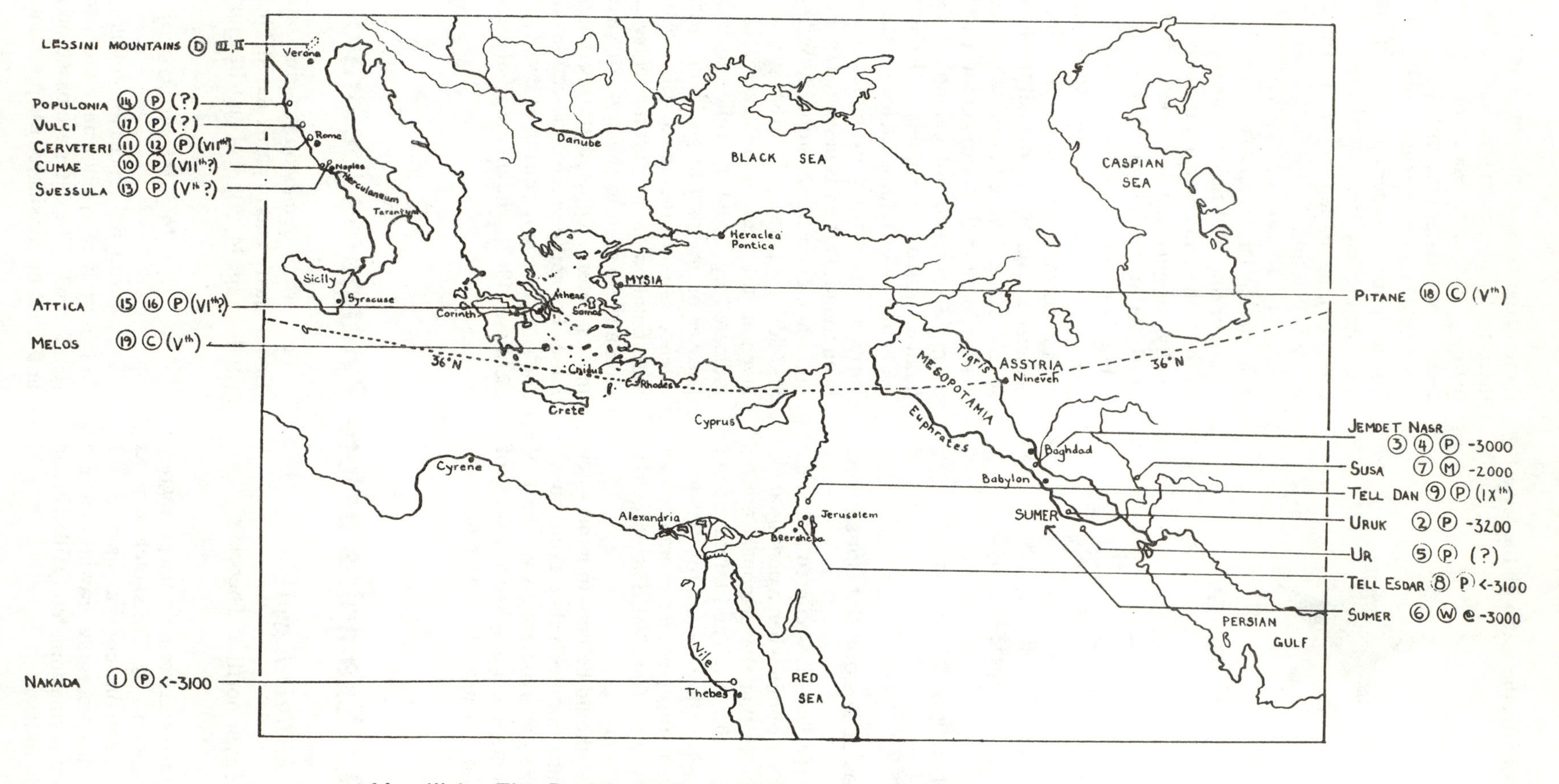

MAP III-1. The Pentagon, Pentagram, and Dodecahedron Before −400

to note Goff's warnings: "the greatest circumspection should be used in drawing inferences about early periods from knowledge of later periods" [p. xxxvi]; sometimes "a symbol is used only because it is reassuring and there may be no explanation in words," or sometimes "a symbol is accompanied by a simple explanation, but the same explanation may be attached to a great variety of symbols" [pp. xxxvii, 9, 77].

(2) Design on tablet from level 4 at Uruk (modern Warka) (Uruk period, −3200) [Goff, 1963, 77, fig. 316].

(3) Design on vase from Jemdet Nasr (Jemdet Nasr period, c. −3000 [Goff, 93, fig. 339]

(4) Design in spindle whorl from Jemdet Nasr (Jemdet Nasr period, c. −3000) [Goff, 1963, 113, fig. 464]

(5) Van Buren [1945, 114] gives references to examples from Ur (date?)

C. *Sumerian and Akkadian Cuneiform Ideograms*

This section deals with the meaning of the pentagram symbol, or the various cuneiform signs derived from it, in Sumerian and Akkadian. The term "Akkadian" here is used to include the Assyrian and Babylonian dialects.

Figure III-2 shows the development of the cuneiform symbol from the Sumerian pictograph to classical neo-Assyrian form. These examples are taken from Labat [1952, 138]. Only one illustration from each period has been used, and I have not shown the Babylonian forms.

Pictograph Uruk period c. −3200

Classical Sumerian

Old-Assyrian

Middle Assyrian

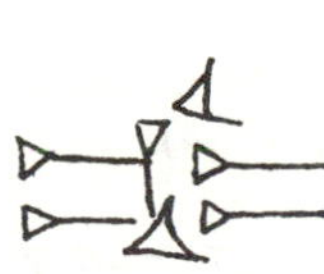

Neo-Assyrian

(6)

Source: Labat [1952, 138]

FIGURE III-2. Development of the "Pentagram" as a Cuneiform Symbol

Since the Akkadian language used the Sumerian cuneiform symbols either phonetically (in our case for *ub*, *up* or *ăr*) or as ideograms in the same sense as the Sumerians—at least for the most part [Labat, 1952, 25]—it is to the meaning as a Sumerian ideogram that we must look if we wish to determine what the pentagram represented to the Mesopotamians. Labat [p. 139] gives the following examples (the Sumerian sign is in uppercase letters, the Akkadian equivalent in italics).

1. UB, *kibrati*: regions, parts of the universe.
   UB-DA-TATTAB, *kibrat erbetti*: the four regions of the world, the universe.
2. UB, *tubqu*: interior.
   UB-DA, *tubqat shachati*: inside and outside.
   UB-LÍL-LÁ, *ibratu*: room.
3. UP-PAD, *challula*: mole-cricket (gryllopalidae).
4. UR, *karmu*: an empty terrain, devastation; *tanittu*, glorification.
   ÁR-NIGIN/NIGÍN-NA, *binati*: members of the body.

Several other examples of the use of the UB(ÁR) symbol have been given by Fuÿe [1934, 23, 24]. The Sumerian texts CTXII and CTXIX, discussed below, were taken by Fuÿe from Thompson [1901*; 1904*]. Rather than giving the entire Sumerian texts and Akkadian translations, I will list some of the meanings of the words presented by Fuÿe which contain the pentagram symbol. The words are indicated by line number within each text. Note that in some cases the meaning is different from that given by Fuÿe and that in other cases a meaning is supplied where Fuÿe has not given one.

TEXT CTXII. (1-4, 6-8) to express accurately; (5,13) utterance; (9,10) to look into, to examine, search; (11) joy; (12) night watchman; (14,15) to know.

TEXT CTXIX. (31) utterance, expression; (35) utterance; (37) derision, mockery; (41,42) place where the gods decide fate.

It should be emphasized that the above meanings are derived from bilingual Sumerian-Akkadian texts and thus may not represent the earliest Sumerian meetings. (See Van Buren [1945, 114] for additional references to the various sources.)

When we look at the uses of the Akkadian word *kibratu* [Chicago Oriental Institute, 1956, vol. 8, 331], we see that the basic meanings given are: regions (referring to the four regions of the inhabited world), edge, shore line. Typical examples of usage are:

1. "who has no rivals among the princes of the four regions";
2. "upon her command she [Istar] subjugated to him [Ammiditana] the four regions of the world";
3. "which are not the regions warmed by the brightness of your light?";

4. "I looked about for coastlines in the expanse of the sea."

i. Fuÿe's Theory

The booklet *The Pythagorean Pentagram* [Fuÿe, 1934] involves a shifting back and forth between various time periods with the motivating factor being the Lucian quotation (Q.2 of Section 11) and its linking of the pentagon with the Pythagoreans and "health." Perhaps indicative of the style and mood of the book is the suggestion [p. 11, fig. 4] that the pentagram was originally an anthropomorphic symbol for Hygeia, the Greek goddess of health (see Rose [*OCD*]), in the sense indicated in Figure III-3. Fuÿe says that this image also recalls representations of the Phoenician goddess Tannit.

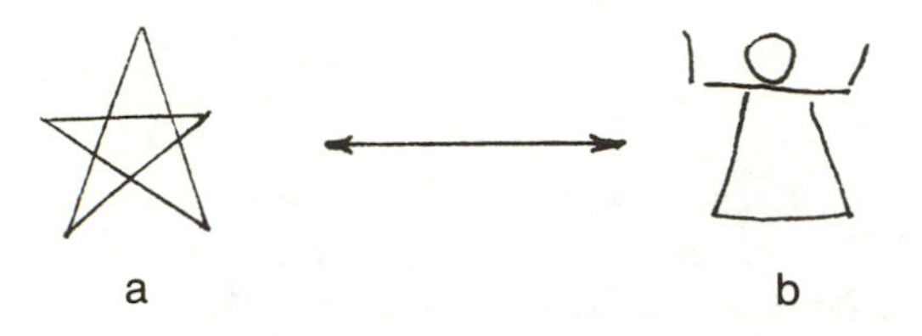

Source: Fuÿe [1934, 12, fig. 4]

FIGURE III-3. Suggested Association of the Pentagram with Hygeia

Fuÿe goes back in time and considers the pentagram as an ideogram and cuneiform symbol. First the author looks at various meanings of the UB symbol that are derivable from bilingual Sumerian texts. I have given some of his examples above.

Next the author considers older Sumerian texts which are written only in Sumerian ("classical" Sumerian in Fig. III-2) and for which it is often the case that the meaning is presently unknown or far from certain. The first batch consists of texts from Fara (c. −2600 ?) and are taken from Deimel [1922*, textes 18,59,60, 69,75]).

Here are some of Fuÿe's interpretations [p. 36] concerning the UB symbol in association with other signs.

(Texte 18—II, 3,5; V,5): mace, mace with 50 [large number of] heads.
(Texte 18—II, 7; V,9): brilliant jewelery or lamps.
(Texte 59—I, 10,11): mammary, wet nurse.
(Texte 69—I, 1): two-edged axe.

The next set of tablets considered [p. 42] are Proto-Elamite documents from Susa (bibliographic source is not clear). In one series of texts, the pentagram is always associated with the number 2 and Fuÿe suggests the possible meanings: ear rings or breast ornaments or perhaps lamps.

Fuÿe now examines [p. 43] some tablets from Jamdet Nasr [Langdon, 1928*]. He states that one "is struck by the importance of the UB sign" and goes on to suggest that it may represent a part of the temple, a sacred object or perhaps a god, in particular a sky god.

Notice that in these suggested meanings there is nothing dealing with health. But now Fuÿe returns to this theme. He points out [p. 53] that on Tablet 46 from Jemdet Nasr there is a figure which he interprets as being a female holding and feeding a snake. This he says is similar to the figure of the Roman goddess of health Salus—who corresponds to the Greek goddess Hygeia—which appears on certain Roman coins bearing the legend "SALUS AUGG." The particular coin illustrated by Fuÿe may be found in Carson [1979, 139, type 1039] where the date 283 is given. I note that the same figure, but with different inscriptions, also appears [pp. 109, 131, 132, types 870, 991, 996] with the dates 262, 263, 275, and 276. In particular on types 991 and 996 we read "SALUS PUBLICA" so that the reference seems to be the general well-being of the population. This is indeed one of the three things that Salus personifies, in addition to health and prosperity (for more details see Schmitz [1849]). Furthermore, there are other coins (see Carson [1979]) in which Hygeia is not holding the serpent, but rather she feeds it while it is coiled around an altar.

There are two major criticisms that one can make with regards to Fuÿe's suggestions. First of all, whatever the importance of the UB sign in the Jemdet Nasr tablets taken as a whole the UB sign does not appear on the same tablet as the figure. Secondly, the association of the pentagram with Hygeia in Figure III-3 is certainly tenuous, to say the least.

D. *A Babylonian Approximation for the Area of the Pentagon*

Bruins and Rutten [1961; see also the summary given in Bruins, 1950a] have published a series of mathematical cuneiform tablets found in 1936 at Suse (Iran) which corresponds to ancient Elam. The authors date the tablets to the end of the first dynasty of Babylon (beginning of 2nd millinaire; see Oates). Of special interest to us is line 26 of Tablet I, Text III [pp. 26,28, plate IV,45] which reads:

(7) 1 40, the constant of the five-sided figure (Fig. III-4).

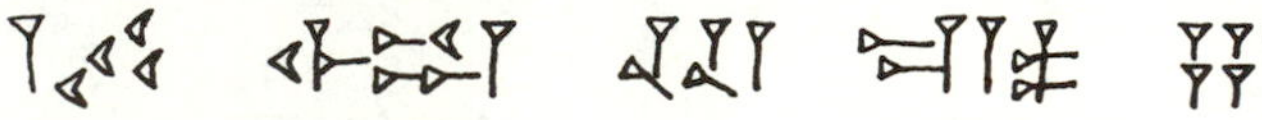

1 40 IGI.GUB šà SAG 5

Source: Bruins and Rutten [1961]

FIGURE III-4. Line 26 of Tablet I

Here we have the numbers 1 40 and 5 written in the standard Babylonian sexegesimal (base 60) manner (see Neugebauer [1957], Van der Waerden [1954], and Bruins [1950b]), that is, without any indication of the absolute value of the numbers in question. From the context the grouping 1 40 is seen to represent 1 + 40/60, i.e., 5/3 in fractional form. According to Neugebauer and Sachs [1945, 132, 170] the Sumerian IGI·GUB BA means literally something like "fixed (or established) fraction" while the Sumerian SAG has the general meaning "head."

Thus the meaning of line 26 [Bruins and Rutten, 1961, 32] is that if we have a regular pentagon side of unit length then the area will be 5/3; a value which may be compared to the true value of 1.720 . . . . It should be emphasized that the text simply gives the formula and does not provide anything in the way of a derivation.

An explanation of how this result was obtained is suggested by Bruins and Rutten [1961, 32] and Bruins [1950a, 1027]. The basic assumption is that the side a_n of the n-gon and the radius r of the circumscribed circle are related by the formula $n \cdot a_n = 6r$. If the side of the pentagon is taken to be $1° = 60'$ then the radius will be $50'$.

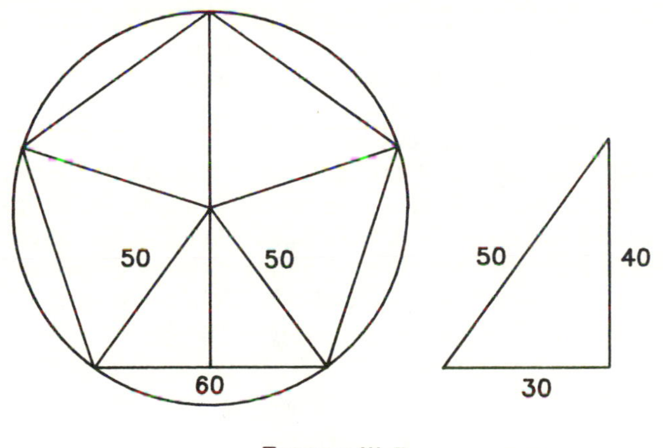

FIGURE III-5

If the pentagon is broken up into five triangles (Fig. III-5), then each of these triangles can be split up into two right triangles with hypotenuse $50'$ and base $30'$. This in turn means that the other side is $40'$ so that the area of each of the five triangles is $\frac{1}{2} \cdot 60' \cdot 40' = 1200' = 20 \times 60'$. Ignoring the $60'$ in the Babylonian manner the area of the pentagon is then $5 \cdot 20' = 100' = 60' + 40' = 1° \, 40''$ which is written as 1 40 in the text.

That the authors of the tablets did indeed use the basic relationship $n \cdot a_n = 6r$ in determining the area of polygons is supported by lines 27 and 28 which deal with the areas of the hexagon and the heptagon and line 29 which deals with the height of an isosceles triangle, all of unit side. The situation in these is somewhat different, however. First of all for the hexagon the relationship $6 \cdot a_6 = 6r$ is exact. Further, whereas for the case of the pentagon the altitude comes out exactly, the case of the hexagon requires an approximation to $R(a^2 - b^2)$. That the approximation in this case was $a - b^2/2a$ (the first two terms of the binomial expansion) is in turn confirmed by lines 27 and 29 and Texte II [p. 24] which also deals with the area of a regular hexagon.

For the case of the heptagon one must both approximate the radius in terms of the side and find bounds on the square roots. Bruins shows that the assumption $7 \cdot a_7 = 6r$ combined with his earlier deductions concerning bounds on square roots fits in with the constant given on line 28 (see Bruins [1950a, 1027; 1950b, 307]).

I remark here that the same constants for the pentagon, hexagon, and heptagon are found in works of the 1st-century mathematician Heron (Section 27,A). But whereas the Babylonian approximation has no relationship to DEMR, Heron, as we shall see, is in full command of its properties.

i. Stapleton's Theory

The article by Stapleton [1958] deals with a variety of cultures, times, and topics involving discussions of the role of the number 5, hands, cubits, etc. In particular there is an attempt to establish a relationship between the Babylonians and the Pythagoreans. With respect to the above calculation of the area of the pentagon he writes: "Geometric study of the pentagon can be traced back at Susa . . . and —apart from its magic association with the Hand—this may have been due to an appreciation that its symmetry is comparable to that of the circle, both being thus superior to the mirror-image relationship of the human hands" [p. 36].

We shall see in Section 12,xii how Stapleton further relates the hands, in a most direct way, to the pentagon.

E. *Palestine*

(8) *Tell Esdar* (Chalcolithic −4500 to −3100)

Kochavi [1978, 1171] illustrates an incised flint scraper (Fig. III-6) from stratum IVb at Tell Esdar (midway between Dimona and Beersheba). He assigns this stratum to the Beersheba culture of the Chalcolithic period (−4500 to −3100). I personally inspected this piece in the Israel Museum and discussed it with the curator, Dr. Ruth Hestrin. It appears that the extension of the side of the pentagram shown in the upper right decreases in depth as the line ends. This, in my opinion, is evidence of quickly executed graffiti.

Another example of a geometric design from the Chalcolithic period is given by the eight-pointed star from Tuleilat Ghassul [Lee, 1978, 1210]. Professor John Hennessy, the current excavator at Tuleilat Ghassul, has informed me that the pentagram has not occurred among the decorations found at that site.

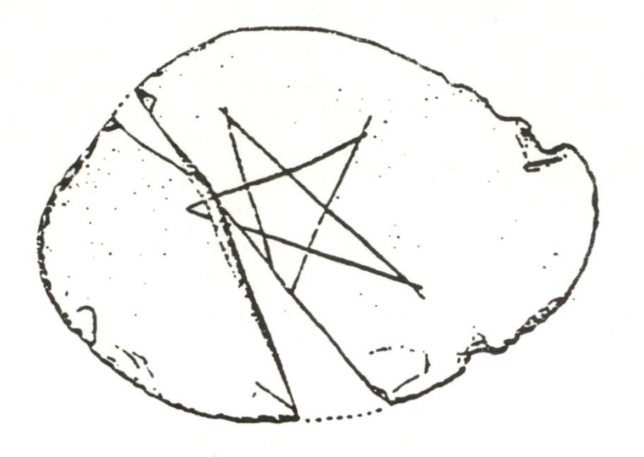

Source: Kochavi [1978, 1171]

FIGURE III-6. Flint Scraper from Tell Esdar

(9) *Tell Dan* (IXth century?)

Avigad [1966, 210] has published a diagram of a bowl (Fig. III-7) found at Tell Dan in northern Israel which displays a pentagram and the writing lṭb[ḥ]y'. Avigad presumes that the language is Aramaic and this enables him to translate the word as meaning "property of the cooks." The date of the bowl is suggested by the known conquest of Tell-Dan by the Arameans (see 1 Kings 15:20) in the IXth century. Further evidence for this date is obtained by a comparison of the *b* and *y* with the same letters on the Mesha stone.

Regarding the pentagram itself, Avigad notes that it was inscribed on the bowl after firing, which would indicate that it was not a trademark of the potter. He presumes that the pentagram had a symbolic or magical meaning. Another pentagram was found on a piece of clay.

Source: Avigad [1966]

FIGURE III-7. Bowl from Tell Dan

The same image (upside down) appears as item no. 123 of the Israel Museum [1972, 130].

Mention should be made here of the seal stamp with a pentagram with the letters y r š l m (Jerusalem) in between the apices of the pentagram [Israel Museum, 1972, 152, no. 155]. It was formerly thought [see Diringer, 1934, 130, plate XVI; Vogel, 1966, 41] that these dated from the IVth, Vth, or VIth centuries. However, they are now securely dated to the Hellenistic period—IIIrd or IInd centuries (see Avigad [1974; 1976, 25]).

Section 10. From Pythagoras until −400

A. *Vases from Greece and its Italian Colonies, Etruria (Italy)*

The names used here are to be interpreted in a cultural sense as much as in the strict political and geographical sense. Indeed, because of the constant cross influence and because in some of the examples of vases it is not known whether we are dealing with local products or imports, I have grouped all these entities here. The term "Italy" refers to the various city states, such as Rome, which lay outside of Etruria.

One notes that most of the vase examples whose source is known come from Etruria, the region which corresponds approximately to modern-day Tuscany. From a political viewpoint, we are dealing with a loose association of city states which expanded and contracted with its zenith being attained c. −500, but with the rise of these city states starting in the VIIth century and their history continuing until the IVth. (For more information on the Etruscans and their culture, see Strong [1968; a chronological table is given on p. 107], Bandinelli [1973]; on early Greece and its colonies, see Cook [1961].)

For a general discussion of vase styles, see Cook [1972], Pryce [*OCD*], and Cook [1961, 43 and plates]. I have used these for an approximate dating—as indicated by ?—of some of the vases. Etruscan vases are discussed by Cook [1972, 146] who concludes that most vases were imitations of Italian and Greek models.

I have not searched through the *Corpus Vasorum Antiquorum* (see Cook [1961, 331]). My thoughts on doing this have been expressed most elegantly in the last two lines of Csörgő and Révész [1981, 19].

(10) *Cumae* (VIIIth?, VIIth?)

Cumae, north of Naples, was the earliest Greek colony in Italy and was founded c. −750 (see Salmon [*OCD* 1]). Gabrichi [1913, 282, plate XLVII,3] has published a diagram of a vase on which we see two pentagrams, a star (flower?), two Xs and two Ts. All these are painted on the vase. (See Cook [1972, 39, 145, 149] for additional comments on vases found at Cumae.)

(11) *Cerveteri* (Etruria, first half of VIIth century)

This vase is illustrated and briefly discussed in Bandinelli [1973, 152, fig. 176, 153]). The pentagram appears here with dots in the apices and appears as background decoration along with an eight-pointed star (four crossing lines) in a scene involving a man and a woman. (Only one side of the vase is shown.) Efforts to obtain more information about this vase, in particular from the museum, have been unsuccessful.

(12) *Cerveteri* (Etruria, c. −650)

This vase is the so-called Aristonothos crater named after the signature on it. Reproductions appear in Arias [plates 14,15], Schweitzer [1955, plates 34,35,40], Bandinelli [1973, 150, 371, figs. 177,438,439]. The pentagram appears twice as background decoration, in a battle scene involving two boats.

Schweitzer [1955] has made a detailed study of this vase and finds [p. 10] that the decoration agrees with the Attic geometric style except for the pentagram. However, he seems to have been unaware of the two vases (discussed in examples 15 and 16) which may come from Greece proper (see also Arias [1962, 274] and Cook [1972, 93]). There does not seem to be agreement on where the vase was made, but it is generally suggested that this is the work of a Greek-speaking immigrant.

(13) *Suessula* (vth century?)

Kuhn [1887, 267] reports that a pentagram appears, along with a feather, as graffitti (presumably meaning scratched in after fabrication) at the bottom of a black cup. Suessula, which is modern Cancello near Naples, was an inland town in Campania. There were Greek settlements in Campania after −750, and these survived until the vth century (see Salmon [*OCD* 2, *OCD* 3]).

(14) *Populonia* (Etruria)

Buonamici [1932, 475, fig. 11] in his survey of Etruscan epigraphical material describes and illustrates a bowl with a pentagram scratched in very roughly—some of the lines are partially repeated—on the bottom. He speaks of many examples but this seems to refer to the general group of stars.

B. *Shield Devices on Vases*

Chase [1902] made an extensive examination of the various devices to be found on shields that are depicted pictorially on Greek coins and vases, etc., or described verbally in literary sources. For the period following −700 Chase distinguishes ten, possibly twelve, classes. The pentagram is assigned [p. 84] to class I, which is "devices purely decorative." Three examples are listed under the heading "pentagram" [p. 117, CCIV] and two of these are vases.

(15) *Attica*(?) (VIth century?)

Jahn [1854, 227, no. 729] describes, without an illustration, an Attic black-figured amphora. Aside from one warrior with a pentagram on his shield there are warriors with an eye and three stripes on their shields. The vase also has paintings of various men, women, and children.

(16) *Attica*(?) (VIth century?)

Gherhard [1830, 219, no. 9, plate XXII,11] describes and illustrates a fragment from an Attic black-figured amphora. This Panathenaic vase—that is, representing scenes from the Panathenaic games [Nilson and Croon, *OCD*] shows Athena (called the Roman "equivalent" Minerva by Gherhard) carrying a shield with a pentagram on it. In addition there is a scene from a foot-race.

The following Etruscan vase, not described by Chase [p. 81], is also of interest.

(17) *Vulci* (Etruscan)

A pentagram with intertwined lines appears on the shield of a warrior, one of several, painted on a Kylix excavated at Vulci in 1837-38 (see Museo etrusco Georgiano [1842, 15, plate LXXXVII]). Attempts to find out more about this vase have been unsuccessful.

Although Chase's study is limited to the portrayal of devices on shields, the following points that he makes are of interest to us:

1. The number of devices used was large; Chase has 268 different categories of which many are purely decorative [p. 84].
2. The subjects chosen for vases apparently followed closely the actual practice of the time [p. 79].
3. There appears to be no correlation between the various gods and heroes and the devices represented on their shields [p. 82].

C. *Coins*

Coins have been a favourite point of reference for those who wish to discuss the pentagon and its possible relationship to the Pythagoreans. Already in 1792 we find Eckhel [1792, 63] mentioning the quotation from Lucian (Section 11,Q.2) in connection with his study of coins from Gaul (there are coins with pentagrams from ist century Gaul; these are illustrated in Deonna [1954, 47, fig. 6; Vogel, 1966, plate 11b]). It turns out, however, that there are only two known coins dating from before −400 on which the pentagram appears. Since most of the literature is out of date, I have used more recent references and information, kindly supplied to me by Dr. Martin Price of the Department of Coins and Medals of the British Museum.

(18) *Pitane* (mid vth century)

Pitane was a city on the Mediterranean located in the country of Mysia in Asia Minor (see Stillwell [1976]). This coin is made of silver and has a pentagon with the name Pitane on the reverse with a head on the obverse. No photograph of this coin appears to be available and the only listing that I have seen is in Head [1910, 537]. Dr. Price informs me that the British Museum now possesses a copy of this. He describes the head on the obverse side as rather nondescript, but certainly not Zeus Ammon (see below) and says that it may possibly be female. Head assigns the coin to the vth century and Dr. Price informed me that the coin cannot be earlier than −475 nor later than −400 and suggests "mid vth century" as a date.

One should be careful not to confuse this coin with a series of bronze (Æ) coins from Pitane which date from the ivth century. These coins are listed in the British Museum catalogue of coins from Mysia [British Museum, 1892, 171, plate XXXIV, no. 5,7,8]; see also Anson [1910, 87, plate XIX] and Plant [1979, no. 2709]. It is these coins which have a picture of the god Zeus Ammon on the obverse. I mention the god Zeus Ammon because of various attempts in the literature to link the pentagon to the cult of Zeus Ammon and then to health (see the Lucian quotation, Section 11,Q.2). However, as we have seen, the pentagon appears on the silver coin of the vth century without Zeus Ammon and as the British Museum catalogue shows Zeus Ammon appears on most of the bronze coins without the pentagram. Thus the association of the pentagram with Zeus Ammon is most unlikely.

The same thing occurs on the coins from other places. An example of this is given by a series of coins of the early iiird century from Cyrene (in the ancient land of Cyrenaica, which is now Barca) in North Africa [Muller, 1860, 57]. Here we find a whole set of bronze coins with Zeus Ammon on the obverse side and a palm tree together with various small subsidiary symbols on the reverse side. On one of the coins [no. 264] we find a pentagon, but on others we find such things as crabs, horns, and serpents. Incidentally, we find Muller [p. 110; see also the references in fn. 5] identifying the pentagon with Hygeia the Greek goddess of health; I have already discussed her in connection with Fuÿe's thesis (Section 9,C,i).

(19) *Melos* (second half of the vth century)

This island, one of the Cyclades, lies in the Aegean Sea approximately halfway between Athens and Crete. The coin in question is a silver "stater" which has most recently been published in Kraay [1964, no. 3, plate 1]; see also Babelon [1901, 854, no. 1357, plate CCXLI, no. 21] and Head [1910, 892]. Melos was conquered by Athens in −416 and Dr. Price suggests a date of c. −450 or a little later (see also Babelon [p. 860]. The obverse side of these coins, which are known only from a single cache, is almost always an apple (? see Kraay [1964, 2 fn. 3]). There are, however, many different reverse types: a circle with six spokes, a square split into eight triangles by the diagonals and a vertical and horizontal line, a crescent, a triskelion with human feet, barley grains, vases, and dolphins. Thus the pentagon is just one design among many designs; some are geometric and some represent real objects.

Several additional remarks are in order. Reference is sometimes made to the coin with a club and dot on the obverse and a pentagram with a dot in the middle which appears on the reverse (listed in British Museum [1873, 59, no. 29, no illustration]). This coin is now dated from −280 to −260 in Thurlow [1979, 40, no. 222, plate 72] (see also the sketch in Boyne [plate IIa, no. 7] and Haeberlin [1910, 167, 169, plate 68]). Also listed in Thurlow [1979, 40, no. 227, plate 73] is another coin in the category "uncertain of Central Italy, c. −280 to −260," again with a pentagon containing a dot in the centre on the reverse but this time with a scallop shell on the obverse. Both of these coins are of the "aes grave" type (cast heavy bronze; see Carson [1970, 107]). Again the pentagon is just one of a large variety of designs that appear on this type (cf. the type index in Bradbury [1979, 44]).

There are many coins with pentagons on them, always apparently as a small subsidiary symbol, from the ivth century on. These include coins minted under Alexander the Great; see Muller [1855, index]; there are specimens in the American Numismatic Society collection, for example, Macedonia-amphipolis 17.10, and from other Greek and Roman cities and rulers. Consult, for example, the indices in Grose [1923], Grueber [1970], and Crawford [1974]. As these indices indicate, the pentagram is only one of a large variety of symbols of all types which served as mint marks or simply as designs.

D. Dodecahedra

In de Stephani [1885] we find the report of the discovery of a more or less regular dodecahedron carved out of steatite (soapstone, a form of talc). The faces of the stone are cut by various lines and seven of the faces have varying numbers (1,1,1,1,2,3,4) of holes (see the plate accompanying the article, reproduced in Lindemann [1896, plates II,III]). The site of the discovery lies in the Lessini mountains which are north of Verona and east of Lake of Garda. While there were fragments of vases nearby [p. 1448], the archaeological indications were not sufficient to determine the date of the dodecahedra and de Stephani reports on the opinions of various people. There was, however, no consensus.

This dodecahedron has become the source of various statements in the literature of the history of Greek mathematics which consider the question of a possible Pythagorean or pre-Pythagorean origin for the mathematics of the pentagon (see, for example, Cantor [1892, I, 175], Sachs [1917, 83], Euclid−Heath [III, 438], and Burkert [1972, 460]). Unfortunately the ultimate source of these reports is not the original article by de Stephani, but the highly speculative article by Lindemann [1896]. In his fourth section Lindemann simply reports on de Stephani's article, but then in his ninth section he tries to determine the age based on the report of de Stephani and people cited by him, and he appears to accept the period from −1000 to −900. Elsewhere in his paper (sections 9,18,19) we find statements about the possibility of learning about dodecahedral shapes from minerals, other examples of dodecahedra, various examples of the appearance of pentagons, and the Aristophones vase, etc. An effort is made to relate all this to the Pythagoreans because of the various statements concerning the Pythagorean discovery of the dodecahedron.

A search of the literature revealed only one mention of this dodecahedron, aside from the references to Lindemann. This appears in Heinevetter [1912, 58] who refers to the "Etrusco-Roman dodecahedron which was published in [de Stephani]." Presumably this designation, which in fact corresponds to a much later but still rather vague date, is based on the comparison with the other examples of dodecahedra and icosahedra. Some of these, which bear Greek letters, have been dated on epigraphical grounds to the IIIrd and IInd centuries. Aside from the references given by Heinevetter, I know of two icosahedra, cited by Alexander [1929; 1937], one of which apparently comes from Egypt and perhaps dates from the IIIrd century. Alexander was unable to find any reference to such objects in ancient works.

After many inquiries I was finally able to locate and observe the de Stephani dodecahedron in the Museo Civico di Storia Naturale (Verona) (inventory no. 3669). The dodecahedron is displayed with other material from the same region, all of which belong to the period from the IVth to Ist centuries. The director, Alessandra Aspes, assigned the dodecahedron itself to the Rhaetic culture in the IIIrd or IInd centuries. I noted that several of the pieces displayed with the dodecahedron have various incisions, lines, or "writing," but apparently nothing is known concerning their possible meaning.

There is another group of dodecahedra that has been referred to in the Pythagorean literature. These are cast out of bronze in such a way that the interior is empty and there are little "pearls" sitting at each vertex. These have been discussed by Deonna [1954] and, according to his article [p. 23], many of the previous articles which dated these as far back as the Bronze Age are erroneous. Some of these examples were cited by Lindemann; Deonna dates these dodecahedra to the period 200 to 400. These dates are also those accepted by Duval [1981] who discusses an accurate mathematical way of describing the dodecahedra—which have holes of various sizes—in the hopes that a serious analysis of dimensions will eventually give a clue to their original raison d'être. Recent photographs may be found in Duval[1982] and Thompson [1970].

E. Additional Material

I have recently come across, or have been informed of, additional material involving inscriptions and game boards from ancient Egypt and prehistoric stone balls from Great Britain which have their surfaces divided by incisions. Lack of time has not permitted sufficient research for a proper examination of the relevant literature and so I will discuss these items in a future article.

Conclusions

It seems to me that the wide diversity of the material that we have seen, together with the scattered sources and enormous spread in time, indicate that the pentagram, like so many other symbols—geometric as well as non-geometric—must, unless some other evidence comes along, simply be considered as a symbol which appeared among many people and at many epochs. It may have indeed served as a sign of the Pythagoreans (Section 11,Q.2) in the same way as we see five-pointed stars being used by different countries, companies, and groups today—as just a decoration without any hidden, deep meaning. It may also be true that the pentagram was somehow associated with health, but it is difficult to believe that a symbol that was far from unknown at the time of Pythagoras and the early Pythagoreans was used as a recognition symbol (see Section 11,Q.3). That somehow the sight of all those pentagrams inspired deep mathematical thought as opposed to simple doodling—something that the pentagram is ideally suited to because it can be drawn without lifting the stylus, knife, brush, pencil, or felt-point marker off the clay, pottery, or paper (indeed we have seen this aspect in several examples)—is of course possible, but in my opinion not very likely.

CHAPTER IV

THE PYTHAGOREANS

In Section 11 of this chapter, I have gathered together the ancient sources which specifically link the Pythagoreans to various mathematical concepts or objects related to DEMR. Section 12, in turn, examines the theories in the literature, many of them based on the quotations of Section 11, which link the Pythagoreans to the origin of DEMR. Before presenting this material, I think it would prove useful to the reader if I first outlined some of the difficulties involved in any discussion of the Pythagoreans and also summarize various names and dates associated with the Pythagoreans and early Greek mathematics.

In commenting on the various theories concerning Pythagoras and the Pythagoreans, Burkert [1972, 9] says, "Pythagoreanism is . . . reduced to an impalpable will-o-the-wisp, which existed everywhere and nowhere." The various historical sources are discussed by Heidel [1940] and Burkert [1972; especially chapter 2 on Pythagoras and chapter 6 on the mathematics of Pythagoras and the Pythagoreans] and Philip [1966, especially Appendix II on mathematics].

This lack of certainty is reflected in the attitude of various historians to Pythagoras himself. Heath [1921, 145, 150] "would not go so far as to deny to Pythagoras the credit of the discovery of [I,47 and would] . . . like to believe that tradition is right and that it was really his" and thinks that it "is certain that the theory of application of areas originated with the Pythagoreans, if not with Pythagoras himself." Neugebauer [1957, 148] writes: "It seems to me evident, however, that the traditional stories of discoveries made by Thales or Pythagoras must be discarded as totally unhistorical."

When we turn our attention to the Pythagoreans the situation is not any clearer. According to Heidel [1940, 29]: "It is one of the singular facts in the history of Greek thought that individual Pythagoreans are rarely mentioned except by later writers whom one has every reason to suspect." Even when we turn to the accomplishments of the Pythagoreans in general the testimony is far from being precise. For example, when Frank [1940, 48] writes, "Ever since mathematics as a science came into the world which was the accomplishment mainly of the Pythagoreans," he is basing his statement on the following quotation from Aristotle:

(Q.1) At the same time, however, and even earlier the so-called Pythagoreans applied themselves to mathematics, and were the first to develop this science; and through studying it they came to believe that its principles are the principles of everything.

(Aristotle, −384 to −332, *Metaphysics* 985b [Aristotle–Tredennick, 31]; this passage is discussed by Heidel [1940, 9] and Burkert [1972, 49, 448, 509])

Aristotle has been vague, and as Heidel [p. 14] points out, there is no reference to Pythagorean geometry proper in Aristotle's work despite some opportunities to do so—for example, with respect to the Pythagorean Theorem (I,47). According to Heidel [pp. 11,18], we can infer from Aristotle that, from the middle of the vth century on, there were Pythagoreans working in number theory and geometry, but that Aristotle cannot be used as a source for the period before the middle of the vth century. This, however, should not be taken as an endorsement by Heidel of the assignment of specific theories to the Pythagoreans, for at the end of the article he writes: "The conclusion to which we are driven by our study is that it is impossible to reconstruct the history of Greek mathematics as one may to a certain extent ... of Greek scientific thought in general by focusing attention upon individual men or groups." However, other writers tend to attribute most of the early development of Greek mathematics to the Pythagoreans; some of these views are summarized by Burkert [1972, 403]. On the other hand, Philip's theory [1966, 205, 207] is that until about −450 Pythagorean mathematics consisted mainly of speculation in number theory (see Philip [chap. 6]; Burkert [chap. 1, chap. 6.2]; Van der Waerden [1954, 96]; Knorr [1975, chap. 5]). Then, following a Pythagorean debacle around −450, there was no activity by the Pythagoreans until about −400 and this "must have been largely dependent on external stimuli, particularly in mathematics which had meanwhile begun to evolve as an independent discipline."

In reading the literature about the Pythagoreans, one should probably distinguish three groups: first, Pythagoras and the Pythagoreans of his epoch; second, the "Early Pythagoreans" of the first half of the vth century; and third, the "Late Pythagoreans" of the second half of the vth century and early ivth century. It is not always clear in the literature which group the author is referring to.

The following is a chronological listing of some of the names of mathematicians who have been associated with the Pythagoreans and who will be mentioned at various points in this book. I have also included, because of his important role in vth century mathematics, Hippocrates of Chios.

i. Pythagoras

Vogel [1966, 24] assigns c. −570 for Pythagoras' birth (cf. Philip [1966, 195]; Burkert [1972, 109]). Freeman [1953, 73] summarizes the various references to Pythagoras in ancient literature.

ii. Hippasus

He apparently lived in the first half of the vth century. If this dating is correct, then Hippasus is the first Pythagorean mentioned in ancient sources who did mathematics. See Q.7 and Q.8 and Burkert [p. 206]. See Freeman [1953, 84] for a complete list of classical commentaries on him and see also Fritz [1945].

iii. Hippocrates of Chios

He appears to have been active c. −430 [Burkert, 1972, 313, 402, 420; Van der Waerden, 1954, 131; Freeman, 1953, 217]; however, this date has been disputed and a more accurate date may be a somewhat later one. The mathematics of Hippocrates has been discussed in Section 4. In his "Catalogue of Mathematicians," Proclus [Proclus−Morrow, 54; Thomas, 1939, 149] says that Hippocrates is the first recorded person to have compiled the *Elements*. Van der Waerden [1954, 135] concludes that Hippocrates was in possession of Books I and II and large parts of the material in Books III and IV of Euclid. Heidel writes: "One rushes to add that all this body [the work of the Pythagoreans] may be certainly referred to a time before Hippocrates of Chios" [1940, 18; cf. also Burkert, 1972, 449].

iv. Theodorus of Cyrene

He was active c. −410 to −390. Theodorus appears in Plato's *Theaetetus*, whose action takes place in −399 (see Sections 7,ii and 18,A). He appears right after Hippocrates in Proclus' "Catalogue of Geometers" [Proclus−Morrow, 54] and is credited with being one of Plato's teachers [Freeman, 1953, 219]. The contributions of Theodorus will be discussed in Section 15. I note that there is a Theodorus of Cyrene listed in Iamblichus' "Catalogue of Pythagoreans" [Iamblichus−Albrecht, 261; Iamblichus−Nauck, 193; Freeman, 1953, 244] but I do not know if this is the same person who is discussed in *Theaetetus*. The same list also mentions a Theodorus from Tarentum, which is the city of Archytas.

v. Archytas

Active at the end of the vth century and the beginning of the ivth century, Archytas was a Pythagorean and a friend of Plato [Freeman, 1953, 233; Burkert, 1972, 78, 92, 198; Van der Waerden, 1954, 149; Heath, 1921, I, 213]. Among his accomplishments was the duplication of the cube (see Van der Waerden [1954, 150]). In the "Catalogue of Geometers" [Section 16,Q.3], we are told that he was a mathematician "by whom the theorems were increased in number and brought into a more scientific arrangement." Burkert [1972, 198] gives a list of some Pythagorean predecessors and pupils of Archytas who are classed as mathematicians. Archytas will be discussed again in Section 22,C.

In my discussion up to now, as well as in the following references, there is no material from or pertaining to Plato. Aside from two passages in the *Republic* (530D, 600A [Plato−Shorey, 189, 439]), there is no specific use of the words "Pythagoras" or "Pythagorean" (see Burkert [1972, 85]). However, there are other passages in Plato which some, but not all, scholars believe have a Pythagorean origin. But as Burkert [p. 83] and Heidel put it: "This is why the question of the nature of historical Pythagoreanism is perhaps hardest of all to answer from Plato alone" and "it is difficult if not impossible for the most part to distinguish what is Platonic and what is Pythagorean" [1940, 7]. What I have done, therefore, is put the several references from Plato which deal with the dodecahedron, and which some authors believe not only reflect a Pythagorean influence but also indicate that certain discoveries are due to the Pythagoreans, in the section (16,D,i) on Plato.

Section 11. Ancient References to the Pythagoreans

The various references that link the Pythagoreans to the pentagram and dodecahedron are presented and discussed in this section. It should be noted that the pentagon as such is never mentioned except in connection with the dodecahedron.

Most of the quotations, at least outwardly, are devoid of any precise association of the pentagram as an object and the mathematical aspects of the pentagram. In other words, it is the pentagram as a symbol or simply as a design which seems to be essential. Some authors, however, have used these quotations in connection with other more mathematically oriented quotations to associate the Pythagoreans with the mathematical aspects of the pentagon and DEMR. For this reason I have tried to be fairly complete.

A. *The Pentagram as a Symbol of the Pythagoreans*

Lucian, 2nd century

(Q.2) The divine Pythagoras chose not to leave us anything of his own, but if we may judge by Ocellus the Leucanian and Archytas and his other disciples, he did not prefix "Joy to you" or "Do well", but told them to begin with "Health to you". At any rate all his school in serious letters to each other began straightway with "Health to you", as a greeting most suitable for both body and soul, encompassing all human goods. Indeed the Pentagram, the triple intersecting triangle which they used as a symbol of their sect [literally, those of the same teaching], they called "Health".

(Lucian, "In Defense of a Slip of the Tongue in Greeting," 5 [Lucian–Kilburn, 177]; [Vogel, 1966, 46; Thomas, 1934, I, 225])

Scholium to Q.2

(Q.3) "the pentagram": because the [symbol] secretly called the pentalpha was a recognition-symbol amongst the Pythagoreans, and they used it in their letters.

(Scholium to Lucian [Q.2], "In Defense of a Slip of the Tongue in Greeting," 5 [Vogel, 1963, 181])

Scholium to Aristophanes, *Clouds*

(Q.4) "First, greetings": They say that Cleon, sending messages to the Athenians from Pylos and Sphacteria, began by writing "greetings"; whence it [the greeting] came into use. ⟦Plato, however, at the beginning of his letters wrote "fare well." And the Pythagoreans [used] "be of good health." And the triply self-entwined triangle, the pentagram, which they employed as a sign among fellow members of their school, was named "health" by them⟧. (Otherwise, it was an old custom to prefix "greetings" in letters. For Cleon was not the first to write in this way, as some say, from Sphacteria to the Athenians.)

(Scholium to Aristophanes, *Clouds*, 609 [Aristophanes–Dubner, 110]. The parts enclosed by ⟦ ⟧ and ( ) indicate variations; consult the critical apparatus.)

This should be compared to the following inscription from IVth century Athens:

(Q.5) Mnesiergus wrote to those at home "greetings and good health."

([Dittenberger, 1898, 1259])

B. *The Pythagoreans and the Construction of the Dodecahedron*

Proclus, 5th century: Pythagoras and the dodecahedron

(Q.6) Following upon these men Pythagoras transformed mathematical philosophy into a scheme of liberal education surveying its principles from the highest downwards and investigating its theorems in an immaterial and intellectual manner; he, it was, who discovered the theory of proportionals [possibly irrationals] and the structure of the cosmic figures.

(Proclus, *A Commentary of the First Book Euclid's 'Elements'*, "Catalogue of Geometers" [Proclus–Morrow, 52]; [Proclus–Friedlein, 65; Thomas, 1939, I, 149; Van der Waerden, 1954, 90])

The cosmic figures here refer to the tetrahedron, cube, octahedron, icosahedron, and dodecahedron as in Euclid, XIII, 13-18 (see Section 16,D,i).

While much of Proclus' summary seems to have been taken from a catalogue written by Eudemus (c. −330), this particular portion is apparently taken from Iamblichus (Q.7). See Van der Waerden [1954, 91]; Burkert [1972, 411]; Heidel [1940, 16]. It is the opinion of these authors that Eudemus gave little or no information concerning Pythagoras and so Proclus filled in the gap with the material from Iamblichus. Thus this evidence from Proclus is not to be trusted.

The next group of quotations comes from the works of Iamblichus, who wrote c. 300. The first two are essentially the same.

Iamblichus: Hippasus and the dodecahedron

(Q.7) And about Hippasus they say, that he was one of the Pythagoreans, and by being the first to draw and publish the sphere from the twelve hexagons [read: pentagons], he perished at sea as irreverent and he took the idea as if he were its inventor, whereas it belongs entirely to "that man"—for so they designate Pythagoras and do not call him by name.

(Iamblichus, *On Common Mathematical Knowledge*, chap. 25 [Iamblichus–Festa, 77])

Iamblichus: Hippasus and the dodecahedron

(Q.8) It is related of Hippasus that he was a Pythagorean, and that, owing to his being the first to publish and describe the sphere from the twelve pentagons, he perished at sea for his impiety, but he received credit for the discovery, though really it all belonged to HIM (for in this

> way they refer to Pythagoras, and they do not call him by his name).
>
> (Iamblichus, *On the Pythagorean Life*, chap. 18, 88 [Thomas, 1939, I, 223]; [Iamblichus–Nauck, 66; Iamblichus–Albrecht, 97])

It is not clear whether "describe" refers to a construction in the rigorous mathematical sense; see Fritz [1945], Thomas [1939, I, 378], and Mugler [1958, 107].

The third text from Iamblichus shows confusion on the matter.

Iamblichus: Pythagorean dodecahedron and incommensurability

(Q.9) And Pythagoras did not think it fitting to speak or write in such a way that [his] ideas would be clear to all who came along, but he is said to have taught this very thing to his disciples first, that being free of all incontinence they should guard in silence the words that they heard. They say that the first [man] to disclose the nature of commensurability and incommensurability to those unworthy to share in the theory was so hated that not only was he banned from [the Pythagoreans'] common association and way of life, but even his tomb was built, as if [their] former comrade was departed from life among mankind.

And some say that even the divine power was angry with those who divulged the [teachings] of Pythagoras; for he who revealed the structure of the icosagon [a twenty-sided figure; perhaps there is confusion with the twenty-faced icosahedron here]—and this was the dodecahedron, one of the five so-called solid figures—to be inscribable within a sphere, perished in the sea like one who has committed sacrilege. But some said that the man who suffered this had spoken out about irrationality and incommensurability.

(Iamblichus, *On the Pythagorean Life*, chap. 34, 246, 247; Iamblichus–Nauck [171]; Iamblichus–Albrecht [239])

One of the difficulties with this text is that the Greek word for "reveal" does not seem to have had the meaning "reveal how" in the sense of "show how it could be done." We would expect that Iamblichus would say something such as "the man who revealed how the dodecahedron could be inscribed in a sphere." Albrecht [Iamblichus–Albrecht, 239] translates this as "who revealed the structure of the icosagon, the fact that the dodecahedron—one of the so-called five solid figures—can be inscribed in a sphere." The sense here would seem to be that "the man revealed that the dodecahedron *could* be inscribed in a sphere" without necessarily showing or perhaps even "knowing"—in a formal sense—how this could be done. As it stands the structure of the sentence makes it difficult to extract the desired meaning.

Burkert [1972, 459] discusses the improbability of some sort of mathematical secret and its betrayal. However, he feels that "the tradition about Hippasus, though surrounded by legend, makes sense.... The dodecahedron may well have been important as a [symbol] in the Pythagorean school, like the pentagram; Hippasus' offense was in analysing the sacred object, publicly, by mathematical means." He refers to a passage in Plato's *Phaedo* 110b (see Section 16,D,Q.6) which describes balls made in the shape of dodecahedron with twelve pieces of pentagonally shaped pieces of leather and feels that it "makes it seem unlikely that the dodecahedron was first made known, outside of Pythagorean circles, by Hippasus" [p. 461 fn. 67]. Further, "the thesis of the Pythagorean foundation of Greek geometry cannot stand, any more than the legend of great mathematics held secret" [p. 456].

Note how the last sentence of Q.9 changes "discovery of the dodecahedron" into "discovery of the incommensurability." For completeness I mention that there are two other sources which speak of a Pythagorean who drowned because he divulged the secret of the irrational; these are by Pappus in a commentary on Book X of the *Elements* [Pappus–Thomson,64] and in Scholium 1 to Euclid's Book X [Euclid–Heiberg, V, 415]. Junge [1948, 336] suggests that the source of the scholium is in fact the commentary of Pappus.

Scholium to Euclid's Book XIII: Platonic figures due to the Pythagoreans and Theaetetus

(Q.10) In this book, that is, the thirteenth, are described the five Platonic figures, which are however not his, three of the aforesaid five figures being due to the Pythagoreans, namely, the cube, the pyramid and the dodecahedron, while the octahedron and icosahedron are due to Theaetetus. They received the name Platonic because he discourses in the *Timaeus* about them.

(Scholium 1 to Book XIII of Euclid's *Elements* [Thomas, 1939, I, 379]; [Euclid–Heiberg, V, 654])

In reference to this scholium, Sachs [1917, 82] states that it is unthinkable that the mathematical construction of the dodecahedron took place before that of the octahedron. She supposes that the Pythagoreans only knew of the dodecahedron in an empirical way, as the sphere approximated by a dodecahedron, the apparent meaning of "the sphere from twelve pentagons" in Q.7 and Q.8. According to Sachs, the tradition of a Pythagorean treatment of the solids is without validity and a later invention (see Section 16,D,i).

Waterhouse [1972] devotes his entire article to this scholium and in particular to the claim that the octahedron came after the dodecahedron. I shall discuss his views in Section 18.

Van der Waerden [1954, 100] considers that the scholium of Q.10 is to be given more credibility than the statement of Proclus in Q.6 precisely because it contradicts the ascription of everything to Pythagoras (see also Burkert [1972, 411]).

C. *Other References to the Pythagoreans*

Proclus: Ascription of the method of application of areas to the Pythagoreans

(Q.11) Eudemus and his school tell us that these things—the application of areas, their exceeding and their falling short—are ancient discoveries of the Pythagorean muse.... Something like this is the method of "application" which has come down to us from the Pythagoreans.

(Proclus, *On Euclid I*, commentary on I,44 [Proclus–Morrow, 332]; [Thomas, I, 187; Heath, 1921, I, 150])

A more specific claim as to the achievements of the Pythagoreans is contained in the following two references.

Scholium to Euclid's Book IV (Book IV is Pythagorean)

(Q.12) In this book it is shown that the circumference of the circle is not three times as great as its diameter, as many believe, but greater than three times, and in the same way, that the circle is not 3/4 of the triangle circumscribed around it. And this book is a discovery of the Pythagoreans.

(Scholium 2 to Book IV of Euclid's *Elements* [Euclid–Heiberg, V, 272])

Contrary to the scholiast, Euclid does not compare the circumference and diameter of a circle in Book IV or, for that matter, anywhere in the *Elements*. The scholiast, however, may be thinking of a proof via the inscription of the hexagon in IV,15.

Scholium to Euclid's Book IV (Book IV is Pythagorean)

(Q.13) And all the theorems of the present book, being seventeen [in number] are discoveries of the Pythagoreans. And the following [story] was transmitted, namely that in the course of time [the name of] Theon also crept into the entire geometry....

(Scholium 4 to Book IV of Euclid's *Elements* [Euclid–Heiberg, V, 273])

This does not mean that the statement about the Pythagoreans only dates from after Theon (4th century). Rather, this particular quotation illustrates the difficulties in using scholia, for even when they can be dated to a certain extent, as in this case, we do not know whether they represent a tradition going back to the IVth or Vth century or they are simply later commentary and "historical overlay." For the source of the various scholia, consult the critical apparatus in Euclid–Heiberg. The scholia are discussed in Euclid–Heath [I, 64].

Scholium to Euclid's Book II,11

The following scholium, which is also of independent interest, is referred to in Section 12 with respect to the Pythagoreans. I present and discuss it here for reference.

(Q.14) Let the whole line AB be cut into 8 and 1/8. Therefore, taking the number contained by the whole, 5+3 and multiplying that sum by 3, it becomes 24: for $3\times 8 = 24$. Taking also the rest of the segments of BH, i.e. the eighth of 8, and adding 24 to this, the [rectangle] contained by the whole and one of the segments is 25. Multiplying in the same way also the number of the other segment by itself, i.e. 5, makes it 25: for $5\times 5 = 25$. So the rectangle contained by the whole AH [AB is meant] and the rest of the segment BH is equal to the square described by the remaining segment AH.

(Scholium 73 to Theorem II,11 of Euclid's *Elements* [Euclid–Heiberg, V, 249])

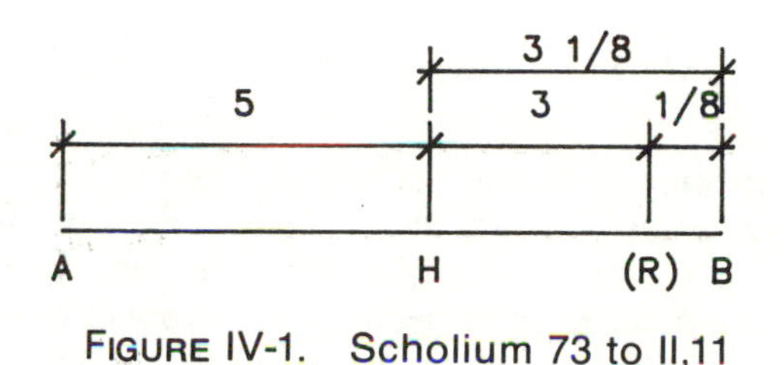

FIGURE IV-1. Scholium 73 to II,11

This text is far from being clear, but it seems to me that the following, in great part due to L. Curchin, is the correct interpretation: The total length of AB is 8 1/8 as stated in the first sentence. The "number contained by the whole" is the integer part of the 8 1/8, that is, 8. AB is divided into two parts at H, but we are not explicitly told what AH and HB are equal to. For instance, since the text speaks of "the number of the other segment AH by itself," it could, a priori, be thought of in the same way as above—that 5 represents the integer part of AH. However, the text speaks of "taking also the rest of the segments of BH, i.e. the eighth of 8." The use of the plural "segments" in addition to the number 1/8 suggests that the scholiast thinks of HB being $3 + 1/8$ units in length and that this in turn can be divided into two segments $HR = 3$ and $RB = 1/8$. Thus we may suppose that $AH = 5$ exactly and that $HB = 3\ 1/8$. We can now write the area of the rectangle as $AB \cdot BH = AR \cdot BH + RB \cdot BH = (AR \cdot HR + AR \cdot RB) + RB \cdot BH = (8\cdot 3 + 8\cdot 1/8) + 3\ 1/8 \cdot 1/8 = (24+1) + 3\ 1/8 \cdot 1/8$. This last number, whose closest integer value is 0, is simply ignored by the scholiast to obtain the value 25. This in turn is equal to $5\cdot 5 = AH\cdot AH$.

Heller [1958, 20] incorrectly translates the first line as "One divides the whole line into eight equal parts" and indicates this in his accompanying diagram. This misunderstanding affects his interpretation of the scholium.

Knorr [1975b, 34], on the other hand, assumes incorrectly that the scholiast first takes the segment *HB* to be precisely equal to 3 and that "the writer, in introducing the fractional error of 1/8 right at the start, is aware that the segment [HB] (=3) will turn out to be too short." Then after writing the arithmetic of the scholium as $5^2 = 3 \cdot 8 + 1 = 3 \cdot 8 + 8 \cdot 1/8 = 8 \cdot 3\ 1/8$, Knorr says, "We thus expect that 3 1/8 will be a better value for the shorter segment than its initial value 3." Knorr is writing about this scholium in the context of his discussion of the ancient use of anthyphairesis and "side and diagonal" techniques and seems to be implying that the scholiast was aware of the "side and diagonal" recursion relationship for the pentagon and its use as an approximating procedure (see Sections 12,ix and 23). It seems to me that it is too much to conclude from this obscurely stated scholium that the scholiast was in the possession of such a procedure.

Note, however, that the ratio of the segments $AH = 5$ and $HB = 3\ 1/8$ is 8/5 and that this in turn implies that the ratio of the whole line *AB* to the larger segment *AH* is 13/8. These are ratios of Fibonacci numbers (Section 32,B) which provide excellent numerical approximations for DEMR. Given the garbled form of the scholium, it is not impossible that the 5 and 3 1/8 values are the result of some crude experimentation. They may, however, represent some more advanced transmitted theory, even the one suggested by Knorr. Unfortunately we have no idea of the date, source, or inspiration of this scholium (see the comments on Q.13).

Section 12. Theories Linking DEMR with the Pythagoreans

Under discussion in this section are the various claims that have been made about the development of DEMR by the Pythagoreans. I begin with statements based on the quotations that have been given. Many of these are widely repeated and I have not tried to be complete in listing who supports each theory. Nor have I attempted to ascertain in each case who was the first to state a theory; nor have I tried to list all the variations.

i. The Pentagram

Many authors point to references Q.2, Q.3, and Q.4, according to which the pentagram was the symbol of the Pythagoreans, and from this they deduce that the Pythagoreans were probably acquainted with DEMR. See, for example, Heath [1921, I, 161], Euclid–Heath [II, 99], and Van der Waerden [1954, 101]. These authors also give possible constructions that the Pythagoreans might have used (see iii below). I have already discussed the work of Fuÿe [1934] in Section 9 where the idea of health is linked to various Sumerian and Babylonian texts. Fuÿe discusses Lucien's remark and believes that originally the pentagram was traced—after much practice—mechanically and without recourse to a circle [p. 8]. After this mechanical stage geometers such as Pythagoras are thought to have looked for a mathematical solution. The diagram (Fig. IV-2) shows Fuÿe's suggestion about how the pentagram may have been constructed by starting with what is now called Heron's method (Section 27) for the division of a line in EMR. Note that the "triple intersecting triangles," *BcE*, *AdC*, and *AbD*, formed in this way correspond to the quotation from Lucien (Q.3).

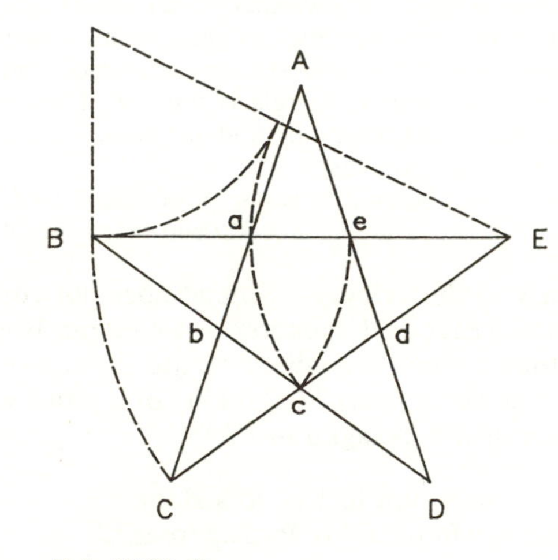

Source: Fuÿe [1934, 9]

FIGURE IV-2

ii. Scholia Assigning Book IV to the Pythagoreans

In Q.12 and Q.13 we saw two scholia that specifically attributed Book IV to the Pythagoreans. Since Book IV contains the construction of the pentagon some authors conclude that the Pythagoreans knew the construction of the pentagon. See, for example, Euclid–Heath [II,97]; Neuenschwander [1972] and also Burkert [1972, 450 fn. 14].

iii. Equations and Application of Areas

I pointed out in Section 5,K,L how II,11 can be interpreted in terms of equations and these in turn in terms of application of areas. This algebraic approach is discussed by Tannery [1882, 399]; Heath [1921, I, 161]; Euclid–Heath [I, 403]; Van der Waerden [1954, 101]. Heath and Van der Waerden both suggest ways in which the Pythagoreans may have obtained the proportions-equations from observations concerning the pentagram (see i above). Both arguments use similar triangles and variations on the theme of XIII,8—that the diagonals of a pentagon divide each other in EMR. The linking of application of areas is usually supported by a reference to Q.11.

iv. The Dodecahedron

Various authors, such as Heath [1921, I, 162], refer to Q.10 as well as to examples of models of dodecahedra that have been found (Section 10) to support their belief that the Pythagoreans were interested in the mathematical construction of this figure. It is not usually supposed, however, that the construction was as elaborate as that of XIII,17.

Allman [1889, 39] accepts the reports about the construction of the solids by the Pythagoreans. He suggests a method which, while not a construction in the sense of Euclid's Book XIII, could in effect be considered a mathematical construction. The starting point would be the observation that the tetrahedron and octahedron, respectively, have three and four equilateral triangles meeting at each vertex. Experimentation with five triangles would show that if they were joined at a common point then the bases would form a regular pentagon and this would lead to the icosahedron. Similarly, since joining three squares at a vertex leads to the cube, one would try to join pentagons at a vertex leading to the "construction" of the dodecahedron.

I shall discuss the dodecahedron again in Sections 16,D,i and 18.

v. A Marked Straight-Edge Construction of the Pentagon

Hofmann [1926, 435] suggests that a marked straight edge and compass construction may have been used by the Pythagoreans to construct the pentagon with a given side. This of course does not meet the Euclidean requirement of using only a compass and unmarked straight edge. The method is set out below.

Using some elementary results concerning both angles inscribed in a circle and isosceles triangles, one could show that $DF = BF = GC = AB = EG = a$.

The line AB is drawn, and with the compass arcs of radius a are drawn about A and B. Next the perpendicular bisector PQ of AB is drawn using the compass and unmarked straight edge. On the straight edge we lay off two points which are a units apart and then we slide and rotate the marked straight edge, making sure that it passes through A, until one of the points falls on the arc about B. Note that there are two possible positions for the point on the marked straight edge, namely, C and F, and these in turn determine, by means of the other point which has been marked on the straight edge, the points G and D. The last vertex E is now determined by the requirement that on line EGB the distance EG is equal to a.

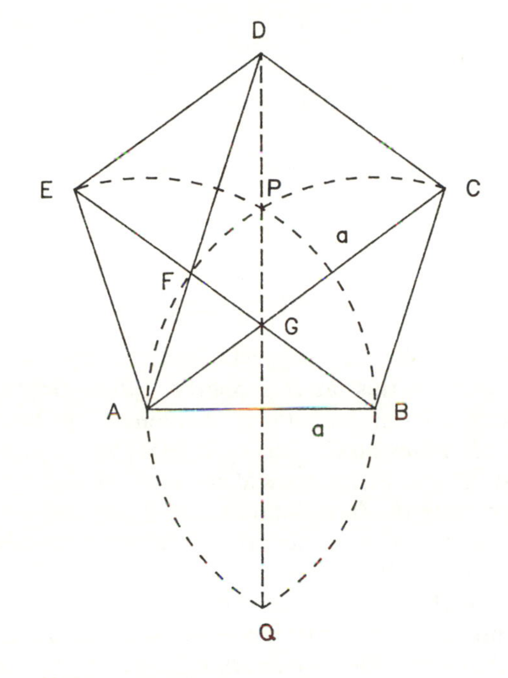

Source: Hofmann [1926]

FIGURE IV-3

Note that the argument does not imply a knowledge of the fact (XIII,8) that diagonals cut each other in DEMR but just that the large segment is equal to the side.

vi. A Gnomon Theory

Suppose that we start with a line AB which has been divided in DEMR at point C with BC being the larger segment, and then form the rectangle $ABDF$ with $DB = BC$. Then from the very definition of DEMR and the equality $DB = BC$ we have that the rectangle $ACEF$, obtained by subtracting the square $CBEF$ from the original rectangle, is similar to the original. Indeed the ratio of the sides of both rectangles is the same as the ratio of the segments of a line divided in EMR. Conversely we can think of starting with the smaller rectangle and adding the square to obtain the similarly shaped larger rectangle. By extending the terminology of Section 3, we could call the square "the gnomon" to the rectangles whose sides have the proportion determined by DEMR. Again from the definition of DEMR one sees that these rectangles are the only ones to have the square as a gnomon.

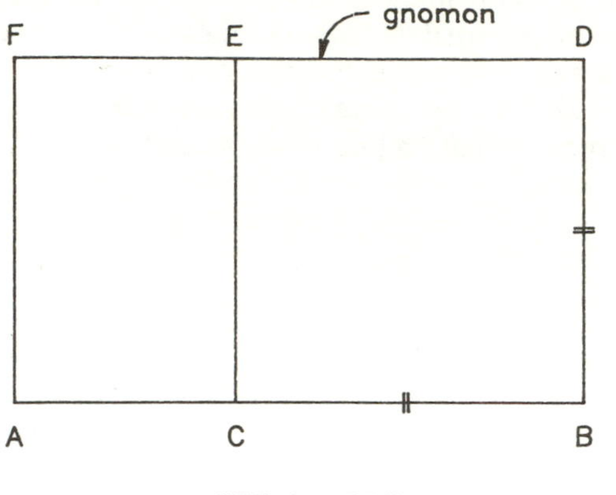

FIGURE IV-4. The Square as a Gnomon

Mugler [1948, 212] suggests that the Pythagoreans were aware of this latter property. Interestingly, Mugler uses non-mathematical "evidence" in his arguments. He relies on Ghyka [1927; 1931] who claims that interest in the mathematical aspects of the "golden number" (see the Introduction,D and Appendix I,C,vi) was engendered by the occurrence of this ratio in nature and art. The only other serious work on Greek mathematics which relies on such statements, for which absolutely no textual support has been offered, is the book by Michel [1950, 591; see Section 20,A,iv] who appears to have been influenced by Mugler, as well as by Ghyka, in this respect.

vii. Allman's Theory: The Discovery of Incommensurability

In Section 7 where the discovery of the concept of incommensurability was discussed, I mentioned that most writers believe that the discovery involved the relationship between the diagonal and side of a square. Allman [1889, 42, 137], on the other hand, believes that it was with respect to the division of a line in EMR that the concept was discovered by the Pythagoreans.

Allman points out that XIII,5 can be used to apply the incommensurability test of X,2 to show that the segments of a line divided in EMR are incommensurable. Theorem XIII,5 says that if we add the larger segment of a line divided in EMR to the original line then the new line is again divided in EMR. But if from this new line we subtract the smaller segment (original larger segment) from the larger segment (original line) we obtain the original smaller segment; or looking at it in another way, the larger segment of a line divided in EMR is itself divided in EMR when we subtract the smaller segment. Thus this process will never end, and so by XIII,2 the larger and smaller segments of a line divided in EMR are incommensurable.

Again in X,1 we always subtract a quantity greater than half of the larger quantity, while XIII,3 can be used to show that the small segment of a line divided in EMR is greater than one-half of the larger segment (on this "bisection principle" in Euclid, see Knorr [1978, Appendix IV]). Theorem X,2 allows us to test for incommensurability by seeing if we can continually subtract magnitudes. Allman says that the relationship between these theorems from Book XIII and Book X explains why X,1 is where it is rather than at the beginning of Book XII (method of exhaustion) to which he thinks it is more closely related. Allman writes: "it seems likely that the writer to whom the early proportions of the Tenth Book are due had in view the section of a line in extreme and mean ratio out of which problem I have expressed the opinion that the discovery of incommensurable magnitudes arose . . . ."

Mugler [1948, 212] seems to support Allman's thesis. He claims in fact the line divided in EMR is the simplest geometric figure for which incommensurability can be shown.

To counter Allman's argument one could point out that the method of X,9 is needed for the irrationality of $R(n)$ because the usual reductio ad absurdum methods [Euclid–Heath, III, 2] only works for $n=2$. This indicates that the case of the diagonal of a square preceded the more complicated one. Furthermore, as Heath [Euclid–Heath, III, 19] points out, the proof for $R(n)$ can be accomplished by a method practically equivalent to that of X,2. Nowhere in the extant Greek literature is there a proof of the incommensurability of the two segments of a line divided in EMR.

viii. Fritz–Junge Theory: The Discovery of Incommensurability

Another approach linking the discovery of incommensurability and DEMR has been advanced by Fritz [1945] who did not seem to be aware of Allman's work.

Fritz [254, 256] argues that the formal proof of the existence of incommensurable quantities that has been transmitted is one that requires a great deal of abstract and strict logical thinking. Thus, he argues, it is probable that incommensurability had already been discovered, probably in a different geometrical context.

In fact the discovery is claimed to have been made by Hippasus in the middle of the 5th century in connection with the pentagon. Fritz notes the two traditions concerning Hippasus (Q.4, Q.5, and Q.6): that he was the first to disclose the construction of the dodecahedron and incommensurability. He recalls the pentagram as a token of recognition among the Pythagoreans (Q.2) and says that this is "the one geometrical figure in which incommensurability can be most easily proved.'' The proposed proof involves a series of nested pentagons and pentagrams (Fig. IV-5).

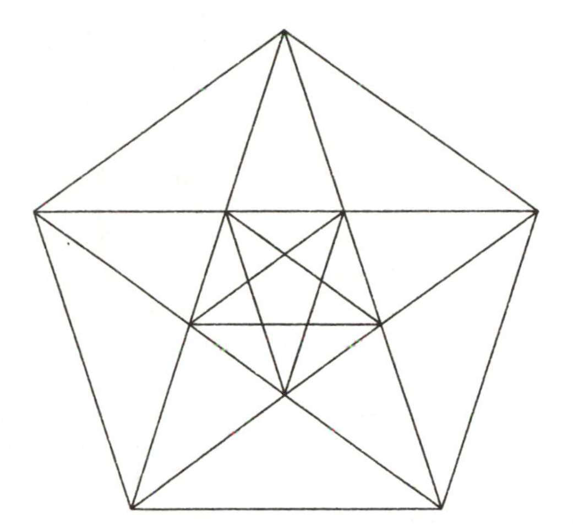

Source: Fritz [1945]

FIGURE IV-5

Fritz claims that geometrical knowledge of the time included both the result that base angles of an isosceles triangle are equal and the calculation of the angles of a polygon. From these he shows how one could find a proof, which was acceptable at that time, of the relations $d_1 - s_1 = d_2$ and $s_1 - d_2 = s_2$. The process continues indefinitely and the relations show that there can be no common measure for the side and diagonal of a pentagon. This reasoning is along the lines of the Euclidean Algorithm, and essentially the same argument is given by Becker [1957, 70].

Junge [1948, 336; 1958] was aware of Allman's work, but he suggests a geometric proof involving the pentagon. This is similar to, but shorter than, that of Fritz. Because of publishing delays due to the war it is not clear to me which article was written first. To substantiate his claim, Junge refers to the various citations involving Hippasus.

Philip [1966, 26, 40 fn. 5, Appendix II] criticizes Fritz's paper on the paucity and unsureness of the information concerning Hippasus, in particular his connection with the irrational where Hippasus' name is never explicitly mentioned. There is a similar criticism of the treatment of Hippasus by Van der Waerden [1954, 106].

Burkert [1972, 459 fn. 62] says: ''To my knowledge the connection of the regular pentagon with the irrational is never emphasized in the tradition.''

ix. Heller's Theory: The Discovery of DEMR

The article by Heller [1958] is entirely devoted to his proposals concerning the historical development of DEMR. It starts off with a discussion of the various references in the classical literature, which is followed by a discussion of Hofmann's marked straight-edge construction (Subsection v). Heller then suggests that what happened next was Hippasus' discovery of the incommensurable via the observation of the nested series of pentagons that one obtains when drawing the diagonals. This statement is based on the paper of Fritz [1945] discussed in Subsection viii above.

Hippasus' discovery of the incommensurability of the diagonal and side of a regular pentagon was followed by the discovery, either by Hippasus or by a disciple, of the incommensurability of the diagonal and side of a square. It is assumed that it was in connection with this that the ''side and diagonal'' numbers (Section 6) was developed.

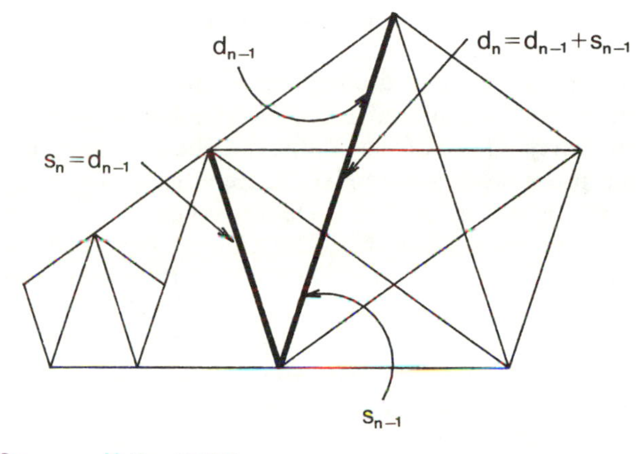

Source: Heller [1958]

FIGURE IV-6

Having gone from the incommensurability of the pentagon to that of the square to the ''side and diagonal'' numbers for the squares, Heller supposes that mathematicians developed ''side and diagonal'' numbers for the pentagon. These are obtained by observing the figure (Fig. IV-6) in which the diagonal—d_{n-1}—of one pentagon becomes the side—s_n—of the next largest pentagon. Further, the new diagonal dn is the sum of the side and diagonal, s_{n-1} and d_{n-1}, of the previous pentagon. Thus we have the recurrence relationships:

$$(1) \qquad s_n = d_{n-1}; \quad d_n = d_{n-1} + s_{n-1}.$$

Recall that for the ''side and diagonal'' numbers of the square (Section 6) one started off with $s_1 = 1 = d_1$. For the pentagon this would lead to the degenerate case as would $s_1 = 1$, $d_1 = 2$, and so the sequence is started off with $s_1 = 2$, $d_1 = 3$. This in turn leads the sequence of $d_n : s_n$ ratios of 3/2, 5/3, 8/5, 13/8 . . . . These are the ratio of successive Fibonacci numbers (Section 32,B); to see why, note that (1) implies $d_n = s_{n+1}$, $s_n = s_{n-1} + s_{n-2}$.

Heller next discusses the area definition (II,11) of DEMR. For this purpose appeal is made to Scholium 73 to II,11 (Q.14). Heller argues, via diagrams and calculations, that a similar process to that of the scholium was carried out with respect to the side and diagonal numbers of the pentagon (1) and that one would discover the following relationships which are analogous to (12) of Section 6:

(2) $$d_n \cdot r_n - s_n^2 = \pm 1; \quad r_n = d_n - s_n.$$

These would have been thought of in terms of rectangles and squares.

The relationships of (2) would in turn lead to the following relationships:

(3) $$d \cdot r = s^2; \quad r = d - s$$

for the diagonal and side of the pentagon. This is the algebraic version of XIII,8, but it would have been thought of geometrically as in II,11.

The development presented up to this point would not have been considered as providing a formal proof of the property XIII,8 of the pentagon. The first step in a formal proof would be to give a geometrical construction of DEMR as in II,11. Heller suggests how the preserved construction of II,11 may have been arrived at. He writes II,6—used in the proof of II,11—in algebraic symbols as $CD^2 = AB \cdot BD + BC^2$ and then suggests that the requirement $AD \cdot BD = AB^2$ was substituted to obtain $CD^2 = AB^2 + BD^2$. Since $BC = \frac{1}{2}AB$, CD could be solved for by the Pythagorean Theorem (I,47; used in II,11).

The next step would be a formal proof that states that if we lay off the side on the diagonal of a pentagon we do indeed obtain XIII,8, the geometric form of (3). Finally this would be followed by the construction of the $72°-72°-36°$ triangle of IV,10 and the pentagon of IV,11. (This theory should be compared to my analysis of Section 2.)

x. *Neuenschwander's Analysis*

Neuenschwander [1972] has analyzed Books I-IV of the *Elements* with the aid of charts that show which axioms and results are used for each theorem and where each theorem is used. He is of the opinion that Books II, III, and IV are Pythagorean in origin. For Book II this conclusion is based on the work of Van der Waerden [1954, 124]; some parts, however, are considered to have been reworked in later times. With reference to Book IV, use is made of the two Scholia to Book IV (Q.12 and Q.13), a discussion of the inscription problem by the Pythagorean Timaeus in Plato's *Timaeus* 54E-55A (Section 16,D,i), and the employment of a related result by Hippocrates in his quadrature of the lunes.

There is a detailed discussion of IV,11, the construction of the pentagon. Neuenschwander says that there are two possible approaches or decompositions that suggest themselves, either the one of IV,11 or the one obtained by drawing lines from each of the vertices through the centre of the circle and extending them to the opposite base, thus forming ten $36°-54°-90°$ triangles. Since the Pythagoreans used the pentagram as a symbol, they would choose the approach of IV,10. There is a difficulty with this argument in that it assumes that either approach is feasible. However, the second one requires the construction of the $36°-54°-90°$ triangle and the way to do it is certainly not a priori clear. Thus, irrespective of any symbolic significance, the method of IV,11 may have been chosen because mathematicians were able to solve the problem in this case.

Neuenschwander goes on to suggest that there was another construction of IV,10 that predated the one given by Euclid. This construction used VI,3 (a variant would use VI,4), which says that since CD bisects the angle at D one has $AC{:}DB = AD{:}DB$. Since $AB = AC$ and $DB = AC$ the point C divides AB in EMR (see also the use of VI,3 by Archimedes and a suggestion concerning his use of DEMR in Section 23). Furthermore, in the present version of IV,10, the proportion version of DEMR—as in VI,def.3 and VI,30—would have been used. Since proportion theory could not be used in Book IV, the original proofs were replaced by the present geometric proofs.

xi. *Stapleton*

In Section 9 I mentioned how Stapleton [1958] linked the study of the pentagon to the human hand. Earlier on in his paper [p. 23], we read how the "earliest Greek Philosophers" may have been led to mathematical definitions, "but it may be useful in understanding the working of the Greek mind in their use of points in the building up of geometrical figures to adopt a similar procedure in discussing how the Greeks may have arrived at their simplest geometric definitions, . . . by substituting, for ink and paper, the simpler process of making depressions with the tips of the fingers on a tablet of damp clay, as was the practice in Mesopotamia. . . . [Now follows an identification of one finger with a point, two fingers with a line, three fingers with a triangle, and four fingers with a quadrilateral.] If the tips of all five fingers are simultaneously pressed on to the clay surface, a five-sided figure called a Pentagon results. If the sides are all equal, the name Regular Pentagon is given to such a figure. . . . With the outlining of a Pentagon with the five finger-tips of a man's hand, primitive geometry must have died unless other means of making geometrical figures could be found."

Stapleton then suggests [p. 37] two ways in which the "early Pythagoreans" may have constructed the pentagram. In the first method, a knowledge of the value of the interior angles of a regular polygon would have led them to the value of 108° for the pentagon and then to 72° for the interior angle. Indeed, "the factors of these two numbers being respectively $2^2 \times 3^3$ and $2^3 \times 3^2$, this could not fail to have been of as great numerological significance to them as 72 was both to the ancient Chinese and the Zoroastrians." The pentagon would then be constructed using a protractor placed at the centre of a circle. The pentagram is obtained by connecting alternate vertices. In the second method, we start with a line BE and then draw, with the aid of the protractor and compass, the lines BD and EC, both equal in length to BE and making a 36° angle with it. Point A is then obtained by taking lines AC and AD to be at 36° to CE and BD.

CHAPTER V

MISCELLANEOUS THEORIES

In this chapter I describe several theories which have not been assigned to any particular mathematician or time period. This description is followed by a listing of the various statements that have been made about XIII,1-5.

Section 13. Miscellaneous Theories

i. Michel

Michel [1950, 570] believes that XIII,1-5 gives a view on the distant past history of DEMR. Because Euclid uses geometrical methods of the type found in Book II, without actually using any of these results, instead of more "algebraic" methods, Michel concludes, he borrowed the proofs from a book which was an "elucidation" of Book II and which Michel considers to be Pythagorean. The style of the proofs is considered "manifestly archaic" in particular because they use the gnomon technique (Section 3). This argument is similar to one given by Allman [1889, 136] (see Section 20,B,ii). Michel disagrees with the view that assigns these results to Eudoxus (Section 20,A,i).

ii. Fowler: An Anthyphairesis Development of DEMR

In Sections 8 and 12,vii I discussed anthyphairesis and various aspects of the literature which linked anthyphairesis and DEMR. In this section I shall discuss possible early relationships between anthyphairesis and the concepts of ratio and proportion and, via these latter, the concept of DEMR.

As noted in Section 1, the definitions of Book V are not completely explicit. In particular, Euclid, V,def.3 gives only a very vague definition of the concept of ratio: "A ratio is a sort of relation in respect of size between two magnitudes of the same kind."

There are places in Euclid and in other Greek mathematical works where, without a precise definition being given, the concept of ratio is discussed and treated. Examples of these are given by Fowler [1979a].

Proportion, however, is well defined by means of an "operational" definition which serves as a powerful tool in later results. In V,def.6 we read: "Let four magnitudes which have the same ratio be called proportional."

This follows V,def.5 which reads: "Pairs of comparable magnitudes A,B and C,D are 'said to be in the same ratio' [i.e., proportional; A is to B as C is to D] if whenever mA is greater, equal to or less than nB, where m and n are integers, then the same relationship holds between C and D." This definition as well as its attribution to Eudoxus are discussed by Van der Waerden [1954, 175].

Several authors (see Knorr [1975, chap. 8, sect. 2]) have argued that before Eudoxus' theory of proportion there existed another theory of proportion, apparently due to Theaetetus and based on anthyphairesis, in which the pairs A,B and C,D were proportional if the anthyphairesis of A,B and C,D were the same. While such a definition could have served for a preliminary theory of proportion, it had several drawbacks—

including the difficulty experienced in proving some seemingly elementary results (see Knorr [1975, chap. 8, sect. 3 and Appendix B especially on Euclid V,9, Theorem 6). Consequently, the proportion theory for plane and solid figures of Euclid's Books VI and XI was based on the Eudoxean proportion theory as developed in Euclid's Book V. Theaetetus's anthyphairesis-based theory in turn remains in Euclid's Book X where it serves in the development of the theory of irrationals.

Fowler has taken the anthyphairesis proportion theory one step further and claims that there was at one time an explicit theory, refined by—if not due to—Theaetetus [1979, 33, 78; 1980, 28], which defined the ratio of two quantities in terms of their anthyphairesis. The argument is based not only on the various, but ill-defined, uses of ratio in Euclid, but also on certain writings of Plato, Archytas, and Aristotle. While this anthyphairetic definition of ratio was superseded by the Euxodean definition of proportion, vestiges still remain in Euclid.

Fowler has continued his discussion of this theme [1980; 1981; 1982a; 1982c] and has also given a generalization of DEMR inspired by these ideas [1982b].

In Fowler [1980] the results of Euclid II are given a new interpretation; they are viewed as the results needed for the study of anthyphairesis. The proofs of the various results dealing with anthyphairesis are linked to the approaches to proofs known as "analysis and synthesis" (see Section 16,B).

Of particular interest to us is the interpretation of the statement of II,11 which, it is suggested, developed out of the following problem. Given a line, divide it so that the segments will have the anthyphairesis [1,1,1,1 . . .]. This could be solved as follows. Given AB we want to find the point H between A and B such that $ANTH(AB,AH) = [1,1, \ldots]$. If this is to be the case, then we must have $ANTH(AB,AH) = [1, ANTH(AH,KB)]$ where K is the point between A and B such that $KB < AH$ and $AB = 1 \cdot AH + KB$. But, since $AK = AB - KB$, we have $AK = AB - KB = 1 \cdot AH = AH$. The only way this can happen is if points K and H coincide. Hence $[1,1,1, \ldots] = ANTH(AB,AH) = [1, ANTH(AH,HB)]$. This means that $ANTH(AH,HB)$ also equals $[1,1,1, \ldots]$ and so $ANTH(AH,HB) = ANTH(AB,AH)$. This in turn implies $AH{:}HB = AB{:}AH$.

In another reference to II,11 Fowler says: "The whole construction is so subtle and ingenious that it is difficult to conceive how it could have been invented without the intervention of some intermediate stage, important enough in itself to justify an independent study." This remark is in support of Fowler's view of Book II which is opposed to the view that Book II deals with "geometrical algebra" (Section 5). The discussion of the role of II,11 continues with the remark that, if the definition of DEMR along with Book II in general does derive from a source based on Theaetetus' work on ratio theory, then the formal construction of the pentagon would be of a rather late date.

iii. Knorr: Anthyphairesis and DEMR

Specific references to the anthyphairesis of the segments of a line divided according to EMR are given by Knorr [1975, 257, 289]: "Likewise, the nature of the anthyphairesis of lines in extreme and mean ratio was certainly known; it is a simple deduction from XIII,5 that the procedure in this case yields an infinite sequence of quotients, each being unity." In connection with earlier remarks, Knorr continues: "we emphasize . . . that, in the context of the early studies of incommensurable magnitudes, anthyphairesis was not used to prove the incommensurability of specified magnitudes, but only to obtain approximations of their ratio."

In connection with this argument it should be remarked that great care is needed in going from statements of the form "could have been known" to statements of the form "was known." In particular I showed in Section 2 that, despite the fact that the result of XIII,8 could have been known and proved at the time that II,11-IV,10 were proved, it seems unlikely that XIII,8 was in fact known. See further comments on this in Section 20,C,iii.

Regarding Knorr's statement, also see his argument in Section 23 which states that Archimedes performed calculations corresponding to the continued fraction expansion of segments in DEMR.

iv. Itard: Theorem IX,15

Theorem IX,15 of the *Elements* shows that if three integers a,b,c are such that they are in "continued proportion" (i.e., $a{:}b = b{:}c$) and no smaller integers will give the same ratios, then any two of the terms added together will be prime to the third.

Itard [1961, 181] has pointed out that this result can be used to show that the segments obtained when a line is divided in EMR are incommensurable (see Section 1 on Book X). Indeed if $a{:}b = b{:}c$ with $c = a + b$, as in the case of DEMR, then of course $a + b$ is not prime to c as stated in the conclusion of IX,15.

Itard does not claim that Euclid had DEMR in mind when he wrote IX,15, but he seems to be rather circumspect in his consideration of the Fitz–Junge theory discussed in Section 12,viii [cf. Itard, 38, 42].

Section 14. Theorems XIII,1-5

Surprising as it may seen, the lowly set of Theorems XIII,1-5, which are used as lemmas for later and much more important results in Book XIII, have been the subject of much commentary and contention. Since my descriptions of these commentaries are scattered in various sections, I shall give a synopsis here, together with an indication of other sections where I discuss them.

i. Bretschneider

This author [1870, 168] assigns XIII,1-5 to Eudoxus; this view is accepted by Sachs [1917, 97]. Heath [Euclid–Heath, III,441] accepts this conclusion, but in a seeming contradiction he also says: "It would therefore appear as though these propositions were taken from an earlier treatise [than Book XIII?] without being revised or written in the light of Book II [which is assigned to the Pythagoreans by Heath]." See Section 20,B,i.

ii. Allman

Allman [1889, 136] assigns XIII,1-5 to a time before Eudoxus and believes that the definition of analysis and synthesis as well as the alternate proofs [Euclid–Heath, III, 442] are due to Eudoxus. See Section 20,B,ii.

iii. Michel

Michel [1950, 570] assigns XIII,1-5 to the distant past history of DEMR. See Section 13,A.

iv. Dijksterhuis and Van der Waerden

Van der Waerden [1954, 173] writes: "Time and again, results are obtained implicitly which had already been obtained explicitly in Book II. . . . The author of Book XIII evidently knew the methods of 'geometric algebra,' but he was not acquainted with its systematic development of Book II. . . ." The reference here is apparently to the direct manipulation of rectangles in XIII,1-5 rather than to the use of theorems from Book II.

v. Lasserre

Lasserre [1964, 178] assigns XIII,1-4 to Leodamas of Thasos and XIII,5 to Eudoxus. See Sections 17 and 20,B.

vi. Fritz

Fritz [1971, 499] concludes from the comments of Heath and Van der Waerden (given above) that Book II is later than, or from a later developmental phase than, Book XIII. This seems to imply that the proofs of XIII,1-5, being more "elementary" than those available if results of Book II had been used, in fact preceded those results.

vii. Knorr

This author [1975, 209 fn. 52] assigns XIII,1-5 to the time of Theodorus. See Section 15.

viii. Heiberg

Although most of Book XIII is assigned to Theaetetus, Heiberg [1925, 19] says that the dates of XIII,1-5 are uncertain but that they predate Euclid.

ix. Herz-Fischler

In Section 2,D I argued that given the elementary nature of XIII,1-5 we should simply consider these results as lemmas which were proved as needed for the later results in Book XIII. Using this and various other indications, I will argue in Sections 20,C,iii, 20,D, and 22 that XIII,1-5 are, along with other parts of Book XIII, due to Theaetetus and his students.

CHAPTER VI

THE CLASSICAL PERIOD: FROM THEODORUS TO EUCLID

The time period dealt with in this chapter covers perhaps a span of 120 years. For the first time we have some actual names of mathematicians with whom theories concerning DEMR have been associated. The most important of these are Theodorus, Plato, Theaetetus, and Eudoxus. Their actual role, if any, in the development of DEMR, however, rests for the most part on a matter of conjecture based on a few isolated historical remarks and some often subtle interpretations of texts.

In this chapter each mathematician is dealt with in a separate section. The order followed is the same as the sequence of names in the "Catalogue of Geometers" of Proclus [Proclus–Morrow, 51] from which I have already quoted and which I will have several more occasions to cite (see Section 16,b,Q.3 for the presentation of the most relevant portion in one block). It should be remarked that this order may give a false picture; for instance, Plato indicates that Theaetetus was doing mathematics in −399 and there is some evidence that Plato's interest in mathematics did not start until a meeting with Archytas around −386. Note also that Speusippus is not mentioned in the list of Proclus.

At the end of the chapter (Section 22) I present my own views on the development of DEMR.

Section 15. Theodorus (fl. c. −410 to −390)

The main source of information about the mathematical activity of Theodorus is the section of Plato's *Theaetetus*, which was given in Section 7,ii; see also the introduction to Chapter III.

i. Knorr

Knorr [1975, 199] argues that Book II "owes its conception to Theodorus." Further, he adds: "save for VI,30 and XIII,6 and 11, all the Euclidean theorems on the extreme and mean section derive from the time of Theodorus, as they employ the area-methods typical of Book II. This is true even of lemmas XIII,1-5 which are normally easily established via Books V and VI . . . . All the fundamental properties of the section were established via these area-methods relatively early; about the time of Theodorus. Only later was the section put into its ratio-form and in fact given its name 'extreme and mean ratio'. Of all the properties of the section which appear in the *Elements*, only one was inaccessible at the earlier time: its placement in XIII,6 and 11 within the classification of the irrational lines" [p. 209 fn. 52]. Knorr also explains [p. 199] how it is possible that later on tradition attributed the application of area results of Book II to the Pythagoreans instead of to Theodorus.

The association of Book II with Theodorus is made for several reasons, the main one being "the utility of that geometry for reconstructing his studies of incommensurables" [p. 197]. In this connection Knorr discusses two possible steps in the development of DEMR which Theodorus might have considered.

Knorr also suggests that rather than coming from the theory of application of areas, II,11 might have its origin in an arithmetic algorithm that produced side and diagonal numbers [p. 200] (see Section 6). In the same way that II,10 can be used to prove the relationship for the side and diagonal numbers of the square, Knorr shows how the diagrams accompanying II,11 can lead to the recurrence relationship—Section 12,ix(2)—for the side and diagonal numbers for DEMR. Heller, however, claimed that (2) was obtained geometrically by looking at diagrams connected with the pentagon rather than from the diagram for II,11. In a note [Knorr, p. 208 fn. 49] a connection appears to be made with the relationship between the side and diagonal of a pentagon (XIII,8) but there is no indication of how this algorithm would fit in chronologically with IV,11 or XIII,8.

The other proof that Knorr suggests [p. 194, Theorem 15] could possibly have been done by Theodorus as part of his development of the irrationals is the result (the first part of XIII,6) that when a line is divided according to DEMR the segments produced are incommensurable. In support of this possibility Knorr provides a proof of the result within the context of Book II.

ii. Mugler

Mugler [1958, 212] says that the infinite reproduction property of XIII,5 was known to Theodorus. This statement is related to the claim—Section 12,vii—that a line divided in EMR is the simplest figure for which incommensurability can be shown.

Section 16. Plato (−428 to −348)

For a short summary of the life and work of Plato, one may consult Robinson [*OCD*] and Field [*OCD*]; there are more extensive summaries in Taylor [1937], Field [1930], Shorey [1933], and Cherniss [1945].

For our purposes the important approximate dates in Plato's life—established from a consensus of the various references mentioned above and elsewhere in this section—are: birth (−428); death of Socrates and the action of *Theaetetus* (−399); visit to Sicily, meeting with Archytas, composition of *Meno* (−388); return to Athens, founding of the Academy (−386); composition of the *Republic* (mid −370s); composition of *Theaetetus* (−368); death (−348).

Plato occupies a unique position in our study of the development of DEMR, as indeed he does in the general history of Greek mathematics, for in addition to any personal contribution of Plato himself as a mathematician we must also consider, first, whether his work contains references to the mathematical work of his predecessors and contemporaries and, second, whether any of the development of DEMR is due to the direction or influence of Plato. In Subsection A, I discuss Plato as a mathematician per se; in B, I consider the influence of Plato on others; in C, I discuss claims concerning the direct involvement of Plato in the development of DEMR; and in D, I discuss passages from the works of Plato which have been interpreted as dealing with DEMR, whether or not they originated with Plato or his predecessors and contemporaries.

As I noted in the introduction, this section does not pretend to give a complete discussion of Plato and the mathematics of his time except where DEMR is concerned. To give an idea of the extent of the literature I note that of the 620 or so pages of the annotated bibliography of Cherniss—which mainly treats the period 1950-1957—more than 15 pages are devoted to mathematics as such [Cherniss, 1960, 396-412]. The period 1958-1975 is covered in Brisson [1977]. Other (pre-1930) references are given in Plato−Chambry [lx, lxxviii].

A. Plato as a Mathematician

There have been many articles and books dealing with the topic of Plato and mathematics. Many of these, however, deal at least in part with the philosophical aspects of Plato's mathematics or the role that mathematics played in the development of his philosophical theories. The following may be consulted: Cornford [1932], Mugler [1948; criticized by Cherniss, 1951], Cherniss [1960, no. 1430], Steele [1951], Brumbaugh [1954; discussed by Gibson, 1955], Wedburg [1955], Hare [1965; discussed by Taylor, 1967], Brown [1967a; 1969], Morrow [1970], White [1975], and Fowler [1985]. Other books and articles will be mentioned elsewhere. The texts of all the mathematical passages are given by Frajese [1963].

A variety of views have been expressed concerning Plato as a mathematician and what he himself really did in the way of mathematics. For instance, Steele [1951, 184] speaks of "a Plato who philosophized from a knowledge of advanced mathematics and not from that of the schoolroom, a Plato whose mathematical inspirations and allusions are too subtle for editors less skilled than Taylor and Burnet, is a Plato who shall be listened to when he speaks on mathematics at large." However, Neugebauer [1957, 152] writes: "It seems to me equally impossible to give any one conclusive 'explanation' for the origin of higher mathematics in the fifth and fourth century in Athens and the Italian colonies. On the negative side, however, I think that it is evident that Plato's role has been widely exaggerated. His own direct contributions to mathematical knowledge were obviously nil."

The main problem in discussing Plato as a mathematician or any aspect of Plato and mathematics is that most of the statements dealing with mathematics are, to modern readers at least, couched in vague language. Indeed in some cases—see, for example, the discussion of the divided line in Section D,ii—we cannot even be sure whether the statement in question can be interpreted in a strict mathematical sense. Consider the statement of Steele [1951, 174]: "The mathematical passages in Plato number more than the half dozen which have trickled wearily from footnote to footnote. Counting by disjunct intervals of standard page and letter, they number at least 121 passages from 25 Dialogs. Slightly fewer than half speak discerningly indeed, but about mathematics rather than from it, while slightly more than half disclose specific knowledge. In these again, references to arithmetic and algebra, in the modern sense of these words, out-number those to geometry in the proportion of nearly four to one." One must be cautious when reading these figures; for the author, as he himself hints, is much too familiar with modern abstract algebra and number theory (see, for example, Steele [174, 176]).

I note that Frajese [1963] gives 101 citations of which the majority are historical, educational, or philosophical.

It seems, however, that when we look at the writings of Plato and his student Aristotle (−384 to −322; he joined the Academy in −367) the emphasis is on mathematics providing methods and examples for use in dialectics. Owen [*DSB*] writes: "It is the mathematics he encountered that impressed him [Plato] as providing a model for any well organized science." Plato himself writes:

(Q.1) the huntsmen or the fishermen hand it over to the caterers, and so it is too with the geometers, astronomers, and calculators—for these also are hunters in their way, since they are not in each case diagram-makers, but discover the reality of things—and so, not knowing how to use their prey, but only how to hunt, I take it they hand over their discoveries to the dialecticians to use properly, those of them, at least, who are not utter blockheads.

(Plato, *Euthydemus*, 290B [Plato−Lamb, 445])

According to the list of Robinson [*OCD*, 842], *Euthydemus* was written in about −385, thus around the time of the opening of the Academy.

The same sort of mathematical-philosophical language can also be found in Books VI and VII of the *Republic*. For references to the mathematical sections, see Plato−Shorey [II, 167], Wedberg [1955, 139], Frajese [1963, 121-53], and Perls [1973, 37-39]. For Aristotle, see Apostle [1952]. See also Isocrates' apparent reference to the Academy: "we gain the power, after being exercised and sharpened on these disciplines [i.e., astronomy and geometry], of grasping and learning more easily and more quickly those subjects which are of importance and of greater value" [*Antidosis*, 265; Isocrates−Norlan, 333].

The very least that Q.1 indicates is that if Plato was a mathematician in some sense of the word, then he was not a chauvinistic one; indeed it is difficult to think of somebody who has actually proved theorems on his own, saying that only dialecticians can make proper use of mathematical discoveries or indeed that the discovery itself was not a valid end in itself.

When we look at those examples in Plato which leave no doubt that they are really straightforward mathematical statements (and not statements, such as those that we will come across in Subsection D, about which there is some doubt that they represent deep mathematics or even mathematics at all), then these examples appear to be at a rather low level. Of course some of them can be interpreted, as Steele [1951] does, as representing a deep result. When we come to Aristotle we find several works that treat mathematics (see Einarson [1936] and Apostle [1952]). Again the level of the examples is rather low, at least compared to what we find in the *Elements* of Euclid which was probably written not more than half a century after Aristotle's death.
half a century after Aristotle's death.

B. *Mathematical Influence of Plato*

I start with two ancient texts which speak of the relationship between Plato and other mathematicians. The first text is probably due to Philodemus (1st century; see Meiggs [*OCD*] and Treves [*OCD*]). Since the text reads "he says," Philodemus is quoting some unknown source. The original manuscript has several lacunae and I have indicated these by { } in which I have placed the conjectural restitutions, suggested in Gaiser [1968, 465] and Herculaneum−Mekler [p. 15]. Since the text speaks of progress in mathematics, the restitutions "atoms" and "Democritus" seem very forced; cf. Gaiser [p. 144] and Section 20,A,v.

(Q.2) and great progress in mathematics [was made] during that time, with Plato as the director and problem-giver, and the mathematicians investigating them zealously. Thus the problems concerning metrology first came into prominence then, and those concerning the {numbers? [Mekler]; atoms? [Gaiser]} and the followers of {Eudoxus? [Mekler]; Theaetetus? [Gaiser]} changed the {archaism? [Mekler]; configuration? [Gaiser]} of {Democri-? [Gaiser]}tus. Also geometry made great progress, for there came into being both solution by analysis and the thesis concerning statements of limits of possibility, and altogether the {...} concerning geometry for much {...}.

(Philodemus?, "Text from Herculaneum" [Herculaneum−Mekler, 15]; [Gaiser, 1968, 465])

The next quotation is from the "Catalogue of Geometers" of Proclus (5th century); we will come across and discuss parts of this catalogue on several occasions. It is usually assumed that the catalogue contains large abstracts from a history of mathematics written by Eudemus, a student of Aristotle and thus a person living shortly after the epoch in question.

(Q.3) Plato, who appeared after them, greatly advanced mathematics in general and geometry in particular because of his zeal for these studies. It is well known that his writings are thickly sprinkled with mathematical terms and that he everywhere tries to arouse admiration for mathematics among students of philosophy. At this time also lived Leodamas of Thasos, Archytas of Tarentum, and Theaetetus of Athens, by whom the theorems were increased in number and brought into a more scientific arrangement. Younger than Leodamas were Neoclides and his pupil Leon, who added many discoveries to those of their predecessors, so that Leon was able to compile a book of elements more carefully designed to take account of the number of propositions that had been proved and of their utility. He also discovered diorismi, whose purpose is to determine when a problem under investigation is capable of solution and when it is not. Eudoxus of Cnidus, a little later than Leon and a member of Plato's group, was the first to increase the number of the so-called general theorems; to the three proportions already known he added three more and multiplied the number of propositions concerning the "section" which had their origin in Plato, employing the method of analysis for their solution. Amyclas of Heracleia, one of Plato's followers, Menaechmus, a student of Eudoxus who also was associated with Plato, and his brother Dinostratus made the whole of geometry still more perfect. Theudius of Magnesia had a reputation for excellence in mathematics as in the rest of philosophy, for he produced an admirable arrangement of the elements and made many partial theorems more general. There was also Athenacus of Cyzicus, who lived about this time and became eminent in other branches of mathematics and most of all in geometry. These men lived together in the Academy, making their inquiries in common. Hermotimus of Colophon pursued further the investigations already begun by Eudoxus and Theaetetus, discovered many propositions in the *Elements* and wrote some things about locus-theorems. Philippus of Mende, a pupil whom Plato had encouraged to study mathematics, also carried on his investigations according to Plato's instructions and set himself to study all the problems that he thought would contribute to Plato's philosophy.
(Proclus, *A Commentary on the First Book of Euclid's* Elements, "Catalogue of Geometers" [Proclus–Morrow, 54]; [Proclus–Friedlein, 66; Proclus–Ver Eecke, 59; Heath, 1921, I, 32; Van der Waerden, 1954, 90; Thomas, 1951, 151; see also Burkert, 1972, 452])

Based mainly on this quotation from Proclus and various statements in the works of Plato, several authors have argued that while Plato himself did not contribute directly to the mathematics of his time his influence and direction played a major role in its development in the early middle part of the IVth century.

Morrow [1970, 319] says: "It is a plausible hypothesis that Plato's greatest contribution to the mathematicians who gathered around him in the Academy was the help he gave them in clarifying and perfecting their methodology. We can at least be sure that the procedure of the mathematicians had great attraction for Plato. It had its analogies in the Socratic method, as he portrays it in the earliest dialogues." Allan [*DSB*, 28], in speaking of Theaetetus' classification of the irrationals and his constructions of the regular solids, feels that "Plato's inspiration may be seen in this effort to systematize." Allan [p. 26] also suggests that it is possibly Plato himself who is referred to in the *Republic* [528B] where there is talk of the need of a director to coordinate research. Burkert [1972, 423] writes: "Even if Plato was not a professional mathematician his philosophy seems to have provided the decisive breakthrough in the establishment of mathematics." Similarly Cherniss states ([1945, 65]; see also [1951, 395] and the quotation from Zordan [p. 396]): "Plato's role appears to have been not that of a 'master' or even of a seminar director distributing subjects for research reports or prize essays, but that of an individual thinker whose insight and skill in the formulation of a problem enables him to offer general advice and methodological criticism to many other individual thinkers who respect his wisdom and who may be dominated by his personality, but who consider themselves at least as competent as they consider him in dealing with details of special subjects"; and [p. 66] "Plato's influence on these men, then, was that of an intelligent critic of method, not that of a technical mathematician with the skill to make great discoveries of his own...."

Against these various statements there are various counter-indications. First of all we must take care when using Q.2, for Proclus was a neo-Platonist and in fact head of the Academy. Thus we might suspect him of exaggerating the role of Plato in mathematics just as he exaggerates the role of the Pythagoreans in the earlier part of history (see Section 5,Q.5); for example, consider his language: "Thus the divine Plato said that geometry is the study of planes" [Proclus–Morrow, 94; Proclus–Friedlein, 117]. Furthermore, as Cherniss [1945, 65] has pointed out, Proclus is hesitant on a later occasion in crediting Plato with the teaching of analysis:

(Q.4) The best aid in the discovery of lemmas is a mental attitude for it. For we can see many persons who are keen at finding solutions but do so without method.... Nevertheless there are certain methods that have been handed down, the best being the method of analysis, which traces the desired result back to an acknowl-

> edged principle. Plato, it is said, taught this method to Leodamas, who also is reported to have made many discoveries in geometry by means of it.
>
> (Proclus, *A Commentary on the First Book of Euclid's* Elements [Proclus–Morrow, 165]; [Proclus–Friedlein, 211])

Note that in Q.3 analysis is mentioned only in connection with Eudoxus and not in connection with either Plato or Leodamas. Recall also that in Q.2 analysis is mentioned, but there it is only said that "there came into being [during the time of Plato] . . . solution by analysis . . . ." Leodamas and the method of analysis will be discussed in Section 17.

Another point that must be taken into consideration is that we must not infer from Proclus' statement in Q.3 that he claims that all the mathematicians mentioned were somehow associated with the Academy. Indeed on four occasions—"Eudoxus . . . [a] member of Plato's group . . . , Amyclas . . . one of Plato's followers . . . , Menaechmus . . . also associated with Plato . . . . These men lived together in the Academy . . ."—Proclus makes a very specific statement about the mathematicians being a "follower" of Plato. However, the text gives the impression that the first five mentioned mathematicians—Leodamas, Archytas, Theaetetus, Neoclides, and Leon—were not directly or at least not greatly influenced by Plato in the field of mathematics. I shall return to this point in Section 22.

Cherniss [1945, 68] also looks at the question of the influence of Plato from the point of view of the works of Aristotle who was associated with the Academy over a long period of time: "It is significant, therefore, that [Aristotle] betrays no knowledge of the higher mathematics which was developed by the specialists associated with the Academy; and hence we are constrained to infer that, while these men may have pursued their investigations in common, as the Proclus summary reports, and while Plato may have discussed their problems and their methods with them, neither these discussions nor the results of their studies were any part of the formal curriculum."

Neugebauer [1957, 152] is even stronger on this: "That, for a short while, mathematicians of the rank of Eudoxus [see Section 20] belonged to his circle is no proof of Plato's influence on mathematical research. The exceedingly elementary character of the examples of mathematical procedures quoted by Plato and Aristotle give no support to the hypothesis that Theaetetus or Eudoxus had anything to learn from Plato. The often adopted notion that Plato 'directed' research is fortunately not borne out by the facts. His advice to astronomers to replace observations by speculation would have destroyed one of the most important contributions of the Greeks to the exact sciences."

C. *Plato and DEMR*

In this subsection I mention several theories concerning Plato and DEMR which are not directly related to his works; the latter will be discussed in Subsection D.

i. In Q.4 from Proclus we read "Eudoxus . . . multiplied the number of propositions concerning the 'section' which had their origin in Plato, employing the method of analysis for their solution." As we shall see in Section 20 many authors associate the word "section" as used here by Proclus with DEMR. If this is so then this is a definite association between Plato and DEMR.

Heath [Euclid–Heath, II, 97; 1921, I, 304] says: "This idea that Plato began the study of [DEMR] as a subject in itself is not in the least inconsistent with the supposition that the problem of Euclid II,11 was solved by the Pythagoreans."

ii. Michel [1950, 570] assigns XIII,6 to Plato's era in contrast to XIII,1-5 (Section 14) which he considers to be older still.

iii. According to Mugler ([1948, 297]; see the review of this book by Cherniss [1951; 1960, no. 1430]): "He [Plato] no doubt knew intuitively that the ratio between these segments [i.e., those obtained by DEMR] was irrational. Perhaps he had even discovered during his researches that these segments were incommensurable . . . ."

The argument goes on to say that Plato would have been particularly interested in DEMR because of his research dealing with polyhedra: "what would be more natural than for Plato to direct his student's attention to this question of such importance to both mathematicians and philosophers . . . and especially to [Eudoxus]."

iv. A search through concordances and thematic indices dealing with Plato and Aristotle [Ast, 1835; Bonitz, 1870; Perls, 1973; Einarson, 1936; Heath, 1949; Apostle, 1952] indicates that neither of the words "pentagon" or "dodecahedron" is ever mentioned by these authors. See, however, Subsection D,i.

In this regard it should be mentioned that Apostle [1952, 112], in his rendition of Aristotle's mathematical works, writes: "Hence of all plane surfaces bounded by straight lines, the triangle is the most elementary; and the quadrilateral, the pentagon, and the rest contain the triangle potentially." This addition of the pentagon seems to be taken from Hicks' comments in *On the Soul* [Aristotle–Hicks, 337]. The original text, however, contains no mention of the word:

> (Q.5) From this it is clear that there is one definition of soul exactly as there is one definition of figure: for there is in

the one case no figure excepting triangle, quadrilateral and the rest, nor is there in the other any species of soul apart from those above mentioned....

The types of souls resemble the series of figures. For, alike in figures and in things animate, the earlier form exists potentially in the latter, as, for instance, the triangle potentially in the quadrilateral, and the nutritive faculty in that which has sensation."

(Aristotle, *On the Soul*, 414 b20, b28 [Aristotle–Hicks, 61]; [Aristotle–Hett, 83])

I shall comment on this passage in Subsection D,i.

D. *Passages from Plato*

We now turn our attention to specific passages from the works of Plato and the comments made on these by various authors. The order of presentation follows the chronological order of composition suggested by Robinson [*OCD*, 842], with the perhaps spurious *Hippias Major* at the end. Because *Timaeus* 55C is related to *Phaedo* 110B I have discussed both of them together.

i. The Dodecahedron in Phaedo *110B and* Timaeus *55C*

Since the precise mathematical construction of the dodecahedron involves DEMR, the mention of this solid is of interest to us in connection with our examination of Plato. The portion from *Timaeus* will lead us into a consideration of texts by other ancient authors which involve the decomposition of the pentagon into various triangles and to a discussion of whether or not these decompositions are related to the origins of IV,10. Furthermore, we will consider the question of whether in fact the various texts are really of Pythagorean origin.

In *Phaedo* 110b Socrates describes the splendour of the true earth:

(Q.6) Well then, my friend, in the first place it is said that the earth, viewed from above, looks like one of those balls made of twelve pieces of leather, painted in various colours, of which the colours familiar to us through their use by painters are, so to say, samples. Up there the whole earth displays such colours, and indeed far brighter and purer ones than these.

(Plato, *Phaedo* 110B [Plato–Hackforth, 176])

This apparently is a reference to the approximation of a sphere by a dodecahedron made out of twelve flexible leather pentagons.

In *Timaeus* (53C-55C) Timaeus discusses the four primary solids and makes a correspondence between them and the four basic elements: cube–earth, tetrahedron–fire, octahedron–air, and icosahedron–water. Concerning the dodecahedron we read:

(Q.7) As there still remained one compound figure, the fifth, God used it for the whole, broidering it with designs.

(Plato, *Timaeus* 55C [Thomas, I, 223])

What makes this section of *Timaeus* especially interesting for us is that, on the one hand, the faces of the first four "Platonic" solids are decomposed into "fundamental" triangles and, on the other hand, no such decomposition is given for the dodecahedron.

(Q.8) Now all triangles have their origin in two triangles, each having one right angle and the others acute; and one of these triangles has on each side half a right angle marked off by equal sides, while the other has the right angle divided into unequal parts by unequal sides....

Of the two triangles, the isosceles has one nature only, but the scalene has an infinite number; and of these infinite natures the fairest must be chosen, if we would make a suitable beginning.

(Plato, *Timaeus* 55C [Thomas, I, 218])

The basic triangles that Plato is speaking of are the isosceles 45°–90°–45° triangle and the 30°–90°–60° triangle, which Plato calls "the fairest." Four of the former can be combined to form the square and then the cube, whereas six of the latter can be combined to obtain the equilateral triangle and then the tetrahedron, octahedron, and icosahedron (Fig. VI-1). Various questions surrounding these decompositions of the square and equilateral triangle—in particular, why the equilateral triangle was not used directly—are discussed by Taylor [1928, 373] and Cornford [Plato–Cornford, 231].

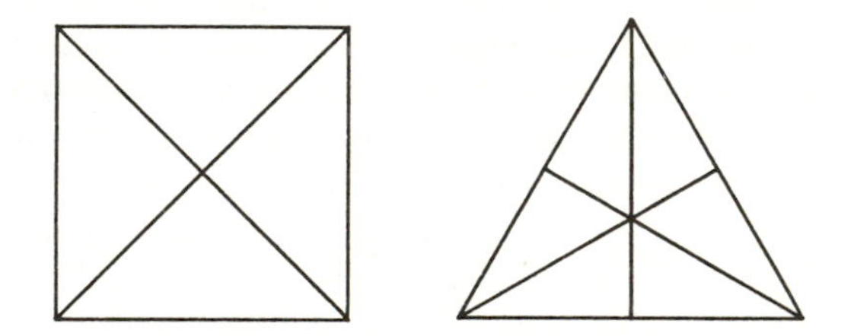

FIGURE VI-1. Decomposition of the Square and Equilateral Triangle

What about the dodecahedron and its decomposition into triangles? Taylor [p. 377] writes: "Timaeus does not describe its construction (Is this just a touch of Pythagorean 'reserve'?)." And then in a footnote he adds: "Perhaps not. His reason for silence may be that he does not know how to construct a pentagon by placing pairs of similar triangles 'diagonally,' a thing which, in fact, obviously cannot be done."

On the other hand, Cantor [1892, II, 176] has pointed out three references in classical literature which speak of the decomposition of the pentagon into triangles.

(Q.9) Did he, as some surmise, associate the dodecahedron with what is spherical, since he said that god employed the former for the nature of the sum of things in tracing the design of this? For, furthest withdrawn from straightness by the multitude of its elements and obtuseness of its angles, it is flexible and like the balls that are made of twelve pieces of leather by being distended becomes circular and circumscriptive, for it has twenty solid angles each of which is contained by three plane angles that are obtuse, since each consists of a right angle and a fifth; and it has been assembled and constructed out of twelve equiangular and equilateral pentagons, each of which consists of thirty of the primary scalene triangles, and this is why it seems to represent at once the zodiac and the year in that the divisions into parts are equal in number.

(Plutarch, 1st to 2nd century, *Platonic Questions* V 10003 [Plutarch–Cherniss, 53])

(Q.10) "Yes," said I, "Theodorus of Soli seems to follow up the subject not ineptly in his explanations of Plato's mathematical theories. He follows it up in this way: a pyramid, an octahedron, an icosahedron, and a dodecahedron, the primary figures which Plato predicates, are all beautiful because of the symmetries and equalities in their relations, and nothing superior or even like to these has been left for Nature to compose and fit together. It happens, however, that they do not all have one form of construction, nor have they all a similar origin, but the pyramid is the simplest and smallest, while the dodecahedron is the largest and most complicated. Of the remaining two the icosahedron is more than double the octahedron in the number of its triangles. For this reason it is impossible for them all to derive their origin from one and the same matter. For those that are simple and small and more rudimentary in their structure would necessarily be the first to respond to the instigating and formative power, and to be completed and acquire substantiality earlier than those of large parts and many bodies, from which classes come the dodecahedron, which requires more labour for its construction.... The generative elements of fire are the pyramid, composed of twenty-four primary triangles, and likewise for air the octahedron, composed of forty-eight of the same.

(Plutarch, "The Obsolescence of Oracles," 427 [Plutarch–Babbit, 441])

(Q.11) But for the Universe the deity made use of the Dodecahedron. Wherefore there are seen [in heaven] the forms of twelve animals in the circle of the Zodiac, and each of them is divided into thirty parts. And nearly so in the case of the Dodecahedron; which consists of twelve pentagons, (each) divided into five triangles, so that, as each consists of six triangles, there are found in the whole Dodecahedron three hundred and sixty triangles, bring as many as there are parts in the Zodiac.

(Albinus, 2nd century, "Introduction to the Doctrine of Plato," 13 [Albinus–Burges, 271]; on the faulty spelling Alcinous see Dodds [*OCD*])

As Cherniss [Plutarch–Cherniss, 53 fn. 9] points out, Plutarch does not state what kinds of triangles are involved in the decomposition. Cantor [1892, II, 177] conjectures that it was the decomposition obtained by dropping a perpendicular from each vertex to the middle of the opposite side and by drawing in all the diagonals (Fig. VI-2). Recalling that the pentagram is mentioned in ancient sources as a Pythagorean symbol (see Section 11,A), Cantor [p. 177] states: "out of the jumble of lines the pentagram distinctly stands out and accordingly this is already evidence for the sought for decomposition of the pentagon into elementary triangles." The diagram also displays the 36°–72°–72° triangle as well as the 36°–54°–90° triangle. Note, however, that the lines connecting the vertices to the midpoints of the side have nothing to do either with the pentagram or with the 36°–72°–72° triangle. Regarding this question of the decomposition of the pentagon into triangles, recall IV,13 where the pentagon is divided into 36°–54°–90° triangles, all radiating about the centre (and my remarks concerning this in Section 2,A).

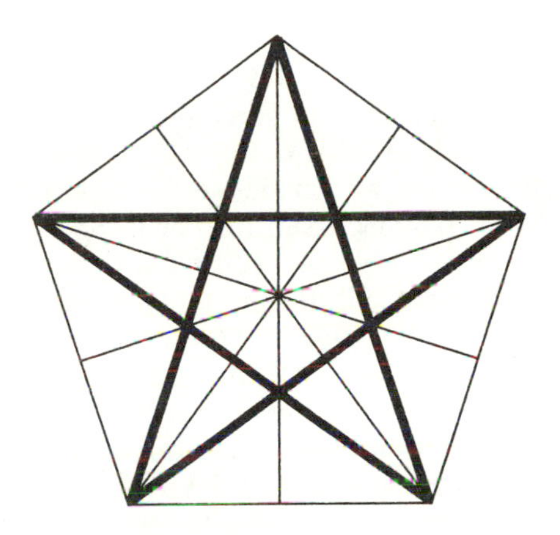

Source: Cantor [1892, II, 177]

FIGURE VI-2. Decomposition of the Pentagon

Combining Cantor's discussion with the assumption that *Timaeus* is an exposition of Pythagorean ideas, Heath [Euclid–Heath, II, 99] concludes: "There seems to be therefore no room for doubt that the construction of a pentagon by means of an isosceles triangle having each of its base angles double of the vertical angle [i.e., IV, 10] was due to the Pythagoreans."

There are, in fact, several Greek texts which link the five element theory to the Pythagoreans. For instance:

(Q.12) Pythagoras, seeing that there are five solid figures, which are also called the mathematical figures, says that the earth arose from the cube, fire from the pyramid, air from the octahedron, and the sphere of the universe from the dodecahedron.

(Aetius, 1st or 2nd century, *Placita*, ii, 6.5 [Thomas, I, 217]. On Aetius, see Ross [*OCD*]. Part of this material appears to be derived from the work of Theophrastus, c. −370 to −288, a student of Aristotle who in turn was a student of Plato.)

As I mentioned in the introduction to Chapter IV, the question of how much of Plato's work is of Pythagorean origin has been surrounded by much controversy, and this is particularly so in the case of *Timaeus*. On the one hand, Taylor [Plato–Taylor, ix] writes: "These considerations long ago suggested to me the necessity of attempting to correlate the *Timaeus* with all that can be learned about the Pythagorean science of the fifth century. It seems probable that careful scrutiny would show that the science of Timaeus is, in the main, pretty much what might be expected from a progressive Pythagorean contemporary of Socrates, and that Plato has, at least, originated very little of it.... If it can establish itself, the *Timaeus* must become a document of first-rate importance for the history of Pythagorean science."

The opposing view is taken by Sachs [1917] who has discussed the entire question in great depth and who gives the texts of the relevant material [pp. 9,52 (table)]. Indeed the whole book is centred around this question and the conclusion [pp. 76,119] is that the stories are not of Pythagorean origin, but of a later invention. Further [p. 82] it is concluded that the Pythagoreans would have had only an empirical knowledge of the dodecahedron. Sachs' views are discussed further in Section 18,iii.

In connection with this discussion, recall Q.5 where we find Aristotle discussing the soul and comparing it to geometrical figures. Note how Aristotle states that "the triangle [exists] potentially in the quadrilateral," but he does not speak of any decomposition of a pentagon into triangles. This would tend to indicate that, despite the quotations from Plutarch and Albinus, the idea of the decomposition of the pentagon into triangles was not of paramount importance—if it was even considered at all—at the time of Plato and Aristotle. Finally, I note Gaiser's introduction of DEMR into his study of ontology in Plato [1963, 144].

The question of Plato and the mathematical construction of the dodecahedron is again discussed in Section 19 in connection with the possibility of transmission of work from the Pythagorean Philolaus to Speusippus.

ii. The "Divided Line" in the Republic *509D*

In Book VI of the *Republic* Socrates is speaking to Glaucon:

(Q.13) "Conceive then," said I, "as we were saying, that there are these two entities, and that one of them is sovereign over the intelligible order and region and the other over the world of the eye-ball, not to say the sky-ball, but let that pass. You surely apprehend the two types, the visible and the intelligible." "I do." "Represent them then, as it were, by a line divided into two unequal sections and cut each section again in the same ratio (the section, that is, of the visible and that of the intelligible order), and then as an expression of the ratio of their comparative clearness and obscurity you will have as one of the sections of the visible world, images."

(Plato, *Republic* VI, 509D [Plato–Shorey, II, 109]; [Plato–Jowett, I, 771])

Concerning this, Brumbaugh [1954, 279 fn. 23, 266, 270] says: "To a reader of his time familiar with geometry, Plato's directions for construction involving the division of segments 'according to the same ratio' would very likely suggest a mean and extreme ratio section."

There is a huge amount of literature dealing with the interpretation of the "divided line." Shorey [Plato–Shorey, xxxii; see also 108 fn. b] writes: "I express the thought that the intermediate place of mathematical ideas on the proportion of the divided line is not to be taken literally...." For other commentary, see Plato–Jowett [p. lxxxvii], Wedberg [1955, 99], and Brentlinger [1963], as well as the bibliography of Cherniss [1960] and Brisson [1977]. Brumbaugh's views are supported by Gibson [1955, 61] but strongly criticized by Cherniss [1960, no. 1440].

iii. Timaeus *31B*

In the part of Plato's *Timaeus* that deals with the making of the "Soul of the World" and the doctrine of the elements we read the following:

(Q.14) Now that which has come into existence must needs be of bodily form, visible and tangible; yet without fire nothing could ever become visible, nor tangible without some solidity, nor solid without earth. Hence, in beginning to construct the body of the All, God was making it of fire and earth. But it is not possible that two things alone should be conjoined without a third; for there must needs be some intermediary bond to connect the two. *And the fairest of bonds is that which most perfectly unites into one both itself and the things which it binds together; and to effect this in the fairest manner is the natural property of proportion.* For whenever the middle term of any three numbers, cubic or square, is such that as the first term is to it, so is it to the last term,—and again, conversely, as the last term is to the middle, so is the middle to the first,—then the middle term becomes in turn the first and the last, while the first and last become in turn middle terms, and the necessary consequence will be that all the terms are interchangeable, and being interchangeable they all form a unity. Now if the body of the All had had to come into existence as a plane surface, having no depth, one middle term would have sufficed to bind together both itself and its fellow-terms; but now it is otherwise: for it behoved it to be solid of shape, and what brings solids into unison is never one middle term alone but always two.

(Plato, *Timaeus* 31B-32A [Plato–Bury, 59])
(My italics.)

The question of interest to us is the meaning of the italicized sentence. Cornford [Plato–Cornford, 44] translates the last part of the sentence as "and it is of the nature of a continued geometrical proportion to effect this most perfectly." The expression "continuous geometrical proportion" is also used by Warrington [Plato–Warrington, 22].

Cornford [p. 45] supports his use of the word "continued proportion" with various references to ancient authors, including Proclus in his commentary on *Timaeus* [Proclus–Festugière]. Cornford gives these references because some ancient authors had tried to read into this passage the other proportions of interest at the time, namely, the arithmetic and harmonic proportions. The view that it is the geometric proportion that is being referred to in Q.14 is also supported by Plato–Taylor [1928, 95], Burkert [1972, 440], Michel [1950, 383], and Mugler [1948, 82].

Note that the next line in Q.14 clarifies the sentence in question and speaks of "numbers, cubic or square" such that the first is to the second as the second is to the third. Thus the reference here is to a sequence of three integers of which the first and third are either perfect squares or perfect cubes. For example, if the two first and third terms are 2^2 and 5^2 then the middle term is the integer 10 and for 4^3 and 9^3 the middle term is 216. There is some problem of interpretation [Plato–Taylor, 98], but this has nothing to do with the requirement that the middle term be an integer.

Contrary to this interpretation, some authors interpret Q.14 as referring to DEMR; see Brunes [1967, 66—the reference is to Marstand] and Neufert [1960, 66]. Taylor [1928, 95, 367] is well aware of DEMR and mentions it in connection with *Timaeus* 53D, but does not even consider the possibility of a relationship of *Timaeus* 31B to DEMR. Incidentally, Brunes [1960, 66, 77], after criticizing Marstand, tries to show that the reference is to his "sacred cut."

It is also interesting to note Michel's [1950] view. After giving the context of the quotation [p. 383] and referring to the geometric mean, he once again refers to the passage [p. 591] and adds: "How much greater would be the harmony in the passage if the sum of two of the terms would equal the third [i.e., if the division were according to DEMR]." See Sections 12,vi and 20,A,iv.

iv. Hippias Major *303B*

In this possibly spurious work (see Plato–Tarrant [129 fn. 79] and Robinson [*OCD*]), we find the following statement by Socrates:

(Q.15) "To which group, then, Hippias, does the beautiful seem to you to belong? To the group of those that you mentioned? If I am strong and you also, we are both collectively strong, and if I am just and you also, we are both collectively just, and if both collectively, then each individually; so, too, if I am beautiful and you also, are we both collectively beautiful, and if both collectively, then each individually? Or is there nothing to prevent this, as in the case that when given things are both collectively even, they may perhaps individually be odd, or perhaps even, and again, *when things are individually irrational quantities they may perhaps both collectively be rational*, or perhaps irrational, and countless other cases which, you know, I said appeared before my mind? To which group do you assign the beautiful? Or have you the same view about it as I? For to me it seems great foolishness that we collectively are beautiful, but each of us is not so, or that each of us is so, but both are not, or anything else of that sort. Do you choose in this way, as I do, or in some other way?"

(Plato, *Hippias Major* 303B [Plato–Fowler, 417])
(My italics.)

Heath [1921, I, 304], in commenting on this passage, offers the suggestion that this is a reference to the fact that if we divide a rational line segment in extreme and mean proportion then each of the segments will be irrational. This view has been disputed by Strycker [1937], including on linguistic grounds. In particular, he argues—following Sachs [1917, 112]; see Section 18,iii—that the irrationality of the segment was discovered by Theaetetus and thus is much later than *Hippias Major*, which Strycker classifies among Plato's earlier works.

Another interpretation that has been proposed for the italicized passage is that it refers to the possibility that the hypotenuse of a right triangle may be rational even though the sides are irrational—for example, in the case when the sides are $R(3)$ and $R(6)$. Strycker also rejects this possibility as well as others that have been suggested.

Strycker's own interpretation of the meaning of the text is based on the use of the Greek word *sunamphoteros* in the italicized passage. He points out, using several examples dealing with the mathematics of music, that this word sometimes has the meaning "multiplication." Thus, it is suggested, the reference is to two irrational numbers, such as $R(3)$ and $R(12)$, whose produce is rational. This argument is supported by Michel [1950, 500, 564], but is criticized by Knorr [1975b, 296 fn. 77]. Mugler [1958, 396] gives various examples of the use of *sunamphoteros*, but only in the sense of addition.

Section 17. Leodamas of Thasos (Period of Plato)

In the "Catalogue of Geometers" of Proclus (Q.3), Leodamas of Thasos is the mathematician who follows Plato in the list. We also saw in Q.4 that Proclus says, although with hesitation, that Plato taught the method of analysis to Leodamas. The same statement is made by another author of the same general period:

(Q.16) He [Plato] was the first to introduce argument by means of question and answer, says Favorinus in the eighth book of his *Miscellaneous History*; he was the first to explain to Leodamas of Thasos the method of solving problems by analysis. . . .

(Diogenes Laertius, early IIIrd century, *Lives of Eminent Philosophers*, III, 24 [Diogenes Laertius–Hicks, 299])

The method of analysis mentioned here refers to the method wherein one starts with what is to be proved and then works backwards until a "true" statement is obtained. For further discussion on the method of analysis, see Heath [1921, I, 371; II, 400], Mahoney [1968], and Hintikka and Remes [1974]. The relationship of this method to Platonic philosophy and Pappus' comments on it are discussed by Klein [1966, 154, 259].

The method of analysis also has a relationship, albeit a tenuous one, to DEMR, for in certain manuscripts of the *Elements* Theorems XIII,1-5 are accompanied by an analysis; see Euclid–Heiberg [IV, 365] and Euclid–Heath [III, 442]. Lasserre [1964, 105, 178, 196] points to the connection that is made in Q.4 and Q.16 between Leodamas and the method of analysis and the appearance of the method of analysis in connection with XIII,1-5, and from these relationships he concludes that XIII,1-4 are due to Leodamas. For reasons that will be discussed in Section 20,B,v, Lasserre assigns XIII,5 to Eudoxus. A list of other theories concerning XIII,1-5 appears in Section 14.

Section 18. Theaetetus (c. −417 to −369)

A. *The Life of Theaetetus*

In the prologue of Plato's *Theaetetus* [142A: Plato–Dies, 156; Plato–Fowler 3, 7] we read that Theaetetus died of dysentery and from wounds he suffered at the battle of Corinth. We also read in the dialogue [p. 143], which takes place shortly before Socrates' death in −399, that Theaetetus was a student, at least to some extent, of Theodorus (Section 14). Theaetetus is also mentioned in a very complimentary fashion in Plato's *The Statesman* 257A [Plato–Fowler 3, 5; see also 266A, 37].

Even assuming that this information from Plato is correct, the questions of when exactly Theaetetus was born and when he died remain controversial points. To confuse matters, there were two battles of Corinth, in −394 and in −369, and Theaetetus would have been of a considerable "advanced" age had he died in −369 and yet had studied with Socrates in −399. The various arguments are discussed by Sachs [1914, chap. 1], Milhaud [1916], and Michel [1950, 278]. Following Sachs [p. 40], I will assume that −369 was the year of his death. Furthermore, assuming that Theaetetus was eighteen years old in −399, I will set −417 as his date of birth.

In the Byzantine collection, *Suidas*, we find the following two entries under the name Theaetetus:

(Q.17) Theaetetus, an Athenian, astronomer, philosopher, pupil of Socrates, taught in Heraclea. And he was the first to describe [or "construct"; see the remark on Section 11,Q.8] the five so-called solids. He flourished after the Peloponnesian wars.

(*Suidas*, article 93, "Theaetetus" [*Suidas*–Adler, II, 689; Thomas, 1939, I, 379])

It should be noted that instead of the meaning "flourish" the Greek might in fact have either the classical meaning of "born" or refer to precisely the fortieth year (see Kent [1905, 155]). Thomas uses the word "lived" [1939, I, 379]. I also note that Sachs [1914, 10 fn. 1] proposes the emendation "and he was the first to describe *Plato's* so-called five solids." This seems both unnecessary and unjustified.

(Q.18) Theaetetus, of Heraclea Pontica, a philosopher, disciple of Plato.

(*Suidas*, article 94, "Theaetetus" [*Suidas*–Adler, II, 689; Allman, 1889, 211])

Before discussing these two texts, I should mention that there is also a Theaetetus mentioned under the year Olympic 85,3 (−438) in Heronymus' translation of Eusebius' *Chronica* (see Sachs [1914, 11]). The Latin version reads: "Theaetetus was recognized as a mathematician"; the Greek version says: "Theaetetus the mathematician flourished." Since the date −438 would imply an age of around sixty years in −399, we can assume either that Theaetetus mentioned by Hieronymus–Eusebius is not the same as Plato's Theaetetus or that somehow Plato's Theaetetus was mistakenly assigned to such a precise but early date.

Returning to Q.17 and Q.18, we see that they present us with certain problems, a situation not surprising in "a work marred by contradictions and other ineptitudes. Many of its sources were already corrupt, and like most works of its kind it has suffered from interpolation" [Forbes, *OCD*]. The main problem concerns the relationship of Q.17 and Q.18. Some authors, such as Sachs [1914, 14], assume that both citations refer to the same person. Allman [1889, 211], on the other hand, writes: "It is much more likely that the second was a son, or relative, of Theaetetus of Athens, and sent by him to his native city to study at the Academy under Plato."

Heraclea Pontica was a city on the Black Sea; see Broughton [*OCD*] and Wilson [1976] as well as the map in Chapter III. Sachs [1914, 64], following Wilamowitz, infers from the mention of Heraclea Pontica in Q.18 that Theaetetus was the teacher of Heraclides Ponticus, a native of that town. This ɪvth century philosopher [Gottschalk, 1980; Lonie, *OCD*] came to the Academy as a pupil of Speusippus and Aristotle, and in fact was in temporary charge when Plato visited Sicily in −361. There are other known connections between Heraclea Pontica and the Academy. In Proclus' "Catalogue of Geometers" (Q.3), we read that Amyclas of Heraclea was one of "Plato's followers." Furthermore, Clearchus [Broughton, *OCD*], a pupil of Plato and Isocrates, seized power in Heraclea in −364.

Sachs [p. 65] also considers that the Greek in Q.18 means "a listener" rather than "a disciple"; "for since he was not properly a student of Plato, he was just like Eudoxus, who when he was a fellow at the Academy was called Plato's listener" (cf. Liddell and Scott [1968, 56] who give two translations: "a hearer of public speakers; hence, a disciple, pupil" and alternatively "a reader"). From this Sachs concludes that Theaetetus "seems to have taught mathematics at Plato's Academy."

It is my opinion that both Q.17 and Q.18 refer to the same person, and I suspect that the reason for the two entries is that the compilers of *Suidas* were unable to reconcile the statement "an Athenian" in Q.17 with "of Heraclea Pontica" in Q.18. It would be too much of a coincidence if the Heraclea of Q.17—there were several cities of this name in antiquity—did not refer to Heraclea Pontica. And it is unlikely that the sources of Q.17 and Q.18 would mention this relatively minor city unless it had played an important role in the life of Theaetetus. While it is not impossible that Theaetetus taught in Peraclea Pontica even though he was not born there (Q.17), the statement of Heraclea Pontica in Q.18 indicates that Theaetetus was indeed a native of that city.

In Plato's *Theaetetus* [142A,C], we read that Theaetetus was "being carried to Athens" but did not wish to stop in Megara because "he was in a hurry to get home. . . ." "Home" does not mean the same thing as "birthplace" nor even as "place of residence while young." Since "philosopher" and "student of Socrates" suggest Plato's play as a partial source of Q.17, it is possible that the compilers of Q.17 misinterpreted the text of *Theaetetus* to mean that Theaetetus was born in Athens.

If Theaetetus studied with Theodorus, Socrates, and perhaps others around −399 when he was about eighteen years old, and if the Academy opened around −386 when Theaetetus was about thirty-one years old, then it is not unlikely that Theaetetus returned to Heraclea Pontica to teach and do mathematical research in the first part of the ɪvth century, only returning to Athens, where he was associated in some way with Plato, at a later period. It should be noted that, according to reports (Section 20), Eudoxus studied with Plato, returned home to Cnidus to found his own school, and then returned to Athens with some of his own students at a later date. It is also important to note that in Q.3 Proclus simply lists Theaetetus as being a contemporary of Plato, while later in the "Catalogue of Geometers" other mathematicians are said to have lived together at the Academy. It may have been that when Theaetetus came to Athens again he remained there. This would account for the text of *Theaetetus*.

As a possible chronology of the life of Theaetetus, I suggest that he was born around −417 in Heraclea Pontica; that he studied in Athens around the turn of the century; that he returned to Heraclea where he founded his own school and did mathematical research, in particular on the theory of irrationals (Section 7, Q.10); that he was associated with Plato at some point, say, after −380 when he had already made a mathematical reputation for himself; and that he spent the last part of his life in Athens and died there in −369.

B. *The Contributions of Theaetetus*

We now turn our attention to the connection between Theaetetus and the domains that are of interest for our discussion: the construction of the regular polyhedra (Book XIII) and the theory of irrationals (Book X).

We have just seen in Q.17 how the *Suidas* associates Theaetetus with the description or construction of the five regular solids. However, we saw in Section 11,Q.10 that Scholium 1 to Euclid's Book XIII attributed only the octahedron and the icosahedron to Theaetetus, with the cube, pyramid, and dodecahedron being attributed to the Pythagoreans. Turning our attention to the theory of irrationals, we can recall the association that was made in Plato's *Theaetetus* (Section 7,Q.10) between Theaetetus and the extension of the theory or irrationals. In the works of Pappus we find another statement connecting Theaetetus with the theory of irrationals.

(Q.19) [I]t was nevertheless Theaetetus who distinguished the powers which are commensurable in length from those which are incommensurable and who divided more generally known irrational lines according to the different means, assigning the medial line to geometry, the binomial to arithmetic, and the apotome to harmony; as it is stated by Eudemus, the Peripatetic.

(Pappus, *Commentary on the Tenth Book of Euclid's* Elements [Pappus–Thomson, 63]; [Gaiser, 1963, 469; see Knorr, 1975, 235])

Based on this, it is generally assumed that it was Theaetetus who developed the theory of the three basic irrational lines, the medial, the binomial, and the apotome (see the discussion of Book X in Section 1). There is, however, a difference of opinion about whether Theaetetus also developed the theory of the other irrationals. While Van der Waerden [1954, 174] credits all of Book X to Theaetetus, Knorr [1975, 273, 287] only credits him with the development of the basic three irrationals, with the minor (X,73) credited to Eudoxus and the rest to later mathematicians; see Section 20,B,vi. Heath [1921, I, 212] credits the other irrationals, aside from the basic three, to Euclid.

Since the construction of the icosahedron and dodecahedron as given in the *Elements* depends on the concept of DEMR, Q.17 of this section and Q.10 of Section 11 are of the greatest interest to us. In particular, we would like to know which of the two references is the more accurate, if indeed they do contain some element of truth. Further, we would like to know in what sense Theaetetus "constructed" the solids. Are we to assume that it was in the rigorous sense of Book XIII? If Theaetetus was involved with the rigorous construction of the icosahedron and dodecahedron, how many of the other results involving DEMR that appear in XIII, or in II, IV, or VI for that matter, can be credited to him? We shall see that the views of various authors differ on these questions.

i. Tannery

Tannery [1887, 101] writes: "The object of Book XIII seems in fact to be less the construction of the regular polyhedra and their inscription in the sphere (problems whose solution can be attributed to Pythagoras) than the determination of the relation that the five polyhedra have with themselves and with the radius of the sphere." As part of the solution of this later problem, continues Tannery, the Greek mathematicians wished to classify the irrationals which were obtained in the course of the geometrical constructions of Book XIII. Thus it would seem that the parts of X that are used in XIII, as well as Book XIII, can be credited to the same author—Theaetetus.

Sachs [1917, 105], following Vogt, criticizes Tannery on the grounds that the exact construction (in the precise sense of Euclid, that is [p. 99]) requires as a prerequisite the actual calculation of the various lengths.

ii. Allman

It was suggested by Allman [1889, 212] that XIII is based on theorems discovered by Theaetetus but contains a recapulation by Aristaeus (Section 24).

iii. Sachs

The work by Sachs [1917] is a study of the history of the five regular or Platonic solids; see also Section 17,D,iii. She rejects the view that they were constructed by the Pythagoreans [p. 76] and assigns their construction instead to Theaetetus [Section 2]. Some of the definitions of the regular bodies in Book XI are also assigned to him [p. 93].

Sachs does not, however, credit Theaetetus with all of Book XIII. She accepts the interpretation (Section 20) that assigns certain of the discoveries—in particular XIII,1-5 [p. 97]—to his contemporary Eudoxus. It is argued [p. 99] that Eudoxus needed Theaetetus' theory of irrationals (see XIII,6,11) and Theaetetus used the work of Eudoxus in his work on the regular solids.

Sachs also says that the construction of the regular pentagon in IV,11 is not due to Theaetetus. The argument goes (cf. Section 2) that the mathematician who developed IV,11 ignored the decagon and in particular the result, stating that the side of the regular decagon is the larger segment of the radius divided in EMR. Knowledge of this would have led to a simpler construction of the pentagon. On the other hand, Sachs concludes that, since Theaetetus needed the decagon for the construction of the icosahedron in XIII,16, he would have obtained this result from XIII,9.

iv. Van der Waerden

Van der Waerden argues [1954, 173] that the author of Book XIII was not acquainted with the systematic development of Book II, but was well acquainted with the classification of irrationals as presented in Book X, for these are used in XIII,11,16,17. Conversely, says Van der Waerden, a study of the sides of the regular polyhedra would lead to a classification of the irrationals. From this follows the conclusion that both Books X and XIII in their entirety can be credited to one and the same author—Theaetetus. Knorr criticizes the ascription of entire books to a single author [1975, 293 fn. 52, 297 fn. 91].

v. Bulmer-Thomas

Bulmer-Thomas [*DSB* 3] notes that "the only use made of Book X in subsequent books of Euclid's *Elements* is to express the sides of the regular solids inscribed in a sphere in terms of the diameter... [XIII,16,17]." As does Heath [1921, I, 162], Bulmer-Thomas credits Theaetetus with being the first to give a theoretical construction of the inscriptions in a sphere, but he assumes that the material in XIII has been arranged by Euclid.

In Section 7,ii,Q.10, I gave part of the conversation involving Theaetetus from Plato's work of the same name. Later on we read:

(Q.20) All the lines which form the four sides of the equilateral or square numbers we called lengths, and those which form the oblong numbers we called surds, because they are not commensurable with the others in length, but only in the areas of the planes which they have the power to form. And similarly in the case of solids.

(Plato, *Theaetetus*, 148 [Plato–Fowler 3, 27]. The translation of this part of *Theaetetus* is controversial; see Van der Waerden [1954, 142, 166] and Knorr [1975, chap. 3].)

About this quotation, Bulmer-Thomas [*DSB* 3] remarks: "It is also significant [regarding the question of the relationship of Theaetetus to Book XIII] that at the end of the mathematical passage in the *Theaetetus* he says that he and his companion proceeded to deal with solids in the same way as with squares and oblongs in the plane." However, Van der Waerden [1954, 166] takes the last sentence as simply meaning: "Line segments which produce cubes whose volume is an integer, but not a cubic number, are incommensurable with the unit length."

vi. Waterhouse

Waterhouse [1972] has examined the meaning of Q.10 of Section 11 and the apparent difficulty in assigning the octahedron to a later date than the dodecahedron. He points out that the crucial concept was the concept of a regular solid. Cubes, pyramids, and dodecahedrons—from nature—may have been studied individually, but the octahedron, being simply two square-based pyramids stuck together, would not have attracted attention as a mathematical object; "only someone possessing the general concept of regular solid would single it out." There is less certainty about the role of the icosahedron. Its construction was perhaps obtained during the search for all regular solids, or perhaps earlier. In support of his argument Waterhouse points to the cube and pyramid having names from common speech and the indications that the dodecahedron was called "the sphere of the twelve pentagons" (Section 11,Q.8). On the other hand, the octahedron and icosahedron apparently never had any other names. Thus Waterhouse credits the general theory of regular solids to Theaetetus.

vii. Neuenschwander

Neuenschwander [1974] discusses the composition and development of Book XIII and attributes the greatest part to Theaetetus, although there were revisions by Euclid; XIII,6,11 and part of XIII,18 may be later interpolations.

I shall return to the question of the contributions of Theaetetus in Section 20,C,iii.

Section 19. Speusippus (c. −408 to −339)

The following citation is of interest to us:

(Q.21) Speusippus, the son of Potone, sister of Plato, and his successor in the Academy before Xenocrates, was always full of zeal for the teachings of the Pythagoreans, and especially for the writings of Philolaus, and he compiled a neat little book which he entitled *On the Pythagorean Numbers*. From the beginning up to half way he deals most elegantly with linear and polygonal numbers and with all kinds of surfaces and solids in numbers; with the five figures which he attributes to the cosmic elements, both in respect of their special properties and in respect of their similarity one to another; and with proportion and reciprocity [perhaps: proportion, continuous and discontinuous].

(Iamblichus, c. 300, *Arithmetical Speculations on Divine Things* [Thomas, 1939, I, 75])

The question of primary interest is what was the precise nature of Speusippus' discussion of the five regular polyhedra. Were they "constructed" mathematically in some sense and if so did the construction involve DEMR? There is no way of knowing for sure, but the title and text suggest that only integers were discussed; how the solids would fit in is not clear. Note also that the fact that his name does not appear in Proclus' "Catalogue of Geometers" suggests that his mathematical contributions were not great.

A secondary question involves the writings of Philolaus. According to various sources (see Freeman [1946, 220]), Philolaus was a Pythagorean active in the latter half of the vth century and a teacher of Archytas (see Section 11) who in turn seems to have influenced Plato. One account says that Plato bought Philolaus' book on Pythagoreanism and copied *Timaeus* from it. Freeman [p. 222] writes: "The question therefore arises, did Plato borrow this from the Pythagoreans, or did he invent it, and was its later ascription to Pythagoras or Philolaus merely an attempt to substantiate the absurd idea that the *Timaeus* was a transcription of the mysterious Pythagorean books." This question is related to my discussion in Section 16,D,i.

Section 20. Eudoxus (c. −400 to −350)

Eudoxus contributed to various fields including astronomy. In mathematics he is usually credited with the theory of proportion and the method of exhaustion (see Toomer [*OCD*], Huxley [*DSB* 1], Mau [1975], and Heath [1921, I, 322]). The estimated date of his birth varies from source to source by eighteen years (see Michel [1950, 234] and Huxley [1963]). As far as his relationship with the Academy is concerned, it seems that Eudoxus arrived in Athens from Cnidus about −377; he returned to Cnidus at some point as the head of a school of his own and then returned to Athens again around −365 (see Huxley [1963]).

The discussions in the literature concerning the relationship of Eudoxus to DEMR revolve around two overlapping issues: the interpretation of the word "section" in Proclus' "Catalogue of Geometers" and the possible contribution of Eudoxus to the development of DEMR.

A. *Interpreting "Section"*

Proclus' "Catalogue of Geometers" (Q.3) contains the following passage dealing with Eudoxus:

(Q.22) Eudoxus of Cnidus, a little later than Leon and a member of Plato's group, was the first to increase the number of the so-called general theorems; to the three proportionals already known he added three more and multiplied the number of propositions [literally, "increased into a large number"; the implication seems to be that there was more than one to start with] concerning the "section" which had their origin in Plato, employing the method of analysis for their solution.
(Proclus, *A Commentary on the First Book of Euclid's* Elements [Proclus–Morrow, 55]; [Proclus–Friedlein, 67])

The Greek *tomos* (cf. the French *tome* meaning "division of a work") which is translated above as "section" has a variety of meanings in ancient Greek mathematics. In Mugler [1958, 425] the entry under *tomos* reads: "A very frequent term which means both the intersection of two geometrical elements in the abstract sense and also the result of this intersection i.e. the precise point, line or surface determined by such intersections." In the course of this section I will discuss how the word "section" is used by Proclus and other sources; further examples are given by Mugler.

The question of the precise significance of the word "section" in Q.22, as well as of the nature of Eudoxus' contribution, has been a matter of controversy since the 1870s. For our purposes the commentators can be split into two classes: those who associate "section" with DEMR and those who give an alternative interpretation.

i. Bretschneider

Bretschneider [1870, 168] argues that, since Q.22 speaks of the "section" and not of sections, we are obliged to ask which particular section is being referred to. His answer is that until Plato the only "section" of importance was that corresponding to the division of a line in EMR.

Against this part of Bretschneider's argument, Tannery [1887, 76 fn. 1] says: "The use of the singular would not prove anything even if the text [that we possess] were better established than it is. Further it is impossible to find a text where one actually speaks of 'a section' in the sense of division in mean and extreme ratio; just above ... Proclus expresses himself quite differently."

The statement of Proclus that Tannery is referring to follows.

(Q.23) The principles of arithmetic and geometry, then, differ from those of the other sciences, yet their own hypotheses are distinct from each other, in the sense mentioned above; nevertheless they have a certain community with one another, so that some theorems demonstrated are common to the two sciences, while others are peculiar to the one or the other.

....

Common to both sciences [i.e., arithmetic and geometry] are the theorems regarding sections (such as Euclid presents in his second book) with the exception of the division of a line in extreme and mean ratio.
(Proclus, *A Commentary on the First Book of Euclid's* Elements [Proclus–Morrow, 48]; [Proclus–Friedlein, 60; Proclus–Ver Eecke, 51])

What is the meaning of this statement? My interpretation is that Proclus is preserving the distinction between arithmetic, where only integers are involved, and geometry, where incommensurable quantities may arise. Consider, for example, Theorem II,5 where we are to cut line AB at the midpoint C and then at an arbitrary point D; these are the "points of section." The conclusion of II,5 will be valid whether the segments are commensurable or incommensurable. On the other hand, II,11, which is the "deviant" result mentioned by Proclus, involves a construction that produces incommensurable segments; in other words, it does not belong to the science of arithmetic.

Returning to Tannery's remark, I presume he means that in Q.23 Euclid's Theorem II,11 is to be considered as just one of the theorems that deals with the sectioning of a line, that is, with the division of a line at certain fixed or arbitrary points of section, and that in no way can Q.21 be thought of as implying that the statement of II,11 defines or specifies a particular point of section.

As Sachs has pointed out, one of the scholia to Book II does use the word "section" in connection with a text involving DEMR [1917, 97] (see also Knorr [1975, 295 fn. 73]).

(Q.24) That the proportion is geometrical is therefore clear: for AB is cut at [H] and the [rectangle] under AB, [BH] was found equal to the [square] of [HA], and this follows the geometrical mean alone; and he says in the sixth book that it [i.e., the straight line] is cut in extreme and mean. But now [i.e., here in Book II], because of our ignorance of anything concerning ratio [i.e., ratio is not defined until Book V], he has not said that it is cut in extreme and mean ratio. And it is not analysed, because he has not defined the section.

(Scholium 70 to Euclid II,11 [Euclid–Heiberg, V, 248])

Are we to interpret "the section" in the last sentence as meaning "the section point defined by DEMR" or perhaps "the sectioning of a line according to DEMR" as in the definition of Mugler? It is not at all certain that such an interpretation would be correct, for we must consider the last sentence as a whole and also inquire about the meaning of "And it is not analysed...." Does "analysed" have to do with the method of analysis that is said to have been employed by Eudoxus and examples of which sometimes accompany XIII,1-5 in manuscripts of Euclid (see Section 17 and below)? If "analysed" does indeed have to do with the method of analysis, then perhaps the scholiast wished to say that we have in II,11 a construction (as opposed to a statement) which involves "a section" in the sense of II,5 and to which an "analysis" could be applied. This possible explanation is weakened by the fact that Euclid himself never does an "analysis." It seems to me that whatever the scholiast meant to say the scholium is so lacking in clarity that, even if it dated from the time of Proclus or before, it is of no help in deciding the meaning of Proclus' statement.

Let us return to Bretschneider's discussion. Not only does he associate "section" with DEMR, but he specifically identified the propositions in question as being XIII,1-5. This association is made because, as pointed out above, in certain manuscripts an "analysis" accompanies XIII,1-5 and because Q.22 says that Eudoxus employed the method of analysis.

Heiberg, however, pointed out in 1881 (see Allman [1889, 136 fn. 29], Tannery [1887, 76], and Euclid–Heiberg [IV, 364 fn. 8]) that the position of these additions varies from manuscript to manuscript—sometimes one per theorem, sometimes grouped after XIII,5. This in turn is taken to imply that they were later additions or interpretations.

Knorr [1975, 277] argues that "the specific theorems which he [Bretschneider] wishes to ascribe to Eudoxus would have been absolutely trivial at this time...."

Burkert [1972, 452 fn. 26] says that "to interpret [Q.22] as referring to DEMR would imply that the report of the scholium [Q.13 of Section 11] on Euclid [IV] was false." I do not see where the contradiction lies.

ii. Tannery

Tannery [1887, 76] disagreed with Bretschneider's thesis and agreed with the opinion held before Bretschneider (the bibliographical data are unfortunately not given either by Bretschneider or by Tannery) that "the section" refers to the section of solids and to the work which preceded the invention of conics. This interpretation has in its favour that such a usage is actually attested to in Greek mathematics (see Mugler [1958, 424]). Indeed later on in his commentary on Euclid's Book I, Proclus himself uses "section" in this sense:

(Q.25) Plato assumes that the two simplest and most fundamental species of line are the straight and the circular and makes all other kinds mixtures of these two, both those called spiral, whether lying in planes or about solids, and the curved lines that are produced by the sections of solids.

(Proclus, *Commentary on the First Book of Euclid's* Elements [Proclus–Morrow, 84]; [Proclus–Friedlein, 104])

In fact Proclus specifically uses "section" in connection with conics.

(Q.26) Geminus [1st century] divides lines first into incomposite and composite, calling a composite line one that is broken and forms an angle. Incomposite lines he then divides into those that make figures and those that extend indefinitely. By those that make figures he means the circular, the elliptical, and the cissoidal; and by those that do not, the section of a right-angled cone, the section of an obtuse angle cone [i.e., the parabola and hyperbola], the conchoid, the straight line, and all such.

(Proclus, *Commentary on the First Book of Euclid's* Elements [Proclus–Morrow, 90]; [Proclus–Friedlein, 111])

Tannery's view is supported by Ver Eecke [Proclus–Ver Eecke, 59]. It has been criticized by Knorr [1975, 277] on the grounds that Eudoxus is never mentioned as having initiated studies of the conics.

iii. Tropfke

Tropfke [1903, 99] suggests that the reference may be related to Eudoxus' presumed development of the theory of proportions as in Books V and VI (see Van der Waerden [1954, 187], Heath [1921, I, 325], and Knorr [1975, chap. 7]). DEMR, of course, appears in VI,def.3 and VI,30.

iv. Michel

Michel's suggestion [1950, 589] about the significance of the passage from Proclus involves the concept of mediety. A mediety can be thought of as a relationship among three quantities or equivalently as a way of obtaining a quantity B which is intermediate between the quantities A and C with $C<A$. For instance, the relationship $(A-B):(B-C)=A:A$ is equivalent to $2\cdot B=(A+C)$ so that B is the arithmetic mean between A and C. The two other classical medieties are the geometric $A:B=B:C$ or $B^2=AC$ and the harmonic $(A-B):(B-C)=A:C$ or $2/B=1/A+1/C$. For a discussion of the medieties as well as historical details, see Heath [1921, I, 85]; Van der Waerden [1954, 231]; Michel [1950, 369]; Mugler [1958, 287]; and Thomas [1939, I, iii].

The particular historical reference of interest to us is the statement by Iamblichus in his *Introduction to Nicomachus' Arithmetic* that three of the medieties were discovered by Eudoxus (see Heath [1921, I, 86], Knorr [1975b, 275], Thomas [1939, I, 120], and Lasserre [1966, 36]).

It should be noted that nothing in the definition of the medieties requires any other relationship to hold among the quantities. Michel, however [p. 576, see also p. 524], investigates the case of "the medieties of partition" where the three quantities in question satisfy the relationship $A=B+C$. In this very restrictive case, there is a great reduction in the number of possibilities; for instance, the geometric mediety reduces to the partitioning of a line in EMR. But just as there is no indication that the problem of DEMR was ever treated by Greek mathematicians as a special case of the geometric proportion, there is no historical evidence that the question of medieties was considered in this special case. Michel [p. 577] considers it "reasonable" that this would be "one of the first problems that presented itself to the spirit of the mathematicians . . . ."

There is, on the contrary, evidence in the Greek texts that the added condition $A=B+C$ was not considered. The counter-indication is the seventh mediety of Nicomachus, which is the same as the tenth mediety of Pappus [Thomas, 1939, I, 123; incorrectly stated in Heath, 1921, I, 89]. This mediety, which takes the form $(A-C):(A-B)=B:C$, automatically implies that $A=B+C$; in other words, the added condition that Michel suggests is itself a consequence of one of the medieties. Michel is aware of this objection, but suggests that the problem of investigating the medieties with the added condition $A=B+C$ could have been posed and solved in the absence of a knowledge of the tenth mediety of Nicomachus.

The interpretation that Michel [p. 589] gives for the meaning of "the section" in the statement of Proclus is based on the supposition that the case $A=B+C$ was indeed considered. He agrees with Bretschneider that "section" refers to the section of a line instead of a section of a conic, but he goes further than Bretschneider and assumes that the reference is to any division of a line according to a mediety and not just a line divided in EMR.

One has the impression that Michel, under the influence of such "golden numberists" as Ghyka [1927, 1931], is trying to force DEMR into early Greek mathematical history. I have already mentioned in Section 16,D,iii his wishful thinking [p. 591] about a passage in Plato's *Timaeus*. I mention also his reprimand [p. 603] of those who have criticized some of the 20th century "golden numberists." In a previous article [Fischler, 1979b], I have pointed out that the quotation from Herodotus (which in turn leads to the "golden number" theory of the Great Pyramid) that Michel refers to [p. 628] has in fact a late 19th century origin.

v. Gaiser

Gaiser [1963, 371 fn. 118, 421 fn. 272] suggests that the statement means that Eudoxus further developed the theory of irrational quantities. This is tied in with Plato, as in Q.22, because of the suggestion that Plato used irrational quantities to clarify his theories of ontology [cf. p. 144, fig. 42—note the appearance of DEMR]. By giving a citation from Aetius where the same Greek word *tomos* is used in the phrase "concerning the division (or cutting) of bodies," Gaiser links the problems of irrationality and the infinitesimal and suggests that Proclus linked both of them (cf. Gaiser's reconstitution of Q.2).

In a similar vein, Knorr [1975, 278] writes: "But all the irrational lines, with the exception of the medial, are also constructed as sectioned lines. Hence, the study of the 'section' might well be suited as a reference to the wider study of irrational lines." In a footnote Knorr mentions the Greek word for division in Book X and notes that "both 'divide' and 'cut' are recognized terms in Archytas' music theory." The implication of the footnote appears to be that even though the word "division" is used in X, the word "section" as used by Proclus is an acceptable substitute. In the list of usages indicated by Mugler [1958, 128] we find two examples from Apollonius where a line is "divided" by a "cut" into two *equal* parts. There do not seem to be examples where a "cut" is used to determine an irrational quantity much less DEMR (cf. Mugler [413, 422, 425] on "cut," "segment," and "section."

vi. Burkert

Burkert [1972, 452 fn. 26] also believes that the reference is to the sectioning of lines, but he takes it to mean the sectioning of lines as the general topic of Book II. Thus he would assign the systematization of Book II to Eudoxus. This view is criticized by Knorr [1975, 295 fn. 75] on the grounds that a systematic theory of Book II was already needed for the earlier work of Theodorus and Theaetetus.

vii. Fowler

Fowler [1980] discusses Q.22 in the light of his theory that there existed an early definition of ratio based on anthyphairesis (Section 8). In this article he shows how Book II can be considered as a study of various properties of squares and rectangles needed to prove that certain anthyphairetic calculations do indeed hold. Then in a discussion of these demonstrations, and the conjectures that preceded them, in the light of an anthyphairetic reinterpretation of "analysis and synthesis" (see i), Fowler shows how an investigation of what magnitudes would have the simplest anthyphairesis, viz. [1,1,1, . . .] would lead to the definition of DEMR (see Section 13,ii).

Fowler feels that the interpretation of "propositions concerning the section" as dealing with II,11, VI,30, and XIII,1-6 might not be valid since they are not sufficiently deep. He then states: "However, by placing the construction II,11 at the beginning of a sequence of increasingly difficult fundamental problems, and setting the discussion within the context of analysis and synthesis, the way is opened up for a new exploration of the possibility."

B. *Contributions of Eudoxus to the Development of DEMR*

Not only is there controversy about the meaning of "section" in Q.22, but even among those who associate "section" with DEMR there is some contention about what propositions were involved.

i. Bretschneider

We saw in Subsection A,i that Bretschneider [1870, 168] associated the propositions with XIII,1-5.

ii. Allman

Allman [1889, 136] believes that XIII,1-5 are older than Eudoxus. For instance, XIII,1-4 involve the gnomon (Section 3) and this is taken to be a sign of their antiquity. Eudoxus' contribution is supposed to have been the definitions of analysis and synthesis given in some manuscripts [Euclid–Heath, III, 442] and the alternate proofs using this method.

iii. Sachs

Sachs [1917, 99] suggests that the construction of the pentagon in IV,11 as well as XIII,8—which states that the diagonals cut each other in DEMR—may be due to Eudoxus.

iv. Van der Waerden

It is suggested by Van der Waerden [1954, 184] that Q.22 refers to VI,30 where proportion theory, generally accredited to Eudoxus, is used. Knorr [1975, 295 fn. 73] objects that this is not a major contribution.

v. Lasserre

Although in some manuscripts all of Theorems XIII,1-5 have associated with them an "analysis," only XIII,5 involves proportion theory (see iv). On the basis of this and Q.22 Lasserre [1964, 105] assigns XIII,5 to Eudoxus.

vi. Knorr

Knorr [1975b, 279] is of the opinion that the reference is to Eudoxus' recognition that the construction of the icosahedron leads to a new irrational quantity, the "minor" of X,76 which is the only non-basic irrational to appear outside Book X; "[XIII,11] was the one result which motivated the development of the entire theory of higher (non-Theaetetean) irrationals" [p. 281]. Part of the argument involves the identification of the terminology "minor" in X,76 and that of "lesser segment" as in the division of a line in EMR as being one and the same (cf. Mugler [1958, 49, 169]).

There are several difficulties with this solution. First of all, it only provides for essentially one result, whereas Q.22 indicates that a group of results is being referred to in Proclus' "Catalogue of Geometers." Secondly, recalling Knorr's criticism of Tannery in Subsection A,ii, we must point out that there seems to be no other evidence to link Eudoxus with the development of the irrationals. Indeed, if we look at the commentary of Pappus [Pappus–Thomson] on Book X we find that there is a very specific assignment (Q.19) of part of the development of the theory of irrationals to Theaetetus; on the other hand there is no mention of Eudoxus, according to the index [p. 185]. Furthermore, even though there are other specific claims in ancient literature concerning the achievements of Eudoxus, irrationality does not appear to be one of them. Finally, one wonders why Proclus would make a connection between Eudoxus and the irrationals and not between Theaetetus and the irrationals, a connection that was made by Plato (Section 7,Q.10).

C. *Commentary*

It is my belief that the association of "section" with DEMR is incorrect. There are several counter-indications to this hypothesis.

i. The description of the era from Plato to Euclid given in the "Catalogue of Geometers" of Q.3 is essentially devoid of specific statements. The only details given in the entire passage are that Eudoxus added three more proportionals to those already known and that Hermotimus wrote something about locus-theorems, if this latter can really be called a detail. The first statement is probably taken from Iamblichus and is at odds with other sources (see Heath [1921, I, 86]). Indeed the whole section is essentially one huge platitude in which the heroes "brought [theories] into a more scientific arrangement," "added many discoveries," "increased the number of general theorems," "made the whole of geometry still more perfect," "produced an admirable arrangement of the elements," "pursued further the investigations," and "carried on . . . investigations according to Plato's instructions." We do not even have the statement to be found in Pappus (Q.19) that Theaetetus found the three basic types of irrationals; rather, we have what seems to be a summary of a joint symposium between philosophers of mathematics and some book editors.

In view of this we may reasonably ask if Proclus really did have anything specific in mind or if this is just another one of those phrases, of which his mentor Plato must be the acknowledged master (see Section 6,Q.6), that intentionally or not serve mainly to occupy and torment the waking hours of historians of mathematics.

ii. Before going beyond the assumption that Proclus is talking in generalities and certainly with no intention of being more specific than in his statements about Leon increasing the number of so-called general theorems and about Hermotimus writing something about locus-theorems, we should see exactly how Proclus uses "section" elsewhere in his commentary. I have already given three examples in Q.23, Q.25, and Q.26—one involving section points of a line, one involving sections of solids in general, and another involving conics.

In his index Friedlein [Proclus–Friedlein, 502] gives the following references to the word "section." I give them in the order in which they appear; the first number indicates the page in Morrow's edition, the second in Friedlein's edition.

(Q.27) [Euclid] divides [the propositions] into problems and theorems, the former including the construction of figures, the division of them into sections . . . . [63; 77]

But it is in imagination that the constructions, sectionings, superpositions, comparisons, additions, and subtractions take place . . . . [64; 78]

[by] making a section [of the cone] from the apex to the base we obtain a triangle, but by cutting it parallel to the base we obtain a section that is a circular plane. [96; 118]

And the spiric sections are three in number corresponding to these different kinds of surface. [97, see fn. 74; 119]

On the plane it is possible to think of straight lines, circles, spirals, and the sections and contacts of straight lines and circles . . . . [97, 220]

(Proclus, *Commentary on the First Book of Euclid's* Elements [Proclus–Morrow]; [Proclus–Friedlein]. Friedlein gives three other references: [65; 79], [134; 170], [251; 322], but these are not of interest to us. Friedlein also indicates that there are other uses in the text of words having the same root.)

Thus while it is not impossible that Proclus used the word "section" in Q.22 in connection with DEMR, the eight uses that I have given all have one of the senses indicated by Mugler (see Subsection A).

iii. Even if we suppose that the statement of Proclus does refer to some specific results, the DEMR hypothesis still remains unsatisfactory for the following reasons. First of all, the Greek phrase "increased into a large number" gives the impression that there were already some propositions concerning "the section" in existence when Eudoxus came along. Secondly, the statement definitely states that Eudoxus added more than one new proposition. Since II,11 and VI,30 are in effect one proposition and since there is no indication that the construction of the pentagon was meant, the already existing theorems concerning "the section" would have to be found in Book XIII if "the section" referred to DEMR.

As I have discussed in Section 2,E, the development of Book XIII appears to have developed, as far as DEMR is concerned, in three stages.

Every indication seems to point to the first two stages being due to Theaetetus (see Sections 18, 22,C and D). Caution must always be uppermost when using a scholium, but Q.10 of Section 11, despite the fact that it presents difficulties, seems to contain some element of fact (see Van der Waerden [1954, 100]). I also assume that Pappus (Q.19) is reliable when he assigns the three basic irrationals to Theaetetus. Further, as my discussion of Section 2 indicates, there is no reason to believe that XIII,1-5 are other than lemmas developed when needed for the first two stages. The same is true of XIII,6 with respect to the second stage. This would only leave the third stage for somebody else.

The third stage consists of only one result, namely (XIII,11) that we obtain the minor line when we inscribe a pentagon in a circle with rational diameter. It is true that XIII,8 is needed for XIII,11, but it is just a simple lemma which was most probably proved especially for XIII,11. Indeed XIII,8 is not used elsewhere and, as I indicated in Section 2,C, it is very unlikely that XIII,8 was known at the time of the original construction of the pentagon. The point is that, as I showed in Section 2,C, the result of XIII,8 *could have* been proved earlier but it was not because nobody "needed" it, and therefore the result remained unknown. It is a "folk axiom" that most basic results in mathematics can be proved once the statement of the result is known; it is the crucial step of realizing that a certain result must surely be true that constitutes the key step, even if the person who realizes this is not technically competent enough to provide the actual proof. I will return to this point and cite ancient evidence relative to it in Section 22.

Thus it would appear that XIII,8 and XIII,11 were developed together by one person, or by a group working together, and that in effect XIII,8 cannot be considered as a new result worthy of note. All this means that, under the assumption that "the section" refers to DEMR, we are left wondering what exactly Eudoxus could have proved.

iv. In view of what I have stated it seems to me that the most likely explanation of Q.22 is that Proclus said, or at least meant to say, that Eudoxus "pursued further" the investigations of various geometrical results that had to do in some way or another with the general concept of "section."

Section 21. Euclid (fl. c. −300)

For a discussion of the little that is known of the life of Euclid, see Bulmer-Thomas [*DSB* 5], Fraser [1972, I, 386; this work provides a description of the work of Euclid and other mathematicians in the context of Alexandrian society], and Euclid−Heath [I, chap. 1]. We simply do not know what Euclid took from other earlier writers and what he rewrote. Various authors, such as Bulmer-Thomas [*DSB* 3], believe that parts of Book XIII were rewritten or rearranged by Euclid himself.

Section 22. Some Views on the Historical Development of DEMR

In Section 2, I examined the *Elements* to see what could be ascertained from the earliest extant text that explicitly mentions DEMR. This was done without any consideration of any other possible historical references and contexts. After that I considered various mathematical topics, historical references, and suggestions that have been made in the literature concerning the historical development of DEMR. This brought us up to Euclid himself and the end of the classical period of Greek mathematics. In this section, I summarize the various theories and present my own views on the subject.

A. A Summary of Various Theories

The principal theories concerning the development of DEMR, some of which overlap, are summarized below. More detailed discussions and criticisms of these theories will be found in the indicated sections.

i. Equations and Application of Areas

According to this theory, which was discussed in detail in Section 5, DEMR developed out of a study of quadratic equations, in particular from a solution of the equation $x^2 = a(a-x)$ by means of the method of application of areas.

ii. Incommensurability

It has been suggested that the development of the concepts of incommensurability and DEMR were linked. See Sections 7 and 12,vii.

iii. Similar Triangles Development Based on XIII,8

The text of Euclid indicates that the result that the diagonals of a pentagon cut one another in EMR was not known at the time the pentagon was first constructed. See Sections 2,C; 12,ix.

iv. Anthyphairesis

This theory assumes that the concept of DEMR developed out of an investigation of the anthyphairesis [1,1,1 . . .]. See Section 13,B.

B. Summary of My Conclusions

My detailed examination of the *Elements* has led me to the conclusion that the Euclidean text accurately reflects the historical development of the concept of DEMR. More specifically, I suggest:

i. The concept of DEMR arose directly out of the task of constructing the pentagon, in the sense that the formulation of DEMR in terms of area (II,11) occurred while attempts were being made to inscribe a pentagon in a circle.

ii. The final approach to the problem via IV,10—possibly after initial attempts with central angles—led via III,32 ("tangent angle equals inscribed angle") to III,36,37 and then II,6. Thus these latter results were most probably originally lemmas for the construction of the pentagon.

iii. The construction of II,11 was inspired by the proofs of III,36,37 and II,6. Thus Theorems 1-10 of Book II are to be looked upon not as being originally one unit but as being, at least in part, a series of lemmas that were needed for various theorems in the *Elements* or perhaps elsewhere. We may suppose that Euclid or one of the earlier authors of "elements" grouped these together because of the outward similarity of their nature and added results such as II,1 for the purpose of "unification."

iv. Theorem XIII,8 ("the diagonals of a pentagon cut each other in DEMR") was not known when the original development took place.

v. The proofs of XIII,1-5 simply represent the work of a mathematician who used the technique of the methods of complements to prove these lemmas from first principles when they were needed for the results of Book XIII.

C. A Chronological Proposal

My starting point for dating the development of DEMR is the question: What indeed was the state of geometry when Plato started his Academy circa −386? It seems to me impossible to obtain a clear general picture by looking at any of the "advanced" geometrical developments that took place before this time and that we know something about with some degree of certainty. When I say "advanced" I specifically exclude all the basic properties of triangles, circles, etc. We simply do not know with what degree of formality and rigour they were stated in earlier periods (I shall return to this point shortly). I also exclude any reference to early theories of proportions or ratios.

When, then, did geometrical developments we know about with some real authority occur? Some sort of theory of incommensurables (Section 7) appeared presumably before −399, the moment of the dialogue in *Theaetetus*; some advances in the field of transformation of areas (Section 4), notably Hippocrates' results on the quadrature of lunes, were made apparently near the end of the vth century; the doubling of the cube with contributions by Hippocrates and a solution by Archytas happened presumably some time not too long before the opening of the Academy.

On the other hand, to credit the Pythagoreans with any mathematics of a conceptual and technical complexity, such as is found in the construction of the pentagon, seems to me to be untenable in view of what is actually known of the state of geometry in the vth and vIth centuries. To link the pentagon to the Pythagoreans because the latter used the former as a symbol strikes me as being part of a rather romantic tradition. As Neugebauer [1957, 147] says, "First of all, it seems necessary to distinguish sharply between the axiomatic style of mathematics, which is the work of Eudoxus and his contemporaries in the fourth century B.C. and the mathematics usually connected with the Ionian and South-Italian schools.... Thales, e.g., is credited with having discovered that the area of a circle is divided into two equal parts by a diameter. This story clearly reflects the attitude of a much more advanced period when it had become clear that facts of this type require a proof before they can be utilized for subsequent theorems. For us, there is nothing to do but to admit that we have no idea of the role which the traditional heroes of Greek science played. It seems to me characteristic however that Archytas of Tarentum could make the statement that not geometry but arithmetic alone could provide satisfactory proofs. If this was the opinion of a leading mathematician of the generation just preceding the birth of the axiomatic method, then it is rather obvious that early Greek mathematics cannot have been very different from the Heronic Diophantine period."

Following is the statement of Archytas to which Neugebauer is referring.

(Q.28) Arithmetic, it seems, in regard to wisdom is far superior to all other sciences, especially geometry, because arithmetic is able to treat more clearly any problems it will... and—a thing in which geometry fails—arithmetic adds proofs, and at the same time, if the problem concerns "forms" arithmetic treats of the forms also.

(Archytas, Fragment D4, attributed to a work entitled "Conversations" [Freeman, 1948, 80]; [Diels, 1934, 438]. Freeman [1948, 81] interprets the term "forms" in this quotation to mean "numerical first principles"; she writes [1946, 235], "This probably means the 'forms' of number i.e. the Odd and the Even....")

Another indication that arithmetic was held in higher esteem than geometry is given by Plato in the *Republic*. Socrates and Glaucon are discussing the curriculum for future rulers.

(Q.29) [Socrates] "It is befitting, then, Glaucon, that this branch of learning [arithmetic] should be prescribed by our law and that we should induce those who are to share the highest functions of the state to enter upon the study of calculation. . . ."

[Socrates] "Assuming this one point to be established . . . let us in the second place consider whether the study that comes next is suited to our purpose."

[Glaucon] "What is that? Do you mean geometry?"

(Plato, *Republic*, VIII, 525B, 526C [Plato–Shorey, 163, 167])

Let us examine the statement of Neugebauer in terms of the two scholia (Section 11,Q.12,Q.13) to Book IV which attribute that book to the Pythagoreans. Not only do they illustrate the uncritical—and therefore for our purposes unusable—approach talked about by Neugebauer, but in addition the scholia will, by the very feature of being uncritical, indicate to us when the concept of DEMR must have been developed.

When the scholiast of Q.12 says, "And this book [IV] is the work of the Pythagoreans," the author is lumping together sixteen theorems which may at first glance appear to be of the same nature, but which in reality must be viewed with a more discriminating eye. In particular, consider those theorems which involve the inscription of regular polygons in a circle. These are IV,2—triangle (in fact the inscription of an arbitrary triangle is shown); IV,6—square; IV,11—pentagon; IV,15—hexagon; and IV,16—fifteen sides. This last theorem presupposes the inscription of the equilateral triangle and the pentagon and is not of direct interest to us. However, there is a fundamental difference between the inscription of the triangle, square, and hexagon and the inscription of the pentagon.

Let us consider what the first three involve. The construction of the triangle depends essentially on the sum of the angles of a triangle being a straight line (I,32). For the square one simply takes perpendicular diameters, connects the ends, and uses the knowledge that the inscribed angle in a semi-circle is a right angle (III,31). For the hexagon one uses the knowledge that each of the six sub-triangles has sides which are all equal to the radius. Thus, while the proofs of the method of inscriptions and the preliminary techniques require some effort, these inscriptions and the preliminary techniques are what may be best described as intuitively obvious. Indeed we have an ancient text from Aristotle that explicitly states this about I,32 and III,31.

(Q.30) Geometrical constructions, too, are discovered by an actualization, because it is by dividing that we discover them. If the division were already done, they would be obvious; but as it is the division is only there potentially. Why is the sum of the interior angles of a triangle equal to two right angles? Because the angles about one point [in a straight line] are equal to two right angles. If the line parallel to the side had been already drawn, the answer would have been obvious at sight. Why is the angle in a semi-circle always a right angle? If three lines are equal, the two forming the base, and the one set upright from the middle of the base, the answer is obvious to one who knows the former proposition.

(Aristotle, *Metaphysics*, IX.ix.4 (1051a) [Aristotle–Tredennick, 467]. The question of proofs of "obvious" facts in Greek mathematics is discussed by Stenius [1978, section 3].)

Thus we should not be surprised by a statement that the facts needed to inscribe the equilateral triangle, square, and hexagon were known, at some undetermined level of rigour, by mathematicians of the late VIth century. We may even suppose that the level of rigour increased, although not necessarily continuously until it reached the level found in Euclid.

The construction of the pentagon, on the other hand, involves some steps which are not in the least bit obvious. For example, there is no intuitive stage which can precede the splitting in IV,10 of AB at C in DEMR in a way analogous to the intuitive laying out of the angles of a triangle along a line as in IV,2. The pentagon can simply not be "constructed" at a low level of rigour.

Thus when our scholiast says, "And this book [IV] is the work of the Pythagoreans," we should perhaps not be overly critical of the author's historical standards, but simply regard the statement as being a short and poetic way of saying: "At an early stage some mathematicians whose names have not come down to us realized how to perform some of the easier constructions of this book. These constructions, of course, are not to be considered as having been at the level of rigour of the text of Book IV in its present state [at the time of the scholiast, that is] but let us give these anonymous early mathematicians credit where it is due because in mathematics the first step is often the most important."

To determine when the pentagon was first inscribed in a circle or, if my conclusion of B,i is correct about when the concept of DEMR first appeared, we must look for a period when mathematics was at the level of rigour where mathematicians would consider working within a "program"; for this is how the construction of the pentagon appears to me, as part of a program to rigorously inscribe the regular polygons in a circle. While the other polygons only required making older intuitive proofs more rigorous, the pentagon required new techniques, new insights, and new lemmas.

This brings us back to the question about the state of geometry at the time of Plato. The view that most of the basic mathematics found in Euclid was already known before Plato is stated by Heath [1921, I, 217]: "There is probably little in the whole compass of the *Elements* of Euclid, except the new theory of proportion due to Eudoxus and its consequences, which was not in substance included in the recognized content of the geometry and arithmetic by Plato's time, although the form and arrangement of the subject-matter and the methods employed in particular cases were different from what we find in Euclid." Presumably we are to interpret this as meaning that the rigour was essentially the same as we find in Euclid. This view, however, has not gone unchallenged. We have already seen how Neugebauer, in citing Archytas, says that before the axiomization of mathematics by Eudoxus and his contemporaries the level of rigour could not be expected to have been very high.

A more precise view of what the level of geometry was at the time Plato started the Academy has been put forth by Brown [1967b]. His argument is based mainly on an interpretation of the dialogue between Socrates and the slave-boy in Plato's *Meno* (81B). Brown dates the composition of *Meno* around −386, close to the time the Academy opened, and shortly after Plato had met, and was influenced by, Archytas on his trip to Sicily. There are two parts to Socrates' questioning of the slave-boy. In the first part the slave-boy "answers" that a square of side four and then a square of side three will double the area of a square of side two. In the second part he "answers" that the square built on the diagonal of the square of side two will result in a doubling of the area. Thus in the arithmetical first part an incorrect answer is obtained while in the geometrical second part a correct answer is given. Nevertheless, argues Brown, Plato disapproves of the second response even though it provides the correct answer: "at least in the case of geometry represented here, geometry and arithmetic stand for important different things in Plato's mind. Arithmetic is for him a model of a rigorous science, a science on secure foundations, whereas geometry models a near-science, a science still insecure in its foundations."

A further discussion of this passage in *Meno* is given by Hare [1965, 25]. While Hare approaches the matter from a philosophical viewpoint rather than from an historical one, we once again find the point made that Plato is finding fault with the rigour of the mathematicians of his day.

In view of these various arguments it seems that we must presume that the inscription of the pentagon took place at some point around and certainly not too long before −386 when we find Plato involved with many things: Italy, Archytas, *Meno*, and the Academy.

Since I have argued that the rigorous development of the inscription problems for the regular polygons and hence the pentagon with its offspring, DEMR, was most probably part of a "program," the obvious place to look is at the Academy itself. In Sections 16,A and B, I discussed the questions of Plato as a mathematician and of his possible influence on the mathematics of others. From that discussion comes the conclusion that Plato seems neither to have done mathematics himself nor to have had a very advanced knowledge of higher mathematics and that no real mathematical research was done in the early days of the Academy.

On the other hand, as I indicated in connection with Proclus' "Catalogue of Geometers" (Section 16,B), not all mathematicians were attached to the Academy and others such as Theaetetus and Eudoxus had either a minor or perhaps an informal attachment to the Academy, or they spent large periods of time elsewhere. Whether the Academy's level of logical rigour directly or indirectly influenced mathematicians elsewhere or whether both the Academy and the rise of mathematics are common by-products of the spirit of the age is a difficult question to answer.

What follows is my own proposal for a chronological plan for the development of the concept of DEMR.

(1) About the time the Academy opened, c. −386, a mathematician or group of mathematicians not directly attached to the Academy embarked on a "program" to rigorously construct the regular polygons (see above).

(2) The inscription of the pentagon led, via the lemmas III,36,37 and II,6, to the area definition of DEMR and to the construction of II,11. All these results were proved more or less as in the preserved proofs (Section 2,B).

(3) A second "program" was started, probably not too long after the completion of the first program or possibly overlapping with it, to inscribe the regular polyhedra. This "program" began before the proportion theory of Book VI was developed (Section 20,C,iii, stage 1).

(4) As part of the second "program," Theorems XIII,3,4,5—in an area formulation—were developed *ab initio*. Theorems XIII,9,10 also date from this period. These results were simply lemmas for XIII,16[a] and XIII,17[a], the constructions of the icosahedron and dodecahedron (Section 20,C,iii, stage 1).

(5) As a sequel to the first "program" and possibly initiated by the same person or group responsible for it, the inscription problem as such was extended to the classification of the irrational quantities obtained from the construction of the pentagon. Lemmas XIII,1,2 and Theorem XIII,6 and the classification part of XIII,17 date from this period (Section 20,C,iii, stage 2).

(6) After a certain period of time during which there was a further development of the classification theory of irrationals—which in turn might have been inspired by the still unsolved classification problem for the icosahedron (see Taisbak [1982, chap. 3])—Theorem XIII,11 and the classification part of XIII,16 were completed. This may have been done before the proportion theory of Book V was

obtained in its final form, by using an older definition of proportion (Section 20,C,iii, stage 3).

(7) After the development of the proportion theory of Book V,def.3 of Book VI and VI,30 replaced II,11. The results of Book XIII underwent minor changes in terminology.

D. *A Proposal Concerning a Name*

To complete my proposal for the development of the concept of DEMR, there remains the task of suggesting a name for the originator of the "programs." It seems that of all the names that are available to us there is only one serious candidate—Theaetetus. More precisely I would associate Theaetetus or possibly some of his students with the first five stages described above and I would assign the sixth stage to a period relatively far removed from the first five. The assignment of the sixth to any particular time or person must be extremely conjectural. If obliged to choose, it might be Hermotimus, about whom Proclus (Section 16,Q.3) says, "[he] pursued further the investigations already begun by Eudoxus and Theaetetus."

The choice of Theaetetus accords well with the chronology that I gave in Subsection C. From the discussion in Section 18 it seems that we can say that Theaetetus was doing active work between −399 and −386; that he was associated with the theory of irrationals and polyhedra; and that he was for some period separated from Athens, but maintained contact with it and in fact returned.

If Theaetetus was in Heraclea before the opening of the Academy, then this would account for Plato's lack of knowledge about advances in geometry around −386 and the impression we have that Archytas was his source (as discussed by Brown). Even if Theaetetus did most of his work away from Athens, the contacts between Heraclea and the Academy would have ensured that his reputation was established in Athens. Furthermore, the proofs of XIII,1-5 suggest the same type of mastery of the "complements" method as is displayed in II,6. This, together with the linking of Theaetetus with the polyhedra, suggests not only that he constructed the pentagon as the first step but that he or his school also followed up with the basic construction of the dodecahedron and icosahedron (stage 1 of the development of the theory of polyhedra). The association of Theaetetus with the basic irrationals also suggests a possible connection, perhaps via his students, with stage 2—the discovery that the side of the dodecahedron is "apotome" (XIII,17[b]).

On the other hand, the apparent distance in the time between IV,11 and XIII,8 (Section 2,C), together with Theaetetus only being linked with the three basic irrationals, suggests that stage 3—the discovery that the side of icosahedron is "minor" (XIII,16[b])—was done by a later mathematician or group. It was left to the latter to put the finishing touches on the theory of DEMR while at the same time terminating that part of the *Elements* that represents the culmination of the discoveries which marked the epic age of Greek mathematics.

CHAPTER VII

THE POST-EUCLIDEAN GREEK PERIOD (c. −300 to 350)

While the classical period discussed in Chapter VI dealt with a span of a 100 or so years, the second phase of Greek involvement with DEMR covers a period of more than 600 years. While some of the material is of the same strictly geometrical nature as the results found in Euclid, we shall see that now the computational aspects, in connection with both trigonometry and area or circumference calculations, come into prominence. It should be kept in mind that some authors (see Section 13) find evidence of an older numerical tradition—masked, so to speak—behind the theoretical geometry which appears to be the overriding interest of the Greek mathematicians of the classical period. Thus the numerical work which is now in plain view would be in part the resurfacing of older work and traditions.

The dates of several of the mathematicians discussed in this chapter are uncertain. For convenience I have grouped together the predecessors of Ptolemy in the field of trigonometry in Section 26,C; Archimedes appears there as well as in Section 23.

Section 23. Archimedes (c. −287 to −212)

For a summary of the works of Archimedes, see Clagett [*DSB*] whose dating I have followed. While most of Archimedes' extant mathematical work is of a geometrical-mechanical nature, it is the surviving fragments of his numerical calculations that are of interest to us.

A. Approximations to the Circumference of a Circle

In his *On the Measurement of the Circle* [Archimedes–Heath, 91] Archimedes gives lower and upper bounds for the ratios of the circumference of a circle to its diameter. These are $6336:2017\frac{1}{2}$ and $14688:4673\frac{1}{2}$, respectively. The method used by Archimedes is as follows.

Consider the right triangle AOC in Figure VII-1 and bisect angle AOC to obtain angles COD and DOA. By VI,3 $CD:DA = CO:OA$ and using V,18 we obtain $CA:DA = (CD + DA):DA = (CO + OA):OA$ or $OA:AD = (CO + OA):CA$. Since A is a right angle we obtain from the Pythagorean Theorem

(1) $\quad OD^2:AD^2 = \{(CO + OA)^2 + AD^2\}:AD^2.$

Thus, given the ratio of the sides of a right triangle with a base angle of 2α, we can find the ratio of the sides, and in particular the ratio of the hypotenuse and opposite side, of the right triangle with base angle α.

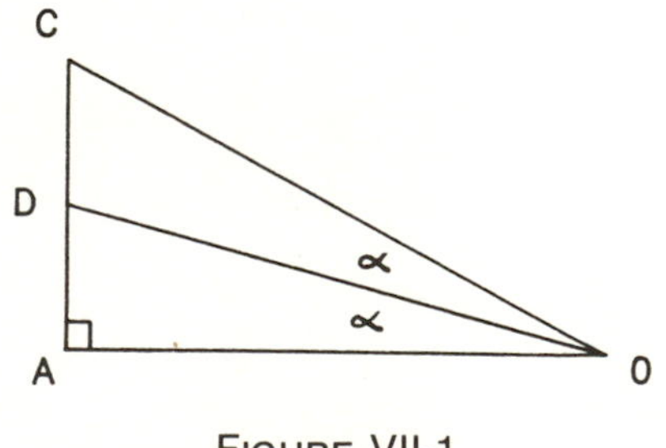

FIGURE VII-1

Archimedes starts with the hexagon circumscribed about a circle with centre O. Thus the starting angle $COA = 2\alpha$ will be 30° so that the ratio of half the side of the hexagon to the radius, namely, $OA{:}AC$, will be—in mixed notation that I will employ where convenient—$R(3)$. Without saying how it is obtained, the text approximates this ratio from below by 265:153. From formula (1) we can now obtain the ratio of side to diameter for the circumscribed 12-gon and, continuing in this way three more times, we obtain the ratio for the circumscribed 96-gon. One works in the same way to obtain estimates for the inscribed 96-gon, but this time $R(3)$ is approximated from above by 1351:780.

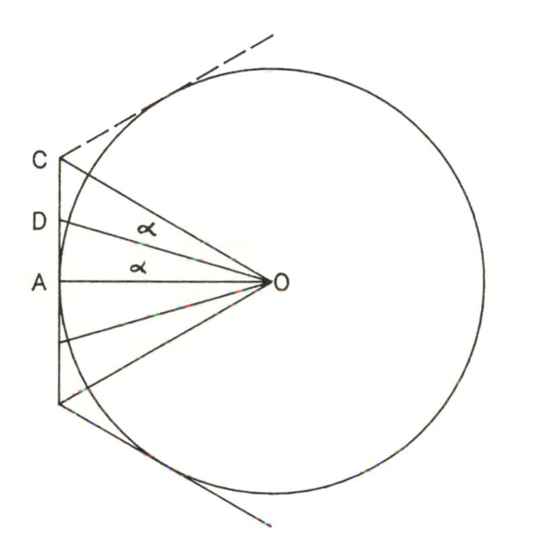

FIGURE VII-2

There is another extant pair of estimates by Archimedes for the ratio of the circumference of the circle to its diameter. This appears in Hero's *Metrica* [Hero–Bruins, III, 144] and reads:

(Q.1) The same Archimedes showed in "On Blocks and Cylinders" that the circumference of every circle has to the diameter a greater ratio than that which 211875 has to 67441; and smaller than that which 197888 has to 62351.

(Hero, *Metrica*, I, XXV [Hero–Bruins, III, 244])

Unfortunately, the text is not satisfactory. Not only is the alleged lower bound really an upper bound, but the alleged upper bound is not even as good as the one given in *On the Measurement of the Circle*. Several commentators have tried to explain and/or emend the text, and it is the recent one by Knorr [1975] that is of interest to us.

Knorr proposes the following emendation: the alleged lower bound 211875:67441 is taken to be the upper bound. The new lower bound is 197888:62991 obtained from the alleged upper bound by keeping the numerator as 197888 and changing the denominator to 62991 from 62351. Knorr notes that 62990 is the smallest integer which will, with the stated numerator, give a lower bound.

The first part of Knorr's argument consists in showing from a photograph of the extant manuscript of the *Metrica* that the scribe has already confused two of the Greek letters which would account for miscopying 62991 as 62351. Next is shown, once again from a photograph of the manuscript, how the reversal of the order of the ratios could be accounted for.

Knorr then turns to the numbers themselves. He notes that the numerators can be factored as follows: $211875 = 5^4 \times 3 \times 113$ and $197888 = 2^8 \times 773$. This leads to the conclusion that 113 and 773 must have been factors of the terms of the ratio of sides of the initial triangles—as in the argument of *On the Measurement of the Circle*. The approximation to $R(3)$ in this latter text apparently came from an anthyphairesis argument (see Section 8) but none of the approximations to $R(3)$ contains a denominator which has 113 or 773 as a factor.

Here is where DEMR enters the argument: for Knorr suggests that, instead of starting with the hexagon as in *On the Measurement of the Circle*, Archimedes started with the decagon. For this case we have from the concepts contained in XIII,9 or IV,10 that $CO/2CA = (1 + R(5))/2$, and so one should look for 113 and 773 as factors of denominators of approximations to $(1 + R(5))/2$. If the sequence of ratios of terms of the Fibonacci sequence (Section 32,B) 1/1, 2/1, 3/2, 5/3, 8/5, 13/8 . . . is formed, then the terms will be alternately proximation becoming better at each successive term. As an indication of the accuracy, I note that my computer gave (in double precision as is the case for all values cited) the value 1.6180340051651 for $(1+R(5))/2$. The sixteenth term is already 1.618034482002258, and starting from the eighteenth ratio, which is 4181/2584, the value given is the same as for $(1+R(5))/2$.

Knorr notes that the term 6765/4181 (the nineteenth term of the sequence) is a lower bound with the property that the denominator can be factored as $4181 = 37 \times 113$. Thus one of the proposed factors mentioned above has been accounted for. In order to account for 773, the other proposed factor, Knorr [p. 128] jumps another five terms in the sequence of Fibonacci ratios to the upper bound 75025/46368. The number 773 is not, however, a factor of 46368, so Knorr proposes that this twenty-fourth term in the sequence of ratios was in turn bounded from above by a ratio that is not in the sequence, namely, 75045/46380 $= (15\times5003)/(60\times773)$. In his Table 2, Knorr shows that if one uses starting values which have been derived, using some rounding-off, from the ratios 6765/4181 and 75045/46380, then the iteration technique of *On the Measurement of the Circle* explained above will lead in six iterations to the emended upper and lower bounds.

There remains the question of how Archimedes would have obtained the Fibonacci approximations stated above; this point is discussed in the appendix to the article. First Knorr returns to the question of how Archimedes obtains the approximations for $R(3)$ used in *On the Measurement of the Circle*. He recalls the side and diagonal numbers for $R(2)$ (Section 6) and points out that an analogous procedure holds for $R(3)$, namely, one forms the ratios $dn{:}sn$ where dn and sn are obtained from the recursive procedure $d_1 = 2$, $s_1 = 1$ and $d_{n+1} = d_n + 3s_n$; $s_{n+1} = d_n + s_n$. The eighth and eleventh term of the sequence of ratios are 265:153 and 1351:780 which are indeed the approximations used by Archimedes.

Despite his obtaining the desired approximations, Knorr suggests that a procedure which provides a more rapid rate of convergence to $R(3)$ was used. This proposed method is as follows: start with the approximation $R(3)>5/3$ and use this to obtain the upper bound $9/5 = 3/(5/3)>3/R(3) = R(3)$. Next for any pair of integers s,d we form the weighted average of 9/5 and 5/3 given by $f(s,d) = (9s+5d)/(5s+3d)$. If we start with the side and diagonal numbers $s=3$ $d=5$, we obtain $f(3,5) = 26/15$ and by iteration we have $f(15,26) = 265/153$; $f(153,265) = 1351/780$. Thus the approximations to $R(3)$ used by Archimedes are now obtained successively in only three iterations. Note that it appears to be an accident that the starting value of 9/5 taken from the square root approximation gives rise to values which are all side and diagonal numbers for $R(3)$.

Having shown how Archimedes could have obtained his approximating values for $R(3)$ in *On the Measurement of the Circle*, Knorr proposes a similar procedure for the initial Fibonacci approximations to $(1+R(5))/2$. First he discusses the side and diagonal numbers for the pentagon (see Section 12,ix) with $s_1 = 1$, $d_1 = 1$; $s_{n+1} = d_n$, $d_{n+1} = d_n + s_n$ whose ratios correspond to the Fibonacci sequence mentioned above. However, since the upper bound 75025:46368 is separated from the lower bound 6765:4181 by three other upper bounds, two of which have "nice" factorizations, Knorr now proposes an iterative scheme which, just like the one for $R(3)$, will skip ratios—in this case every fifth one. The defining function is $g(s,d) = (8d+5s)/(5d+3s)$, and we obtain $g(3,5) = 55/34$, $g(34,55) = 610/377$, $g(377,610) = 6765/4181$, $g(4181,6765) = 75025/46368$.

In a previous article [Curchin–Herz-Fischler, 1985] I have indicated that there are certain difficulties with Knorr's argument. The major weakness appears to be the replacement of both the numerator and the denominator of the term 75025/46368 in the sequence of ratios of Fibonacci numbers to obtain 75045/46380 which is not a term in the sequence. The latter was chosen by Knorr so that he would have the desired factors, but one can ask if somebody who did not know the final result would have come up with both these changes.

Again in the proposed iterative scheme it should be noted that in the case of $(1+R(5))/2$ the d has been associated with the numerators 8 and 5, whereas in the proposed scheme for $R(3)$ the d term has been associated with the denominators 5 and 3. The choice of the decagon also seems strange; since Archimedes' extant development involves the hexagon, we would expect the octagon—involving $R(2)$—to appear in a better approximation rather than the decagon. The existence of approximations involving $R(2)$ is attested to before the time of Archimedes, but this is not the case for $(1+R(5))/2$. The use of a rather sophisticated weighted average procedure to skip terms would also imply an advanced knowledge of anthyphairetic related techniques which may be compared with the techniques that appear to have been used in trigonometry (Section 26,C).

It should be remarked that several other interpretations of Hero's text have been given; some of these are discussed by Knorr (see also the criticism of Knorr's analysis by Bruins [1979, 9; 1976]).

B. *Broken Chord Theorem*

We shall see in Section 28 how the 10th century mathematician al-Biruni used the second Broken Chord Theorem to construct the decagon. Since the first Broken Chord Theorem is explicitly ascribed to Archimedes it is possible that Archimedes was involved with the second and the construction of the decagon.

C. *Trigonometry*

A discussion of this aspect of Archimedes' work can be found in Section 26,C.

Section 24. The Supplement to the *Elements*

A. *The Text*

Before considering the historical aspects of the first Supplement to the *Elements* (often referred to as Book XIV), I shall comment on some of the theorems. The text [Euclid–Heiberg, V, 1-67] does not number the theorems nor does it use the names that I have adopted. The text of Peyrard [Euclid–Peyrard, III, 481-507] deviates from that of Euclid–Heiberg. A condensed version of the Supplement which gives the proofs in modern notation appears in Euclid–Heath [III, 512-19]. All solids are assumed to be inscribed in the same sphere.

The first results—which I have called the Ratio Lemma and XIV,**—are of a special nature. The first appears in the text after Theorem 8 and before the Summary while the second is implicitly used in Theorems 2 and 7. Various indices have led me to believe that XIV,** was explicitly stated and proved in at least one of the Greek versions of the Supplement (see Subsection C) and that the proof has survived, partially in the Arabic *Elements* and essentially completely in a Latin manuscript, Bibliothèque Nationale (Paris) Latin 7373. (For a complete discussion of the texts as well as my arguments, see Herz-Fischler [1985].)

RATIO LEMMA. If two lines are divided in EMR [then the ratio of the entire line to the larger segment is the same in both cases].

THEOREM XIV,**. Let a regular hexagon and a regular decagon be inscribed in the same circle. If we divide the side of the hexagon in EMR, then the larger segment is the side of the decagon.

This last result should be compared to XIII,9.

THEOREM 1. If a pentagon, hexagon, and decagon are inscribed in a circle, then the perpendicular drawn to the centre of the side of the pentagram equals one-half the sum of the sides of the hexagram and the decagon.

Remarks: The proof has nothing to do with DEMR. The result should be compared with XIII,10 which relates the squares of the same quantities and is intimately related to DEMR.

THEOREM 2. If we inscribe an icosahedron and a dodecahedron in the same sphere and consider a triangular face of the former and a pentagonal face of the latter, then the circumscribing circles for these two plane figures are identical.

This is the key proposition of the Supplement. As we shall see, the extant text indicates that at least three different proofs of this result had been given. Only the one due to Hypsicles is preserved in the text of the Supplement, but for the purposes of our historical discussion in Subsection B, I have given numbers to these proofs.

Proof 1 (due to Aristaeus): unknown, possibly preserved by Pappus.

Proof 2 (due to Apollonius): existence uncertain, possibly preserved by Pappus.

Proof 3 (due to Hypsicles): in text.

LEMMA FOR PROOF 3. Inscribe a pentagon in a circle of radius r. Then $S(a_5) + S(d_5) = 5 \cdot S(r)$.

Note on the Lemma: Whereas XIII,10 involves the side of the decagon as well as the side of the pentagon inscribed in a circle, this lemma only involves the pentagon. In the proof the side of the decagon is eliminated from the statement of XIII,10 by the use of the Pythagorean Theorem applied to the right triangle whose sides are the side of the decagon and the diagonal of the pentagon and whose hypotenuse is the diameter. In Section 25,C I shall discuss a passage from the works of Hero which deals with the volumes of the icosahedron and dodecahedron and I will point out how this lemma, in connection with other results used in Theorem 2, could have been used to perform the numerical computations. The result was in fact used by Fibonacci (Section 31,B,ii).

Discussion of Proof 3: We recall from the note on XIII,16[a] that the construction of the icosahedron is based on the construction of the "defining" pentagons which form the bases of the two caps of the icosahedron. The proof proceeds (Fig. VII-3) by looking at the side (KL) of this defining pentagon in two ways: (i) as the side of the equilateral triangle of the icosahedron, which in turn can be related to the radius (r_I) of circumscribed circle of the triangle; (ii) as the side of the defining pentagon of the icosahedron which can be related to the diameter of the sphere and then via a series of comparisons to the radius (r_D) of the circumscribed circle of the pentagon of the dodecahedron. Let $d_S = AB$ be the diameter of the sphere.

a) Since KL is the side of the equilateral triangle of the icosahedron (Fig. VII-3b), we have by XIII,12 that $S(KL) = 3 \cdot S(r_I)$ or

(1) $$5 \cdot S(KL) = 5[3 \cdot S(r_I)].$$

Thus we have easily obtained an expression involving r_I. To find an expression involving r_D requires more work.

b) Considering KL as the side of the defining pentagon of the icosahedron and letting MN be the radius and MO the side of the decagon, we have by XIII,10 that $S(KL) = S(MN) + S(MO)$ or

(2) $$5 \cdot S(KL) = 5 \cdot S(MN) + 5 \cdot S(MO).$$

We next proceed to relate both MN and DG to d_S and then to one another.

c) By XIII,16[a],corollary(i), we have that $5 \cdot S(MN) = S(d_S)$. On the other hand, the corollaries to XIII,17[a] and XIII,8 tell us that DG, the diagonal of the pentagon of the dodecahedron, is the side of the cube. Thus by XIII,15, $3 \cdot S(DG) = S(d_S)$ and so

(3) $$5 \cdot S(MN) = S(d_S) = 3 \cdot S(DG).$$

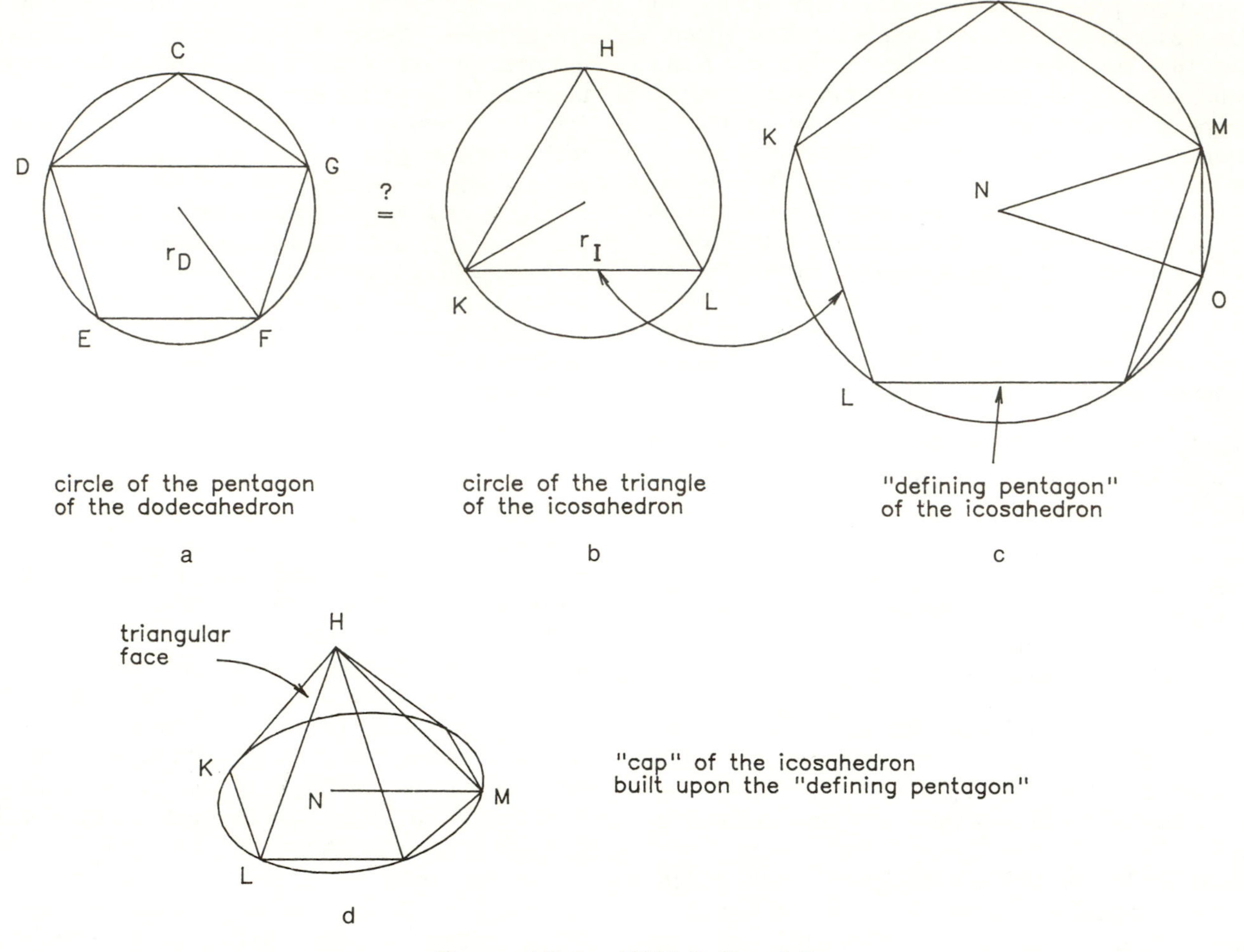

FIGURE VII-3. XIV**,2, Proof 3

This takes care of the $S(MN)$ term in (2); it is to deal with the $S(MO)$ term that the Ratio Lemma and XIV,** are brought in.

d) Consider once again DG and MN. When we divide DG, the diagonal of a pentagon, in EMR, then XIII,8 tells us that the larger segment is the side of the pentagon CG. On the other hand, if we divide MN, the radius of a circle, in EMR, then by XIV,**—used implicitly—the larger segment will be the side of the decagon MO. Thus by the Ratio Lemma—also used implicitly—$DG{:}CG = MN{:}MO$. But in c) we saw that $5 \cdot S(MN) = 3 \cdot S(DG)$, and because of the proportion just stated, we will also have the same relationship between MO and CG, that is,

(4) $$5 \cdot S(MO) = 3 \cdot S(CG).$$

e) Substitution of (3) and (4) in (2) gives $5 \cdot S(KL) = 5[S(MN) + S(MO)] = 3[S(DG) + S(CG)]$. But DG is the diagonal of a pentagon and CG the side, so the lemma for this theorem states that $S(DG) + S(CG) = 5 \cdot S(r_D)$. Thus finally we have

(5) $$5 \cdot S(KL) = 3[5 \cdot S(r_D)].$$

A comparison of (1) and (5) now shows that $r_D = r_I$ as required.

Note: Theorem 2 also appears as Theorem 48 of Book V, Part 3 of the *Collection* of Pappus (Section 27,B). The second proof given by Pappus is the same as Proof 3. The first proof given by Pappus is much more involved. Theorem 2 is also proved in an extremely elegant way as a by-product of Pappus' construction of the icosahedron and dodecahedron in Theorem 58 of Book III, Part 4 (Section 27,A). I shall return to this point in Subsection B.

THEOREM 6. $S_{12}{:}S_{20} = e_6{:}e_{20}$.
The text includes two proofs:
Proof 1 (due to Hypsicles?; see Subsection B).
Remark: In addition to Theorem 2, which itself uses properties of DEMR, and XIII,17[a],corollary, which

tells us that the diagonal of a pentagonal face of the dodecahedron is just the edge of the cube, the proof uses XIII,9. The proof also uses Theorems 3, 4, and 5 which involve integer comparisons of geometrical quantities as well as Theorem 1.
Proof 2 (due to Apollonius?; see Subsection B).
Remark: This second proof uses Theorem 2 and XIII,17[a],corollary in the same way as Proof 1, but does not use any other property of DEMR. The rest of the proof is entirely based on integer comparisons of geometrical quantities. Proof 1, via XIII,9 and its approach, relies more heavily on the properties of DEMR.

The following result, which forms part of the second proof of Theorem 6, does not involve DEMR and in fact involves nothing deeper than the fact that the area of a triangle is 1/2 base·altitude. I state it here for reference because it was used by later authors, such as Fibonacci (see Section 31,B,i), to actually compute the area of a pentagon. The wording follows the computational format of these authors rather than the geometrical language in which it is stated in the Supplement.
Preliminary to Theorem 6, Proof 2: The area of a pentagon is $5/6\, d_5 \cdot 3/4\, d$ where d is the diameter of the circumscribed circle.

THEOREM 7. The relationship between the edge of the cube (e_6) and the edge of the icosahedron (e_{20}) is as follows: take any line which is divided in EMR, then $S(e_6):S(e_{20}) = [S(\text{line}) + S(\text{larger segment})]:[S(\text{line}) + S(\text{smaller segment})]$.

THEOREM 7,[COROLLARY]. Take any line which is divided in EMR, and let L_1 be another line such that $S(L_1) = [S(\text{line}) + S(\text{larger segment})]$. Let L_2 be a third line such that $S(L_2) = [S(\text{line}) + S(\text{smaller segment})]$. Then $e_6:e_{20} = L_1:L_2$.
Notes: 1) The proof of the theorem itself only deals with $S(e_6)$ and $S(e_{20})$ and this is what I have given as the statement. The actual statement in the text [Euclid–Heiberg, V, 25, 29] is what I have called the corollary, but nothing more than "ergo" is said in the proof. It is the corollary that is needed in (iv) of the following Summary to compare V_{12} and V_{20}.

2) It is only in reading the proof that we see that there is a particular line behind this result, namely, the radius a_6 of the circle which circumscribes the pentagon of the dodecahedron and the triangle of the icosahedron. Just as in the case of Theorem 2, the result I have called XIV,** is used implicitly to obtain that when we divide a_6 in EMR the larger segment will be a_{10}. By the Ratio Lemma, which is not referred to explicitly, whatever relationship is shown to hold between a_6 and a_{10} will also hold for the line divided in EMR.
Outline of the Proof: By Theorem 2 the circle of the pentagon of the dodecahedron is the same as the circle of the triangle of the icosahedron. Thus in the following all the terms refer to the same circle and it is this fact which enables us to pass from the dodecahedron in b) below to the icosahedron in c).

a) By XIII,17[a],corollary, we know that e_6 is equal to d_5, the diagonal of the pentagon of the dodecahedron. Also by XIII,8 and XIV,**, d_5 and a_6 have a_5 and a_{10}, respectively, as larger segments when divided in EMR. As a result of these two statements and the Ratio Lemma, we can produce the following string of equalities:

$$(6)\qquad S(e_6):S(a_5) = S(d_5):S(a_5) = S(a_6):S(a_{10}) = 3\cdot S(a_6):3\cdot S(a_{10}).$$

b) By XIII,10 the first term of (6) equals $S(e_6):[S(a_6) + S(a_{10})]$.

c) Consider the last term of (6). By XIII,12 the numerator $3\cdot S(a_6) = S(a_3)$. But a_3, the side of the triangle, is identical with e_{20}, the edge of the icosahedron, so that the numerator of the last term of (6) is equal to $S(e_{20})$. As for the denominator of the last term, XIII,4 informs us that, since a_{10} is the larger segment, it is equal to $[S(a_6) + S(\text{smaller segment})]$.

d) Combining the conclusion of b) and c) we have $S(e_6):[S(a_6) + S(a_{10})] = S(e_{20}):[S(a_6) + S(\text{smaller segment of } a_6)]$ and this is equivalent to the statement of the theorem.

THEOREM 8′. $V_{12}:V_{20} = S_{12}:S_{20}$.
Proof (contained in the proof of Theorem 8): This is immediate from Theorem 2, for if the pentagon of the dodecahedron and the triangle of the icosahedron have the same circumscribing circle, then the pyramids with vertices at the centre of the sphere and the faces of the solids as bases have the same altitude. Thus the volumes of the pyramids will be in the same ratio as the bases.

THEOREM 8. $e_6:e_{20} = V_{12}:V_{20}$.
Proof: This is just a combination of Theorems 6 and 8′.

SUMMARY
(i) Theorem 7,[corollary].
(ii) Theorem 6.
(iii) Theorem 8′.
(iv) $V_{12}:V_{20} = L_1:L_2$ where L_1 and L_2 are in Theorem 7, [corollary].
Proof: Statement (iv) simply puts together the equalities (iii), (ii), and (i).
Comment: Since L_1 in (iv) is greater than L_2 we have that the dodecahedron will have a greater volume than

the icosahedron inscribed in the same sphere. This result is not stated in the text.

I note here that the so-called Book XV [Euclid–Peyrard, III, 481; a summary of the contents without an indication of the proofs is given in Euclid–Heath, III, 519] does not explicitly involve DEMR even though the icosahedron and dodecahedron appear.

B. *Questions of Authorship*

Now let us consider the historical aspects of the Supplement. Since the meaning of parts of the text are far from certain I shall use my analysis of the results to try and draw some conclusions. I will leave the question of dates until Subsection C.

The text of the Supplement, which is assumed to be due to Hypsicles, starts off with an introduction of an historical nature and then gives the statement and proof of Theorem 1. This is followed by the statement of Theorem 2 and some more remarks of an historical nature. Here Hypsicles tells us that Theorem 2 had been proved by Aristaeus in his work *Comparison of the Five Figures*. He then states that Apollonius, in the second version of a work which dealt with the comparison of the dodecahedron and icosahedron, had proved Theorem 8′. Immediately following this Hypsicles sketches the same proof (based on Theorem 2) of Theorem 8′ which is given in the text of the Supplement. We immediately run into difficulties here for the proof of Theorem 8′ follows in one or two lines from Theorem 2 and it is difficult to imagine a new treatise based on this. Furthermore, from the title of Aristaeus' book we would suspect that if he did prove Theorem 2 then he too proved Theorem 8′. Several possibilities suggest themselves. Perhaps Aristaeus gave one proof of Theorem 2 and Apollonius gave another, or perhaps Apollonius extended Aristaeus' work by proving Theorem 6 and/or 7, either using Aristaeus' proof of Theorem 2 or giving his own proof.

Next we note that Hypsicles says immediately after his statement about Apollonius that "it is right that I too should prove [Theorem 2] and for this I need the following [lemma to Theorem 2]." We may suppose then that the extant proof of Theorem 2 is due to Hypsicles. But Theorem 6 has two proofs and, as I noted, the first one is more geometrical than the second in that it relies more heavily on properties of DEMR. It is this first proof that is closest to Hypsicles' proof of Theorem 2. What about the second proof? Either Hypsicles also discovered it or it came from someplace else; the likely source would be the work by Apollonius.

There is another historical point of interest in the text. For we read in the introduction that Hypsicles' father and Basilides of Tyre came across and corrected a tract by Apollonius dealing with the comparison of the dodecahedron and icosahedron, "that is to say what ratio they bear to one another." Furthermore, we learn that Hypsicles came across a corrected version by Apollonius (presumably the same as the second version mentioned above in connection with Theorem 2), and that Hypsicles dedicated what he "deem[ed] to be necessary by way of commentary. . . ." The statement about the ratio between the dodecahedron and icosahedron seems to imply that something more than Theorem 2 was involved in Apollonius' work, most probably Theorem 7. Further, the statement about commentary suggests that Hypsicles himself did not prove anything new, but rather gave different proofs. Hypsicles' father and Basilides seem to have limited themselves to patching up Apollonius' proof.

I would like to suggest, then, the following stages in the development of what I have called the Supplement (see also Langermann and Hogendijk [1984] on the possibility that a larger treatise may be involved):

1. Proof by Aristaeus of Theorems 2 and 8′.
2. Proof by Apollonius of Theorem 6 (Proof 2) and of Theorem 7. The original version was corrected by Hypsicles' father and by Basilides. A second corrected version was also written by Apollonius. Possibly Apollonius gave a new proof of Theorem 2.
3. A new proof (the one that appears in the extant text) of Theorem 2 was given by Hypsicles, who also gave Proof 1 of Theorem 6 as well as the proofs of Theorems 3, 4, and 5, which are, in effect, lemmas for Proof 1 of Theorem 6.

In connection with Proofs 1 and 2, recall that I noted with regard to Theorem 2 that Pappus gave two proofs of that proposition and that the second proof is the same as that of the Supplement. Pappus does not give any reference for this second proof and furthermore the first proof is more involved. This situation would at least suggest the possibility that the first proof given by Pappus is related to the proof by Aristaeus or, if he indeed gave one, that of Apollonius. In this respect, recall that Aristaeus is said by Hypsicles to have written a book entitled *Comparison of the Five Figures* and this subject is precisely what Book V, Part 3 of the *Collection* is all about. Furthermore, as we shall see below, Pappus does mention an Aristaeus elsewhere in the *Collection*, although we do not know if the two people called Aristaeus are one and the same.

C. *Chronology*

For the purpose of dating the development of DEMR we would like to know the dates of Aristaeus, Apollonius, and Hypsicles. Unfortunately very little of our information is precise.

Apollonius, known for his contributions to the theory of conics and astronomy, seems to have lived from the second half of the IIIrd century to the early IInd (see Toomer [*DSB* 1]).

Bulmer-Thomas [*DSB* 1] suggests Hypsicles' father was an older contemporary of Apollonius. His reasoning is based on the statements in the introduction of the Supplement that Hypsicles' father saw the first version of Apollonius' tract while Hypsicles himself saw a widely circulated second version. The reasoning appears to be that Apollonius himself produced a corrected version in the meantime. Toomer [*DSB* 1] suggests −150 as a central date. It is possible, however, that a later date in the IInd century is correct.

What about the Aristaeus mentioned by Hypsicles as having written a work entitled *Comparison of the Five Figures* in which he proved Theorem 2? The only other mention of an Aristaeus in Greek mathematical literature appears to be the references in Book VII, "Treasury of Analysis," of the *Collection* of Pappus:

(Q.2) It [the "Treasury of Analysis"] is the work of three men, Euclid the writer of the *Elements*, Apollonius of Perga and Aristaeus the elder, and proceeds by the method of analysis and synthesis.

. . . .

Apollonius, who completed the four books of Euclid's *Conics* and added another four, gave us eight books of Conics. Aristaeus who wrote the still extant five books of *Solid Loci* supplementary to the *Conics*, called the three conic sections of an acute-angled, right-angled and obtuse angled cone respectively . . . . Now Euclid regarded Aristaeus as deserving credit for his contributions to conics, and did not try to anticipate him or to overthrow his system . . . .

(Pappus of Alexandria, *Collection*, Book VII [Pappus–Ver Eecke, I, 476, 507], [Thomas, II, 597, I, 487])

From the last sentence we learn that the Aristaeus of Pappus was a predecessor or contemporary of Euclid. He is not, however, mentioned in Proclus' "Catalogue of Geometers" (Section 16,Q.3) even though this summary lists names until Euclid's time [Proclus–Morrow, 56]. Note also the mention of the method of analysis and synthesis which was discussed in Section 17. The Aristaeus mentioned by Pappus is discussed by Vogel [*DSB* 2] who assigns c. −350 to c. −330 as his flouraison.

The question, then, is whether or not the Aristaeus mentioned by Pappus in connection with loci is the same as the Aristaeus mentioned by Hypsicles in connection with DEMR. Furthermore, if they are the same what can be said about the chronology of *Comparison of the Five Figures* with respect to that of Euclid's *Elements*. We shall see that three views have been expressed: Aristaeus is the same person in both these cases and *Comparison of the Five Figures* predates the *Elements*; Aristaeus is the same but *Comparison of the Five Elements* postdates the *Elements*; Aristaeus is not the same person in both these cases.

Bretschneider [1870, 171] assumes that Aristaeus mentioned by Pappus is the same as the Aristaeus mentioned by Hypsicles. He concludes that the work by Aristaeus on the comparison of the regular solids was the newest and last on the subject written before Euclid and thus that Book XIII represents at least a partial recapitulation of the work of Aristaeus.

Allman [1889, 202] supports Bretschneider's conclusion by examining which results are used in Theorem 2. According to Allman, "This supposition of Bretschneider receives, I think, great confirmation from the above examination [of the proof of Hypsicles], which shows that the principal propositions in Book XIII of the *Elements* are required for the demonstration, as given by Hypsicles, of the theorem of Aristaeus. This theorem, moreover, goes beyond what is contained in the *Elements* on this subject."

Allman also points out the possible relationship between the *Comparison of the Five Figures* and the comparison results in Book V, Part 3 of the *Collection* and further compares the relationship of the analysis and synthesis of Aristaeus as mentioned in Q.2 and the use of analysis and synthesis by Pappus in his inscription of the five solids (this is in Book III, Part 4; see Section 27,A).

Sachs [1917, 107] continues the argument developed by Bretschneider and Allman. She appears to believe that the two references concerning Aristaeus are to the same person, but that the *Comparison of the Five Figures* came after Euclid's elements.

The key point in her argument is the use by Hypsicles of XIV,** in Proof 3 of Theorem 2. According to Sachs [p. 99], XIV,** was not known to Euclid; if it had been he would have used it to simplify the construction of the pentagon (see Section 2,A,i).

The second point raised by Sachs [p. 112] involves XIII,18. In this result, Euclid proves that the side of the icosahedron is greater than the side of the dodecahedron inscribed in the same sphere. But this result is an immediate consequence of Theorem 2, and so Sachs argues that if Euclid had known this result of Aristaeus he would have used it in XIII,18. On the other hand, the article by Bulmer-Thomas on Hypsicles [*DSB* 1] supposes that the Aristaeus of Hypsicles is not the same as that of Pappus.

It seems to me that the nature of the Supplement suggests that it postdates the material from Book XIII; it seems reasonable that the construction of the polyhedra would precede their comparison. Thus Bretschneider's conclusion that Book XIII contains some of Aristaeus' work does not seem reasonable. Further, we do not know the date of Book XIII (see Sections 21 and 22).

There are also objections that can be raised against Allman's arguments. First, the transmitted proof of Theorem 2 is due to Hypsicles and not to Aristaeus, and since Hypsicles lived well after Euclid it is not surprising to see his proof use results from Book XIII. Second, the connection between Pappus and Aristaeus is very tenuous, and in any case it seems irrelevant to the identification of the two people named Aristaeus. In Q.2 Pappus does not mention *Comparison of the Five Figures*, but this again means nothing since Pappus is talking about conics. Similarly, the use of the term "Aristaeus the Elder," while suggesting that there were two people known by this name, does not tell us very much.

The arguments of Sachs also do not seem sufficient, for if Euclid used older material he may have simply left the older proofs alone, just as he did elsewhere (see the various discussions in Section 2).

It would seem, then, that absolutely nothing is certain or even probable in the way of the chronology of the Supplement either at the beginning or at the end.

Section 25. Hero (fl. c. 62)

The mathematical works of Hero are a mixture of theoretical and computational-practical material. Parts of his work appear to be a transmission of older traditions, some of which seem to go back to Babylonian times. For a discussion of these questions as well as the dating of Hero, see Mahoney [*DSB*], Drachmann [*DSB*], and Neugebauer [1957, 46, 146], as well as the various comments by Bruins [Hero–Bruins]. I have not considered the scholia; some at least are from the xvth century [Hero–Bruins, I, x; see also the reference to Pappus in III, 306].

A. Approximations for the Area of the Pentagon and Decagon

There are several formulae of interest scattered throughout the manuscripts of Hero's work. We shall see that Hero used a variety of techniques involving varying levels of rigour and that in some cases he gave numerical results without an indication of how these results were obtained. I have grouped the various references by topic rather than by the order of the manuscripts—which, in any case, do not represent a true chronological order. In certain instances I have put parts of the same section of the text into different topic headings. I have also omitted texts dealing with the topics but which are not related to DEMR. Because of the overlap of techniques and suggestions concerning them, I have reserved the discussion of the various techniques as well as a comparative table for the end. The summary of texts has been modernized—angles are presented in degrees, etc.—but aside from integer ratios I have kept the ratio terminology of Hero. Where Hero is speaking in strictly geometrical terms, I have employed (as up to now) a geometric notation. However, when numerical quantities are involved I present the material in modern algebraic notation and in particular I use a hybrid ratio notation. (The same will be true in later sections.)

i. The Area of the Pentagram

a) A detailed analysis of the area of a pentagon whose side is 10 units and a numerical approximation of this area appears in *Metrica*, I, 17, 18 [Hero–Bruins, III, 224].

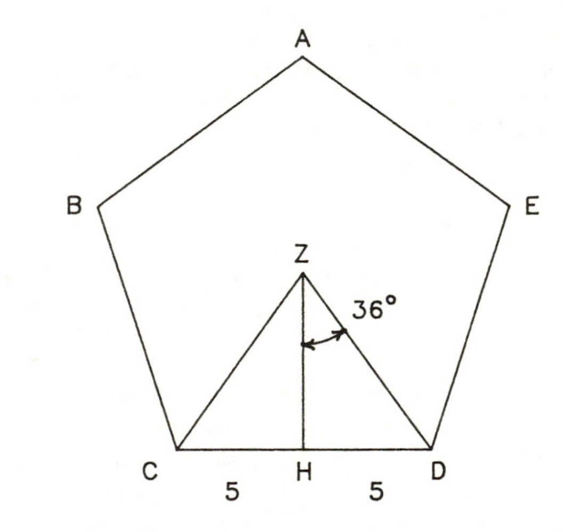

Source: Hero–Bruins [I,151]

FIGURE VII-4

To find the area of the pentagon $ABCDE$ (Fig. VII-4) whose centre is Z, one first finds the area of the triangle CZD and then multiplies by 5. The area of the triangle in turn is one-half of the product of the base CD and the altitude ZH. Since the base CD is given to be 10 units, one need only find the altitude ZH.

Note that the central angle of the pentagon is 72° so that in the right triangle CZH the vertex angle CZH is 36°. This is where Hero's ingenuity comes in; his simple proof should be compared with the rather involved summary given by Heath [1921, II, 327]. He remarks in a lemma that not only is 36° one-half of the central angle of the pentagon but it is also the angle *FGJ* between the side *FG* of a pentagon and its diagonal *JG* (Fig. VII-5). (For clarity I have not reused *ABCD* as Hero does and have drawn Figure VII-5 which actually places the triangle *JFG* in a pentagon.) Then, since XIII,8 tells us that *JG* and *FG* are in the relationship of a line divided according to EMR, we have by XIII,1 that $S(FG+KG) = 5 \cdot S(KG)$. Thus in the original pentagon:

$$S(CZ + ZH) = 5 \cdot S(ZH). \tag{7}$$

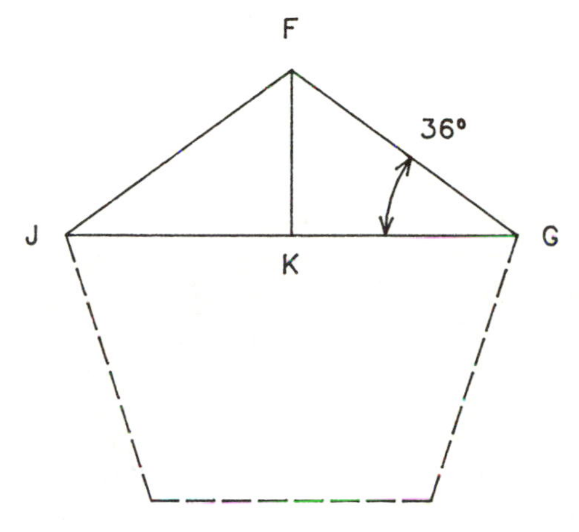

FIGURE VII-5. Lemma (not in text)

Now Hero begins his numerical computations and approximations. If we take, for ratio purposes, $ZH = 4$, then $5(ZH)^2 = 80$. But this is not a perfect square so 80 is approximated by 81 and (7) gives $CZ + ZH = 9$ or $CZ = 5$. We now have that *CZ* and *ZH* are in the approximate ratio of 5 to 4. But since triangle *CHZ* is a right triangle and thus a 3−4−5 right triangle, we have that *ZH* and the half-base *CH* are in the approximate ratio of 4 to 3. The rest of the argument and numerical example amount to saying that the area of triangle *CZD* is $\frac{1}{2}[CD][(4/3)(\frac{1}{2}CD)] = \frac{1}{3}CD^2$ and that therefore the ratio of the area of the pentagon to the square of the side is 5/3.

Recall that this ratio 5/3 also appeared in the Babylonian text discussed in Section 9,D. Of course the Babylonian method did not involve the concept of DEMR.

For some historical and pseudo-historical consequences of Hero's development, see Curchin and Fischler [1981] and Curchin and Herz-Fischler [1985].

b) The same formula as above appears elsewhere in Hero's works but without a demonstration (see Hero−Bruins [III, 51, 172, 279] and Hero−Heiberg [p. 382]).

c) At the end the passage from *Metrica*, I, 18—discussed in a)—Hero writes, "And if we should take another square being nearer to the fivefold of another square, we shall find the area more precisely." However, there is no indication of what other number(s) might be involved or even if Hero had some other specific number(s) in mind.

d) In *Geometry* [Hero−Heiberg, 382] we are given, without proof or commentary, the approximation, area pentagram:square of side = 12/7, and this is applied to a pentagram of side 35.

ii. The Area of the Decagon

a) In *Metrica*, I, 23 [Hero−Bruins, III, 241], Hero finds an approximate formula for the area of a decagon in terms of the side. The development uses the ratios already obtained in the pentagon case, although this is not explicitly referred to.

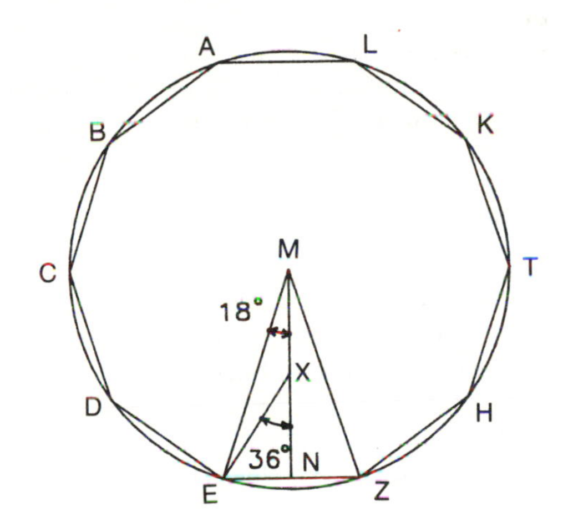

Source: Hero−Bruins [I,155]

FIGURE VII-6

Let *M* be the centre of the circle circumscribed about the decagon *ABCDEZHTKL* whose sides are 10 units (Fig. VII-6). If *N* is the perpendicular bisector of *EZ* then $\sphericalangle EMN = 18°$. We now pick *X* on *MN* so that $\sphericalangle MEX$ also equals 18°. This means that $\sphericalangle NXE = 36°$ and thus we have the same angles as in the case of the pentagon. Thus $EX{:}NX = 5/4$ and $EX{:}EN = 5/3$. But $MX = EX$ so that $MN{:}EZ = (MX + XN){:}2 \cdot EN = 9/6 = 3/2$ (Fig. VII-7). The rest of the argument and numerical example amount to saying that the area of triangle $EZM = \frac{1}{2}[EZ][(3/2)(EZ)] = 3/4\ EZ^2$ and that therefore the ratio of the area of the decagon to the square of the side is $30/4 = 15/2$.

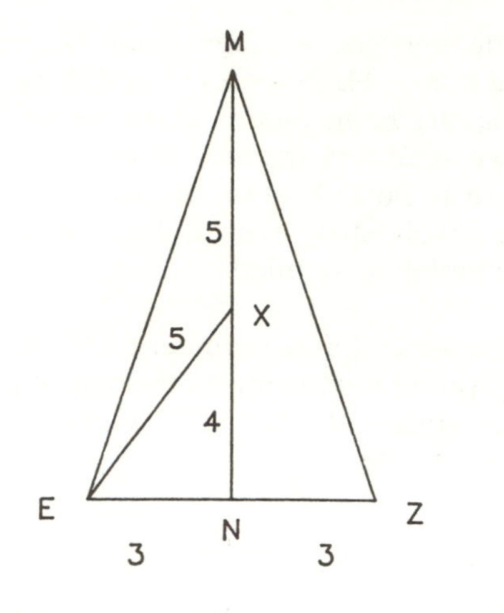

FIGURE VII-7. Enlargement of VII-6

b) The same formula as above appears in "Methods of Polygons," but without a demonstration (see Hero–Bruins [III, 58]).

c) In the same passage as in b) we are also "given" the ratio 38/5 but without any details about its derivation. We are simply told, "This method is [very] accurate."

iii. The Diameter of the Circumscribed Circle of a Pentagon

a) In "Measurements of Pyramids" [Hero–Bruins, III, 163], Hero uses XIII,10 to find the diameter of the circle circumscribed about a pentagon of side 12. Written in modern notation this becomes

(8) $$a_5^2 = a_6^2 + a_{10}^2.$$

Although a_{10} and a_6 can be related via XIII,9, Hero uses the approximation $a_{10} = (a_5)/2$. Since $r = a_6$, the calculation becomes $r^2 = 12^2 - (12/2)^2 = 108$. Then by approximating the square root, $r = 10\ 2/5$.

b) In the "Method of Polygons" [Hero–Bruins, III, 51] we are told, without an explanation, that to obtain the diameter of the circle circumscribed about a pentagon of side 10 we must multiply by 17/10. This ratio may have been obtained by the method of a) or perhaps some other method was used.

iv. Commentaries

As we have seen, Hero uses and combines various methods. Furthermore, there are still other approximation techniques not related to DEMR which appear elsewhere in his works. This means that we cannot be at all sure how the figures cited above, which are given without explanation, were obtained. Following is a table of values for the ratio of the area of a pentagon to the square of the side as obtained by various methods.

a) Bruins [Hero–Bruins, III, 59] has pointed out that the ratio 38/5 for the decagon as given in ii,c corresponds to assuming $R(5) = 38/17$ whereas the assumption $R(5) = 9/4$ would correspond to the ratio 15/2. For the first case Bruins shows this by writing (7) in the form $(CZ+ZH){:}ZH = R(5)$, solving for $CZ{:}ZH$ (see below), then for $ZH{:}HC$ via the Pythagorean Theorem and some square root approximations, and finally rounding off the answer for the area. Note that for the second case one is essentially doing the same thing as in ii,a, but there one is thrown back to the pentagon case of i,a. While these methods are mathematically equivalent it is perhaps worthwhile pointing out that Hero never speaks of square root approximations as such, but rather uses (7) directly to obtain his approximation for $CZ{:}ZH$. Now it turns out that $38^2 = (5 \times 17^2) - 1$ (see Niven and Zuckerman [1960, 159] for an explanation of this phenomenon in terms of Pell's equation; this is also shown by geometrical methods by Bruins [III, 63]). That is, the same situation holds as for the computation of i,a. Bruins [p. 63] thus suggests that the improved value for the ratio for the pentagon alluded to by Hero in *Metrica*, I, 18 (see i,c above) was also determined by assuming $R(5) = 38/17$. In Table VII-1 I have given the ratio for this case using two approximations for the square root involved.

Assumption	Area:(Side)^2	Deviation
Exact (5/[4 tan 36°])	1.7205	0
Hero (i,a)	5/3 = 1.667	.0538
$R(5) = 38/17$; $R(152) = 12\,1/3$ (iv,a)	255/148 = 1.7229	.0024
$R(5) = 38/17$; $R(152) = 12$ (iv,a)	85/48 = 1.7708	.0503
Unknown (i,d)	12/7 = 1.7143	.0062
Ptolemy (chord 72° = 70:32,3 (iv,e)	1.7200	.0005

Area of a Pentagon:(Side)^2 Under Various Assumptions

TABLE VII-1

We note two things: first, even though the ratio $CZ{:}ZH$ turns out to work out nicely from (7), as in i,a, the ratio $ZH{:}HC$ no longer works out so nicely. Secondly, one wonders why Hero did not give the better value if he actually had one in mind.

b) One wonders how Hero obtained the value of 12/7 for the pentagon as mentioned in i,d. It has been noted by D. Fowler (in private communication) that both the ratios 5/3 and 12/7 happen to be convergents (Section 8) of the continued fraction expansion of the ratio of the area of the pentagon to the square of the side. This may be an accident since 12/7 turns out to be a "good" approximation and involves a very small denominator (see Hardy and Wright [1960, 161, Theorem 181]).

c) One might ask if the number 12/7 was not arrived at by trial and error. To check estimates one could work backwards to see if (7) is satisfied, and if we do this with 12/7, then we see that $ZH{:}CH = 48/35$ and $CZ^2 = 3529 = 59^2 + 48$. However, since this gives $CZ = 59 + 48/118$, it is difficult to see how 12/7 would have been hit upon.

d) In Hero's computations for the area of an ennagon and hendecagon [Hero–Bruins, III, 239, 242] we find references to a "book of chords." It is of course possible that the value 12/7 also came about from computations with chords. If this were so and if the same method was used as for the ennagon and hendecagon, then the table of chords would probably have provided a value for the ratio of the diameter of the circle and the side of the inscribed pentagon which would correspond to sin 36° (see also the discussion in Neugebauer [1975, 300] and Section 26,c).

e) Another way of using a table of chords in connection with the pentagon is to work directly with half of the central angle (36°) and then the triangle CZH of Figure VII-4. If we do this using the value that appears in Ptolemy's table of chords (Section 26,A), then we obtain the ratio 1.7200.

B. A Variation on II,11

The Arab commentator al-Nayrizi (fl. c. 900; see Murdoch [*DSB* 2, 441-58]) has preserved for us a commentary by Hero on II,11 in which he presents us with a variation of the method used in Euclid II,11 to divide a line in EMR (al-Nayrizi–Curtze, 106; the text is from a 12th century Latin translation by Gerard of Cremona).

In Hero's version the line $ga = \frac{1}{2}ab$ is drawn perpendicular to ab, the given line. Line gb is then drawn and point d is obtained with $gd = ga$. Finally point e is obtained on ab with $db = be$. This may be compared with the method of II,11 which is illustrated in the same format in Figure VII-8b.

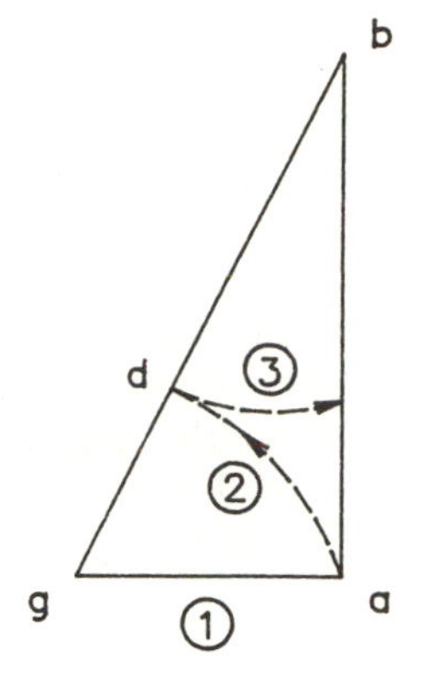

a. Hero (dotted lines added)

Source: Al-Nayrizi–Curtze [p. 107]

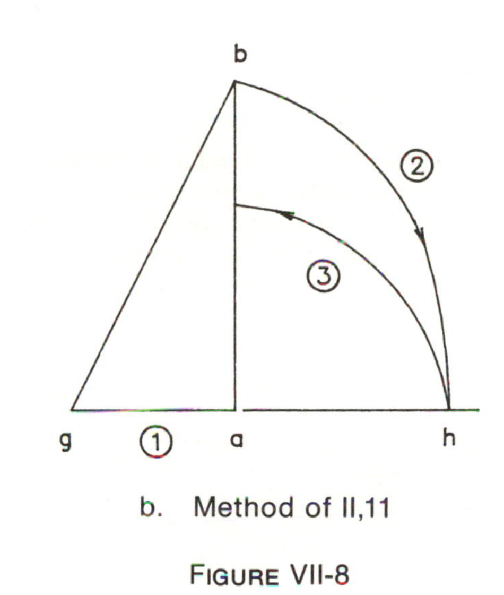

b. Method of II,11

FIGURE VII-8

The proof proceeds as follows: $S(bg) = S(gd + db) = S(gd) + S(db) + 2R(gd, db)$ using II,4. But another expression for $S(bg)$ is given by the Pythagorean Theorem, namely, $S(bg) = S(ag) + S(ab)$. If we equate the two expressions for $S(bg)$ and use $gd = ag$ and $db = be$ then we obtain $S(ag) + S(ab) = S(ag) + R(2ag, be) + S(be)$. This, together with $2ag = ab$, gives $S(ab) = R(ab, be) + S(be)$. On the other hand, by II,2 $S(ab) = R(ab, ae) + R(ab, be)$. Equating these expressions for $S(ab)$, we have $R(ab, be) + S(be) = R(ab, be) + R(ab, ae)$ so that $S(be) = R(ab, ae)$ which is just the statement of II,11.

C. The Volumes of the Icosahedron and Dodecahedron

i. The Text

In *Metrica*, II, 18, 19 [Hero–Bruins, III, 304, 311] Hero wishes to find the volumes of an icosahedron and a dodecahedron whose edges are 10 units. In both cases one is to take one-third the product of the area of a face and the altitude of the pyramid whose vertex is the centre of the solid which gives the volume of the pyramid; this volume in turn is multiplied by the number of faces.

Now in order to complete the solution it is necessary to know the relationship between the edges of the solids and the altitudes. Without the slightest indication of how the results are obtained, we are told that the ratio side:altitude equals 127/93 for the icosahedron and 8/9 for the dodecahedron.

ii. Commentary

The question arises as to how the values of the text were obtained. It turns out that a very direct manner of finding the relationship between the altitude h and the edge a of the icosahedron or dodecahedron is suggested by the extant proof of Theorem 2 of the Supplement to the *Elements* (Section 24).

ICOSAHEDRON. Let a be the edge of the given icosahedron and imagine it to be inscribed in a sphere of radius r. In the defining pentagon of the icosahedron the side will also be a. Let s be the radius of the corresponding circumscribed circle and let d be the diagonal of the pentagon. Let x be the radius of the circumscribing circle of the triangle. Then we have the following relationships for the pentagon (Fig. VII-9a):

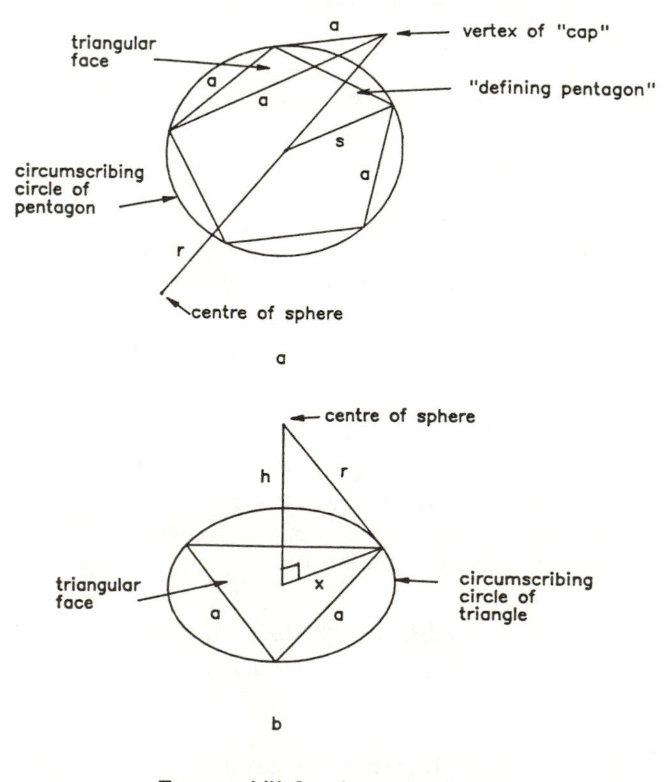

FIGURE VII-9. Icosahedron

(9) $(2r)^2 = 5s^2$ or $4r^2 = 5s^2$ (XIII,16[a],corollary,i)

(10) $d^2 + a^2 = 5s^2$ (lemma to Theorem 2 of the Supplement).

From (9) and (10) we have

(11) $r^2 = (d^2 + a^2)/4.$

Turning to the equilateral triangle (Fig. VII-9b) we have

(12) $a^2/3 = x^2$ (XIII,12).

If we combine (11) and (12) with the relationship $h^2 = r^2 - x^2$ obtained from the Pythagorean Theorem we have

(13) $h^2 = d^2/4 - a^2/12$ or $(h/a)^2 = 1/4(d/a)^2 - 1/12.$

Since d/a is just the numerical value of the ratio defined by DEMR the value h/a (i.e., altitude/side) can be approximated as closely as desired. The exact value is $h/a = .7557$ or $a/h = 1.3232$. The value given in the text is $a/h = 127/93 = 1.3656$.

Note that it is the ratio h/a, and not the ratio a/h, which is needed for the computations of the text. The fact that the value for a/h is given in the text may indicate that another method was used to obtain the text value (see also Hero–Bruins [p. 306] and the method of Fibonacci in Section 31,B,ii).

In Section 27 we shall see that, for the case of the icosahedron, Pappus (Proposition 43) gives $R(12/5) = 1.5491$ as an upper bound for the ratio, side:altitude (i.e., a/h). I note that Pappus uses XIII,10 and XIII,15,corollary in his result just as does the proof of Theorem 2 of the Supplement and that he too (Proposition 48) states and gives the same proof of Theorem 2.

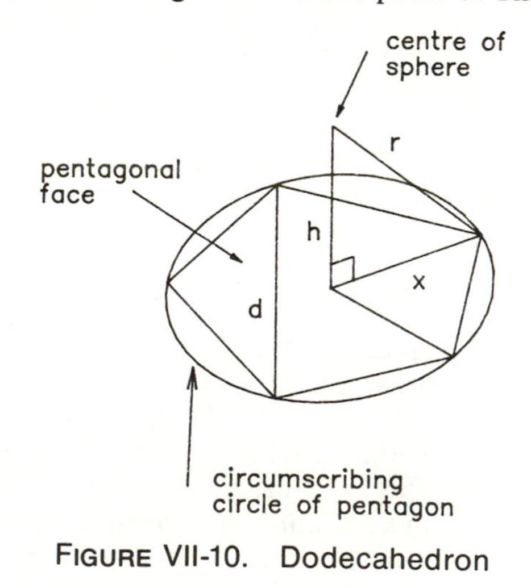

FIGURE VII-10. Dodecahedron

DODECAHEDRON. Let a be the edge of the given dodecahedron and imagine it to be inscribed in a sphere of radius r. Let x be the radius of the circumscribing circle of the pentagon and let d be the diagonal. Then we have the following relationships (Fig. VII-10): $(2r)^2 = 3 \cdot (\text{side of cube})^2 = 3d^2$ (XIII,15,17[a],corollary, 8) or

(14) $$r^2 = 3/4\ d^2.$$

Also $d^2 + a^2 = 5x^2$ (lemma to Theorem 2 of the Supplement) or

(15) $$x^2 = (d^2 + a^2)/5.$$

Combining (14) and (15) with the relationship $h^2 = r^2 - x^2$ obtained from the Pythagorean Theorem, we have $h^2 = 11/20\ d^2$ or

(16) $$(h/a)^2 = 11/20\ (d/a)^2 - 1/5.$$

Since d/a is just the numerical value defined by DEMR the value h/a (i.e., altitude/side) can be approximated as closely as desired. The exact value is $h/a = 1.1135$ or $a/h = .8991$. The value given in the text is $a/h = 8/9 = .8889$.

Again as in the case of the icosahedron, the fact that the value for a/h is given in the text may indicate that another method was used (see also Hero–Bruins [p. 310] and the method of Fibonacci in Section 31,B,ii).

Section 26. Ptolemy (c. 100 to 179)

A. The Chords of 36° and 72° in Almagest

In Book I, chapter 10 of his work on astronomy, entitled *Almagest*, Ptolemy constructs a table of chords [Ptolemy–Manitius, 24] which corresponds to the modern trigonometric table.

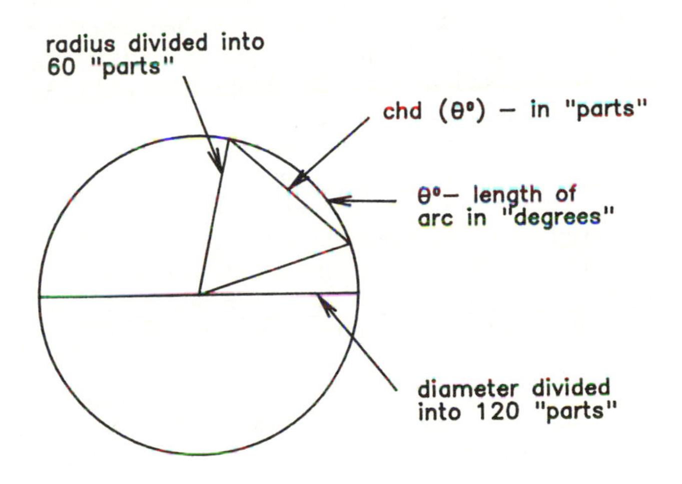

FIGURE VII-11. The Concept of a Chord

As opposed to the modern trigonometric functions which are independent of the value of the radius, the chord of an arc of length $\theta°$—written as chd($\theta°$)—will have a value that depends on the radius of the circle. In Ptolemy's work the diameter is divided into 120 parts, but in Section 29 we will come across other divisions. All calculations follow the sexagesimal system and thus, for example, $a;b,c$ stands for $a + b/60 + c/3600$. (For a discussion of Ptolemy's methods, see Pederson [1974, 56] and Neugebauer [1975, 22].)

Of special interest to us is the calculation of the chords of 36° and 72° which Ptolemy recognizes as being the sides a_{10} and a_5 of the regular decagon and pentagon inscribed in a circle of radius $r = 60$.

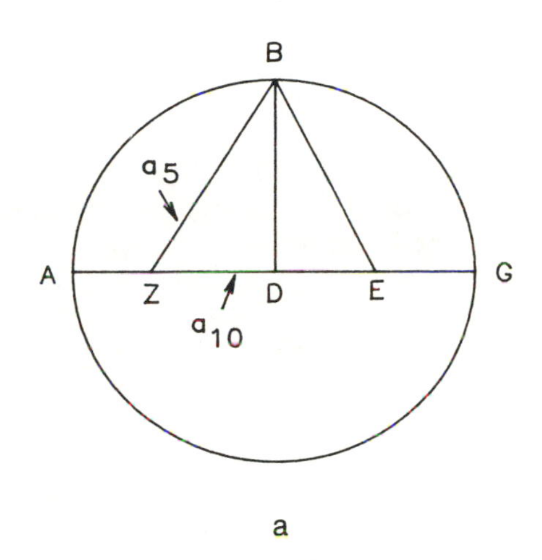

a

Source: Ptolemy–Manitius [p. 25]

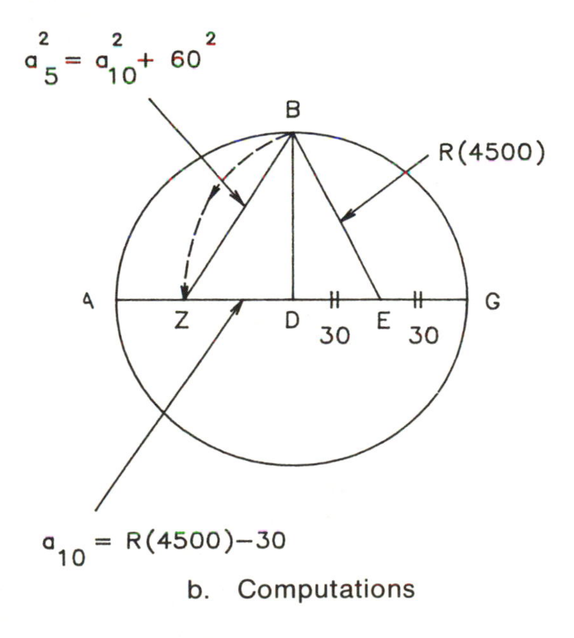

b. Computations

FIGURE VII-12

In the construction (Fig. VII-12a) E is the midpoint of DG and $EZ = EB$. This is, in effect, the construction of II,11 turned on its side with the point Z of the present construction corresponding to the point F of II,11. There is, however, no point here that corresponds to point H of II,11. The reason for this is that Ptolemy is not dividing the radius DG in EMR; rather he is extending DG to Z so that D divides ZG in EMR. This construction is what I referred to as II,11′ in Section 2,C and indeed the first part of Ptolemy's proof is identical with that of II,11′.

Ptolemy invokes XIII,9 which says that $a_6 + a_{10}$ is a line divided in EMR. He then states that, since DG, the radius, equals a_6, and since D divides AG in EMR, it must be that $ZD = a_{10}$. Note that it is not XIII,9 itself that is invoked but rather a "converse." Since Ptolemy does not comment further he may have thought that it was obvious from XIII,9, but in view of Theorem XIV,** (Section 24,A) I do not believe that we can safely assume this (see Herz-Fischler [1985, sections 3G and 4E] concerning this "converse" and its appearance in Arabic and medieval texts).

To show that $BZ = a_5$ Ptolemy uses XIII,10, which states that $S(a_5) = S(a_6) + S(a_{10}) = S(BD) + S(ZD)$, as well as the Pythagorean Theorem which implies that the last sum equals $S(BZ)$.

All the above was brought in by Ptolemy for theoretical purposes only and not for computation. To actually compute a_{10} Ptolemy goes directly to the geometry of the situation (Fig. VII-12b) and only uses the Pythagorean Theorem and subtraction. The former gives $EZ^2 = (60)^2 + (30)^2 = 4500$ so that $ZE = BE = R(45000) = 67;4,55$. In turn $ZD = ZE - DE = 37;4,55$. For comparison I note that in decimal notation this equals 37.0819, whereas the exact value for the side of the decagon inscribed in a circle of radius 60 is $2(60 \sin 18°) = 37.0824$.

In Sections 28 and 29 we will see how Arabic and Indian mathematicians handled the computation of a_{10}.

Using the above value for ZD and the Pythagorean Theorem applied to triangle BDZ, Ptolemy obtains $a_5 = 70;32,3$. In decimal notation this equals 70.5342 whereas the exact value for the side of the pentagon inscribed in a circle of radius 60 is $2(60 \sin 36°) = 70.5348$.

B. *Chord(108°)/Diameter in* Geography

There is another calculation of interest in Ptolemy's *Geography* I,20 [Toomer, 1973, 23]. It so happens that the 36th parallel passes through the Isle of Rhodes (see the map on page 54, Section 9) and that Hipparchus (fl. mid-IInd century) made observations there (see Hipparchus–Dicks [2,5,176,193] and Toomer [*DSB* 3, 208]. Ptolemy is concerned with the ratio of the parallel to the equator (Fig. VII-13). This ratio is equal to chd(180°−2·36°)/diameter = chd(108°)/diameter and is stated by Ptolemy, without explanation, to be 93/115. This ratio whose value is .8087 may be compared with the true value of $(2r \cos 36°)/2r = \cos(36°) = .8090 = (1+R(5))/4 = (\frac{1}{2}$ diagonal pentagon)/side.

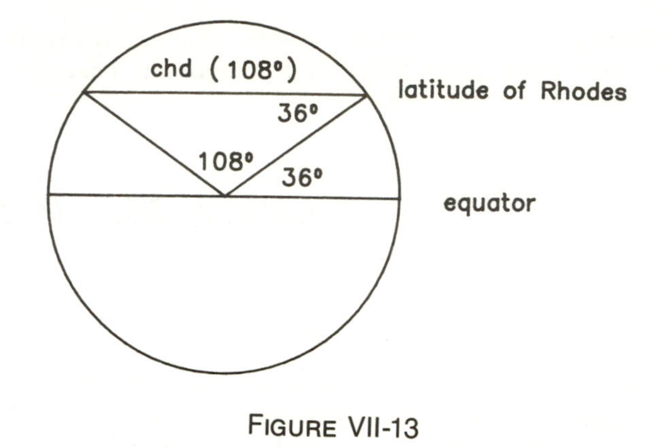

FIGURE VII-13

Note that the denominator 115 is not the same as the value 120 which is the number of parts in the diameter of the *Almagest* tables. Toomer explains this by assuming that Ptolemy used an older table of chords by Hipparchus (see Subsection C) which gave a denominator which was very convenient insofar as the calculations for Thule were concerned. Toomer has reconstructed Hipparchus' table of chords and has concluded that it was based on accurate calculations for multiples of $7\frac{1}{2}°$ with linear interpolation being used for values between these points. The value 93/115 then would have come from interpolation between the values for 105° and $112\frac{1}{2}°$.

Note, of course, that the appearance of 36° in *Geography* is entirely accidental and has nothing a priori to do with DEMR, but the existence of this computation provides a value with which other values may be compared.

C. *Trigonometry before Ptolemy*

Since the chord table of *Almagest* has values for chd(36°) we wish to know if any previous tables of chords dealt with 36° or related angles. The history has been discussed by Toomer [1973]; see also *DSB* 3 and Neugebauer [1975, 209] who may be consulted for the following references to the older literature.

The problem is that Ptolemy acknowledges that much of his astronomical data is taken from Hipparchus (first quarter of IInd century to after −127) but does not say anything precise about the table of chords that Hipparchus used. While some scholars assumed that the table of Ptolemy was essentially the same as that of Hipparchus, Toomer refutes this and states: "the plane trigonometry of the *Almagest* owes much to Ptolemy, and that the trigonometry of his predecessors, notably Hipparchus, was a great deal cruder..." [1973, 20]. Indeed, Toomer, using data from Hipparchus that appear in Ptolemy as well as other evidence including Indian sine tables, argues that the table of Hipparchus was based on a basic division of the circumference into units of $7\frac{1}{2}°$ instead of the half degree division of Ptolemy, and that the radius used by Hipparchus was 3438 "minutes" instead of the 60 "parts" used by Ptolemy. Toomer also notes that his reconstruction would only require a knowledge of $R(2)$, the Pythagorean Theorem, and the half-angle formula.

If Toomer is correct then there would have been no direct calculations involving 36° or 72°—and thus DEMR—in the Hipparchus table (cf. the values involving 108° in Subsection B).

There have also been suggestions in the literature that chord tables were compiled by Archimedes (Section 23) and Apollonius (Section 14). Again Toomer rejects this and states [*DSB* 3, 209] that, while particular problems in trigonometry had been solved before Hipparchus, he was the first to construct a table of chords and thus provide a general solution of trigonometrical problems. Regarding Archimedes, it should be recalled that in his *On the Measurement of the Circle* he essentially calculated the chords of 30°,15°, . . . and that if the conclusion of Knorr (Section 23) concerning the text of "On Blocks and Cylinders" is correct then Archimedes essentially calculated the chords of 36°,18°, . . . but not by the same method as Ptolemy. Recall also our mention of al-Biruni's use (Section 28,D) of Archimedes' lemma to calculate the side of the decagon or equivalently chd(36°).

Two more people must be mentioned. First, Menelaus (fl. c. 100) is said to have written a work on chords but no trace has survived (see Toomer [1973, fn. 1] and Heath [1921, II, 257]). Secondly, we saw (Section 25,A,4,d) that Hero (fl. c. 62) also spoke of a table of chords and that either this table contained values for (1/9) 360° and (1/11) 360° or the values for these angles were obtained by interpolation. If the actual values appeared then it is not unreasonable to assume that values for (1/10) 360° = 36° also appeared. In other words, DEMR would have been treated numerically for trigonometric purposes at least before Hero and thus before Ptolemy.

Section 27. Pappus (fl. first half of 4th century)

There are two parts of Pappus' work that are of interest to us: first, the construction of the icosahedron and dodecahedron by a method fundamentally different in spirit from that of Euclid and which as a by-product gives an elegant new proof of the main result—Theorem 2—of Book XIV; and, secondly, the comparison of the volumes of regular polyhedra whose surface area is the same, by means of a series of lemmas which involve properties of DEMR and numerical estimates obtained from these properties. All references in what follows are to the *Collection* in Pappus–Ver Eecke.

A. *Construction of the Icosahedron and Dodecahedron*

In Book III, Part 4 Pappus gives constructions for the icosahedron and dodecahedron, but his approach is quite different from that of Euclid. There are two things immediately noticeable about the constructions. First of all, whereas Euclid balances them on a vertex and edge, respectively, Pappus rests both solids on a face. Secondly, Pappus avoids the numerical lemmas at the beginning of Book XIII. Thus for the icosahedron Pappus uses XIII,8,10,12 and XIV,** (Section 24), whereas Euclid uses XIII,3,9,10. For the dodecahedron Pappus employs XIII,8,12 and XIV,** as opposed to XIII,4,5 in Euclid.

PROPOSITION 57 [p. 112]. To inscribe an icosahedron in a given sphere.

Sketch of Proof: Whereas Euclid (XIII,16) balances the icosahedron on a vertex, Pappus lays it flat down on one of its faces so that the upper triangle ABG will be horizontal (Fig. VII-14a,b). With the icosahedron oriented in this manner two types of relationships appear.

First, in Figure VII-14a, the defining pentagon $AEHZG$ which corresponds to the "cap" with vertex B will be slanted; horizontal line AG in Figure VII-14a,b,e will be a side of this defining pentagon and ZE in Figure VII-14a,c,e, which is also horizontal, will be a diagonal.

Secondly, all the points of the icosahedron lie on four horizontal circles. The top circle (Fig. VII-14b) contains the triangle ABG. The second circle (Fig. VII-14c) will contain the points E,Z,D which correspond to the vertices of a fictitious equilateral triangle whose sides are equal to the diagonal of the defining pentagon. Triangle HKT (Fig. VII-14d) will lie on a circle of the same radius as that of EZD, but will be oriented in the opposite direction. Because of this last relationship the points Z,E,K can be thought of as the vertices of a fictitious, slanted right triangle (Fig. VII-14e,f). The hypotenuse

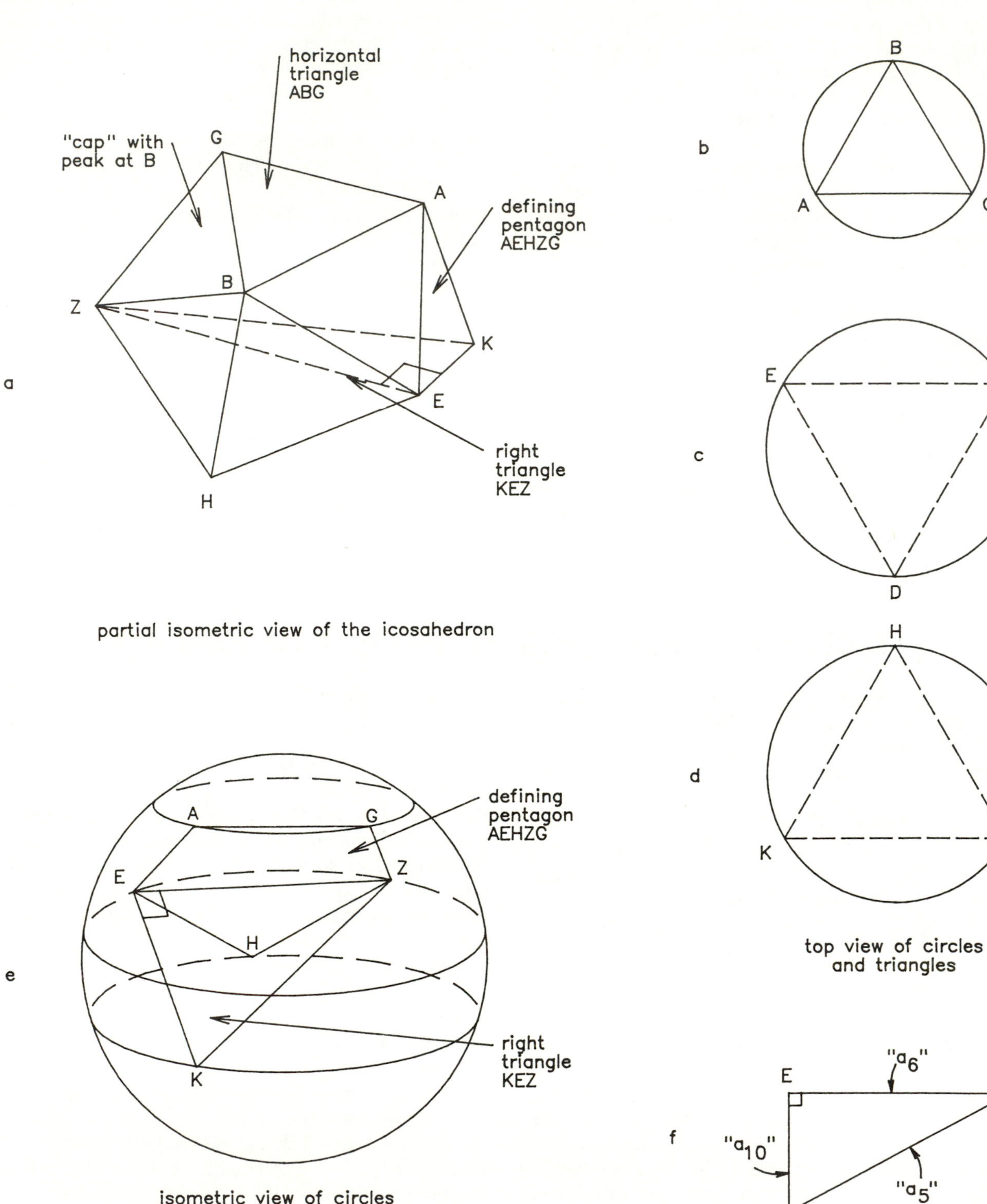

FIGURE VII-14. Pappus' Construction of the Icosahedron

of this right triangle is ZK, the diameter of the sphere. One leg is EZ, the diagonal of the defining pentagon, and the other leg EK is an edge of the icosahedron and thus is equal to the side of the defining pentagon.

Because EZ and EK are equal, respectively, to the diagonal and side of the defining pentagon, their ratio is, by XIII,8, that of EMR. But XIV,** of Section 24,A (which, as we shall see, is explicitly stated and proved by Pappus as Proposition 47 of Book V) tells us that, when we divide the side of the hexagon of a circle in EMR we obtain the side of the decagon. So what Pappus does is think of EZ as "a_6" and EK as "a_{10}" in some fictitious circle (Fig. VII-14f). But XIII,10 tells us that $S(a_6) + S(a_{10}) = S(a_5)$, and since we have by the Pythagorean Theorem that $S(EZ) + S(EK) = S(ZK)$, the diameter ZK corresponds to "a_5" in the fictitious circle (compare with the construction of Ptolemy in Section 26,A). Thus we have

(17) $$ZK:EZ = a_5:a_6.$$

Since EZ and EK correspond to a_6 and a_{10}, respectively, in the fictitious circle, we also have $ZK:EK = a_5:a_{10}$. But since $AG = EK = e_{20}$ we then have

(18) $$ZK:AG = a_5:a_{10}.$$

Thus given the length of the diameter $ZK(=D)$ of the sphere we can find the lengths of the horizontal lines AG and EZ by using proportions (17) and (18). Furthermore, since AG and EZ are not only the sides and diagonal of the defining pentagon but also the sides of equilateral triangles inscribed in the horizontal circles, the relationship $S(a_3) = 3 \cdot S(r) = 3 \cdot S(a_6)$ of XIII,12 enables us to find the radii of these horizontal circles. In addition, the relationship of (17) implies $S(ZK):S(EZ) = S(a_5):S(a_6) = S$(side pentagon in larger circle): S(side hexagon in larger circle) so that $S(D):S$(side pentagon) $= S(EZ):S$(hexagon) $= 3$; that is, $S(D) = 3 \cdot S$(side pentagon inscribed in the larger circle). Note that the radius of the circle of the defining pentagon, which is the key to the Euclidean proof, does not enter the picture.

PROPOSITION 58 [p. 116]. To inscribe a dodecahedron in a given sphere.

Sketch of Proof: Whereas Euclid (XIII,17) balances the dodecahedron on an edge, Pappus lays it flat on one of its faces so that the upper face $ABGDE$ will be horizontal (Fig. VII-15a). With the dodecahedron oriented in this manner, the vertices will lie on four horizontal circles with the pentagon $ABGDE$ lying on the top circle. Then on the second circle we will have the points L,K,T,H,Z which correspond to the vertices of another regular pentagon whose sides are equal to the diagonal of the pentagonal faces of the dodecahedron. The pentagon $LKTHZ$ is oriented in the same direction as the upper pentagon $ABGDE$. The pentagon $MNUOP$ will lie on a circle of the same radius as that of $LKTHZ$ but will be oriented in the opposite direction.

Pappus now shows that certain pairs of diagonals such as GL and UY (Fig. VII-15c) are both parallel to a common edge—KN in this case—and therefore are the sides of a square $LGUY$ (Fig. VII-15b). This square is just the face of the inscribed cube of Euclid's construction which corresponds to the edge KN; Pappus never mentions the cube. Then ZU will be the diameter of the sphere and the diagonal UL of the square will be perpendicular to the horizontal element ZL ($UOZL$ is simply the rectangle obtained by slicing the cube in half diagonally).

A double application of the Pythagorean Theorem gives $S(UZ) = S(UL) + S(LZ) = 2 \cdot S(LG) + S(LZ) = 3 \cdot S(LZ)$. But this last relationship between UZ and LZ is, by XIII,12, just the relationship between the side of an equilateral triangle and the radius of the circumscribing circle.

Now let a_3, a_5, a_6, a_{10} be the sides of the triangle, pentagon, hexagon, and decagon inscribed in the circle of pentagon $LKTHZ$. Then, since LZ is equal to a_5, the above relationship between UZ and LZ implies $UZ:a_5 = a_3:a_6$ or

(19) $$UZ:a_3 = a_5:a_6.$$

This relation between the diameter of the sphere and the side of the triangle of the bigger circle is precisely the same as (17) obtained for the icosahedron in Proposition 57. Thus, as Pappus notes, the larger circles will be the same for both the icosahedron and dodecahedron inscribed in the same sphere.

Turning our attention to the diagonal ZL and side DE of the faces we have, by XIII,8 and XIV,**, that $ZL:DE = a_6:a_{10}$. Furthermore, if we let b_3 be the side of the triangle inscribed in the upper circle, since ZL is the side of the pentagon of the larger circle and DE the side of the pentagon for the smaller circle, then $a_3:b_3 = ZL:DE = a_6:a_{10}$. Combining this with the relationship of (19), we have $UZ:b_3 = a_5:a_{10}$. This relationship between the diameter of the sphere and the side of the triangle inscribed in the smaller circle is precisely the same as that of (18) obtained for the icosahedron in Proposition 57, so that the small circles will be the same for both the icosahedron and dodecahedron inscribed in the same sphere.

Since the smaller circle is the circumscribing circle for both the triangle of the icosahedron and the pentagon of the dodecahedron we have a new proof of Theorem 2 of the Supplement to the *Elements* (Section 24). Pappus, as we shall see, also gives two proofs of this result in Proposition 48 of Book V. As was the case with the icosahedron, the various relationships can be used to find the side of the pentagonal face and the radii of the various circles for a given diameter.

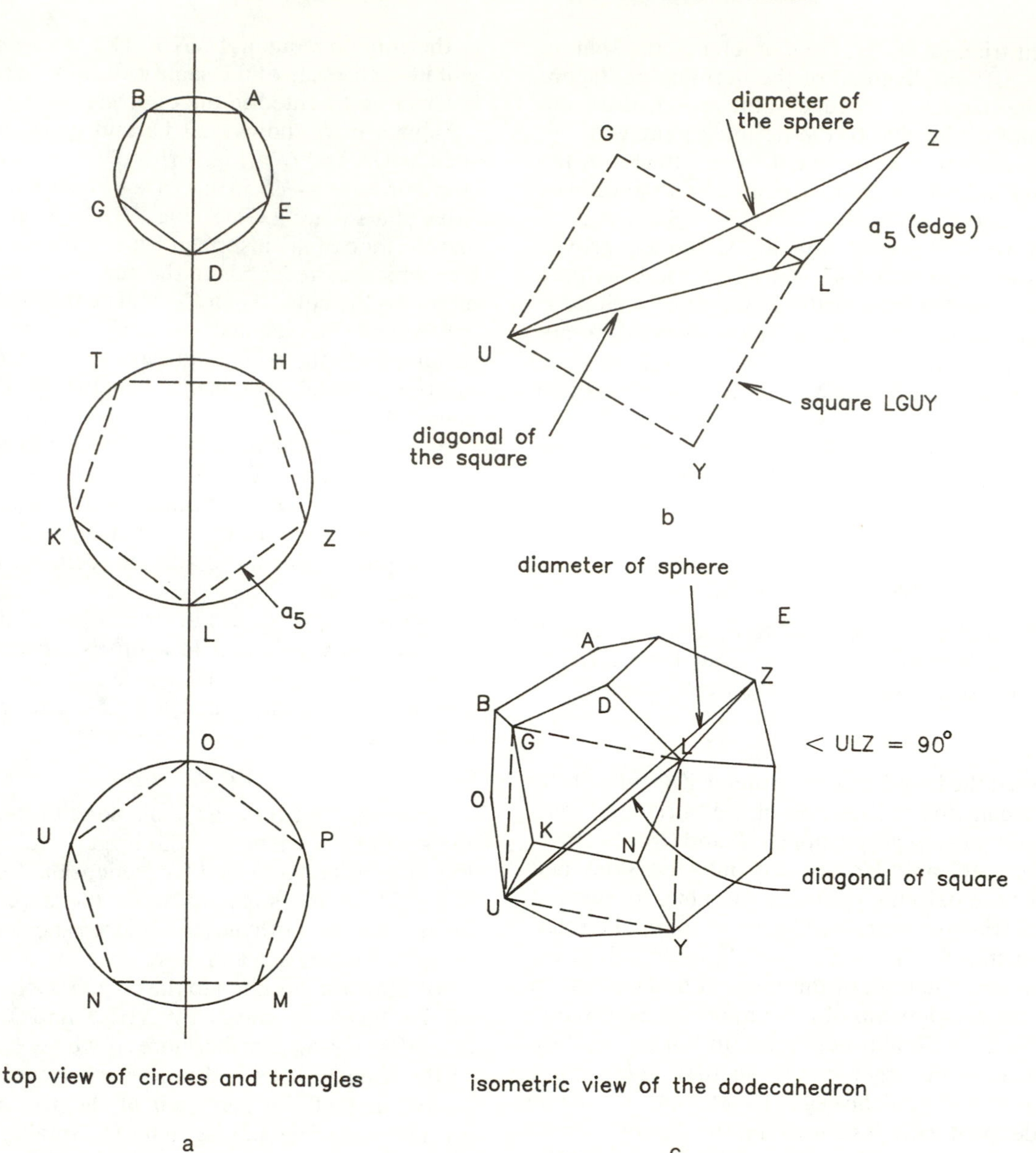

FIGURE VII-15. Pappus' Construction of the Dodecahedron

B. *Comparison of Volumes*

In Book V, Part 3, Pappus again considers the five regular polyhedra, but this time with the aim of comparing them from the viewpoint of volume for a given surface area. This problem should be compared with the problem of computing the volumes for a given edge (Hero in Section 25,C; Francesca in Section 31,C; Bombelli in Section 31,F) or for a given diameter of the sphere in which the polyhedron is inscribed (Fibonacci in Section 31,B).

Several of the preliminary lemmas deal with DEMR. I shall only give sketches of some of the proofs and statements of others. I use geometric terminology—$S(\cdot)$, $R(\cdot)$—whenever Pappus does even though he mixes geometric and ratio terminology.

PROPOSITION 41 [p. 321]. See Herz-Fischler [1985, Section 2] for the statement, an outline, and a discussion of the proof which contains as a corollary—although it is not stated—Theorem XIV,** of Section 24,A.

PROPOSITION 42 [p. 322]. Let AB be cut in EMR at G, with GB being the smaller segment, then $S(AB){:}5\cdot S(GB)>4/3$.

Remark: This rather long proof uses XIII,4. I note that the exact value of the ratio on the left is 1.371.

PROPOSITION 43 [p. 326]. If an icosahedron is inscribed in a sphere, then twelve times the square of the perpendicular from the centre to the middle of one of the faces is more than five times greater than the square of the side of the icosahedron.
Remark: The proof uses Propositions 41, XIII,10,16 [a],corollary. The estimate is used in Propositions 54 and 56 to compare the icosahedron, octahedron, and dodecahedron. I have mentioned this result in Section 25,C in connection with Hero's determination of the volume of an icosahedron whose edge is given.

PROPOSITION 44 [p. 330]. This is the Ratio Lemma of Section 24,A and the proof is the same as that found in the Supplement to the *Elements*. Pappus uses it in his Propositions 45,46,47 and explicitly in both proofs of Proposition 48.

PROPOSITION 45 [p. 332]. Let AG be the diameter of a circle with centre E.
(i) Pick the vertical line BD so that $AG = 3 \cdot GD$. Form $\triangle ABG$.
(ii) Let T divide BG in EMR with BT being the larger segment.
(iii) Pick Z so that $S(EG) = 5 \cdot S(EZ)$. Then $S(BT)$: $S(GZ) = 5/3$.

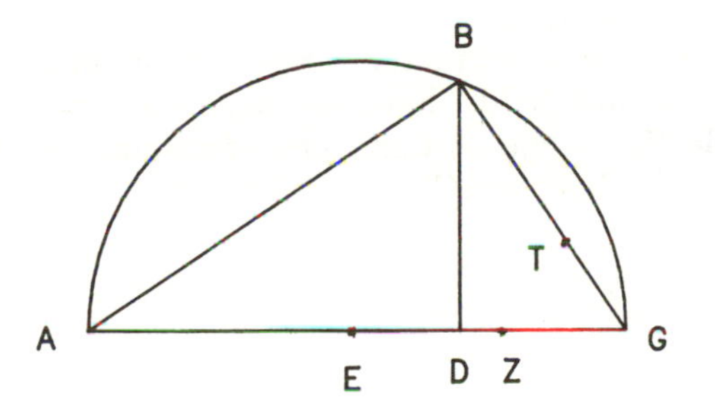

Source: Pappus–Ver Eecke

FIGURE VII-16. Proposition 45

Remark: XIII,2—see condition (iii) above—XIII,5, and Proposition 44 are used. Compare conditions (i) and (ii) with XIII,18; BG is the side of the cube and BT is the side of the dodecahedron. Proposition 45 is used in Proof 1 of Proposition 48 (XIV,2).

PROPOSITION 46 [p. 334]. Let AG be the diameter of a circle with centre Z.
(i) Pick the vertical line BE so that $S(GZ) = 5 \cdot S(EZ)$.
(ii) Let D divide BG in EMR with BD being the larger segment. Then $S(GB) + S(BD) = 5 \cdot S(EG)$.

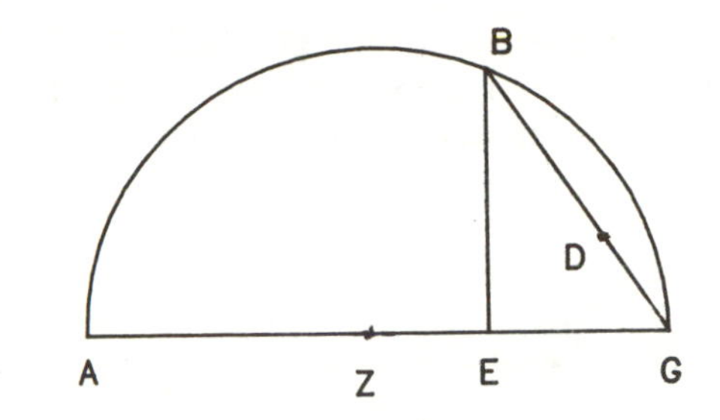

Source: Pappus–Ver Eecke

FIGURE VII-17. Proposition 46

Remark: The proof uses Proposition 44; XIII,5.

PROPOSITION 47 [p. 336]. This is Theorem XIV,** of Section 24. For reasons discussed in Herz-Fischler [1985], I believe that the proof found in Pappus is not the original one, but a shortened version inspired by the original.

PROPOSITION 48 [p. 336; second proof p. 340]. This is Theorem XIV,2 of the Supplement to the *Elements* (Section 24). The second proof is the same as that found in the Supplement; however, it uses XIV,**—Proposition 47—explicitly. The first proof is longer and more complicated and uses Propositions 44,45,46, XIV,** (that is, Proposition 47 explicitly) and XIII,10,15,16,17.

As discussed in Section 24 it is possible that the first proof is due to Aristaeus or Apollonius. A third proof of XIV,2 was discussed in Subsection A in connection with the construction of the icosahedron in Proposition 58.

PROPOSITION 51 [p. 346]. Let DE be one of the equal sides of the $72°-54°-54°$ isosceles triangle EDB, and let KE be the side of the equilateral triangle EKG whose area is the same as that of triangle EDB. Let MN be any line which is cut in EMR at Q with QN being the smaller segment. Then $S(KE):S(DE)<S(MN):5 \cdot S(QN)$.
Remark: The proof uses Proposition 42 which estimates the term on the right. The true value of the ratio on the left is 1.098, while on the right the value is 1.371.

The above results are used to compare the volumes of the icosahedron and octahedron (Proposition 54 [p. 354]); the icosahedron and dodecahedron (Proposition 55 [p. 356]); and the dodecahedron and octahedron (Proposition 56 [p. 358]); where in each case the surface areas of the compared polyhedra are assumed to be the same.

PROPOSITION 55. If $S_{20} = S_{12}$ then $V_{20} > V_{12}$.
Sketch of Proof: The volume of a regular polyhedron is the number of faces times the volume of the pyramid with vertex at the centre of the sphere and whose base is one of the faces or equivalently:(altitude to one of the

faces)·(1/3 surface area of the polyhedron). Since it is assumed that the dodecahedron and icosahedron have the same surface area, one must show that the altitude h_I of the icosahedron is greater than the altitude h_D of the dodecahedron.

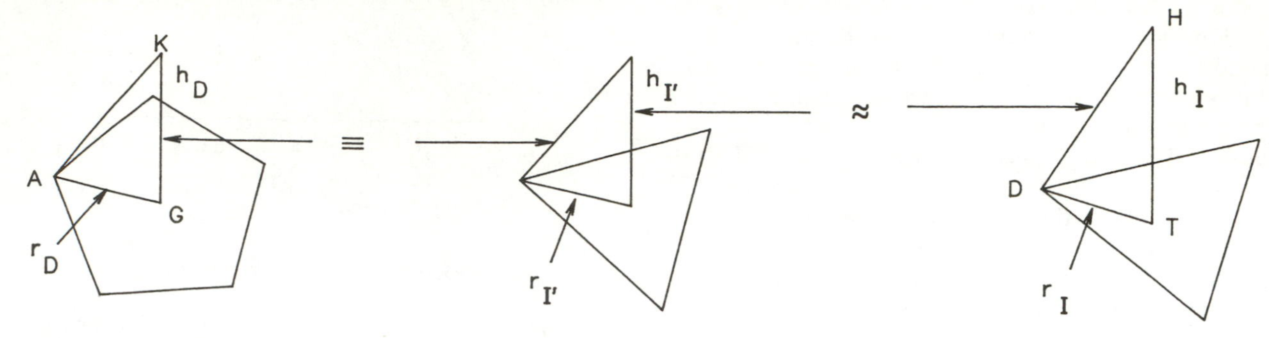

Source of D and I: Pappus–Ver Eecke [p. 356)

FIGURE VIII-18. Proposition 55

Thus consider (Fig. VII-18) right triangles AKG and DHT built upon a pentagon of the dodecahedron D and a triangle of the icosahedron I. But by Proposition 48 (XIV,2) the circle about the pentagon is also the circle about the triangle of the icosahedron I' inscribed in the same sphere as the dodecahedron.

In Proposition 49 Pappus used a series of estimates that did not involve DEMR to show that if a triangle and a pentagon are inscribed in the same circle then $12 \cdot A_5 > 20 \cdot A_3$. This means that $S_{12} > S'_{20}$. Our hypothesis is $S_{20} = S_{12}$ so that $S_{20} > S'_{20}$. This in turn implies that $r_{I'} < r_I$ and further, since the triangles for I and I' are similar, that $h_{I'} < h_I$. Finally, since the circle for D and I' are the same and both D and I' are inscribed in the same sphere, we have $h_D = h_{I'}$, so that $h_D < h_I$ as required.

Comment: Proposition 55 can be derived immediately from XIV,2 and XIV,6,7[corollary] without using Proposition 49. Indeed we have $S_{12}:S'_{20} = e_6:e_{20} = L_1:L_2 > 1$. Thus, if $S_{20} = S_{12}$, then $S_{20}:S'_{20} > 1$, which implies that $(\text{diam})_I > (\text{diam})_{I'} \equiv (\text{diam})_D$ and subsequently $h_I > h_D$ as in the Pappus proof.

I do not know if Pappus just overlooked this simpler proof or if perhaps his version—which as we saw in Proposition 48 contained XIV,2—did not contain Theorems 6 and 7 (see the discussion on Proposition 48 and, in Section 24,C, the stages in the development of the Supplement).

Pappus also constructed some semi-regular polyhedra, but the details have not been preserved. It is possible that some of these constructions may have involved DEMR; see Clagett [1978, 399].

CHAPTER VIII

THE ARABIC WORLD, INDIA, AND CHINA

The major part of this chapter deals with Arabic mathematics and will involve algebraic and trigonometrical as well as purely geometrical techniques. The extant work from India, which is trigonometrical in nature, all dates from the 12th through 17th centuries, but may represent in part earlier and foreign sources. The material of interest from China appears to be very limited but is presented here for completeness.

Section 28. The Arabic Period

The period discussed here extends from the early part of the 9th century to the mid-10th century. The adjective "Arabic" is used here in a wide sense and is applied to mathematicians, Arabic or not, of the Near and Middle East who were part of the great intellectual developments which followed the rise of Islam and the Arabic nation.

There would seem to be a gap in the material which has passed down to us. For example, in Section B,i I will discuss why it is not unlikely that there was a work predating Abu Kamil's *On the Pentagon and Decagon* which presented some more elementary problems. Similarly, the *Handbook* of Ibn Yunus (which will be discussed in Section D) seems, according to the estimate of King [*DSB*], to have been only one of about 200 such medieval Arabic works; unfortunately many are lost. It is possible that some of these works contained intermediate stages between Ptolemy and/or Abu Kamil and the 11th-century writers Ibn Yunus and al-Biruni. In the next section we will see some Indian material which may reflect some of the Arabic work.

i. Authors Consulted

The following is a list of the authors and works I have consulted. For more details on the mathematicians discussed here, one may consult various articles in the *Dictionary of Scientific Biography*, *Encyclopedia of Islam*, Suter [1900], Steinschneider [1893], and Sezqin [1974], and the surveys by Kapp [1934], Youschkevitch [1976], Anbouba [1978], and Berggren [1985, 1986].

a. Al-Khwarizmi (before 800 to after 847); see Section A,i.

b. Ibn Turk (9th century); see Section A,ii.

c. *Mishnat ha Middot* (2nd century?; 9th to 12th centuries?); see Section A,ii.

d. Banu Musa (fl. middle of 9th century). The work of these three brothers, *The Book on the Measurement of Plane and Spherical Figures* [Banu Musa–Curtze; Suter, 1902], is mainly concerned with the area of a circle. It is shown that the surface area of any regular polygon circumscribed about a circle is (1/2 diameter) · (1/2 perimeter), a result also found in al-Khwarizmi's work on mensuration (see Section A,ii). However, this is proved in connection with the estimate of the area of a circle and it is not applied to the pentagram.

e. Thabit ibn Qurra (836-901). This author's "algebra" [Thabit ibn Qurra–Luckey; mentioned in Section 5,G] contains nothing concerning DEMR. Thabit ibn Qurra is said to have written a commentary on Books XIV and XV; unfortunately no manuscripts of this work are known (see Kapp [1934, vols. 23 and 65, no. 86].

f. Abu Kamil (c. 850 to c. 930); see Section B.

g. Abu'l-Wafa' (940-998); see Section C.

h. Al-Samaw'al (al-Maghribi) (12th century). The *al-bahir* [al-Samaw'al-Ahmad, Rashed] contains nothing of interest.

i. Ibn Khaldun (10th century). The part of the work [Ibn Khaldun–Woepcke] that has been published contains nothing of interest.

j. Al-Karaji (10th century). The extracts of *al-Fakhri* given by Woepcke [al-Karaji–Woepcke] contain nothing of interest.

k. Ibn Yunus (died c. 1009). See Section D.

l. Al-Biruni (973 to after 1050); see Section E.

m. Al-Khayyami (1048 to 1131). In the *Algebra* [al-Khayyami–Woepcke, 41, case 1], one of the cubic equations in a certain case leads to the construction of a parabola and to the point $\left(c(1+R(5))/4, c\left(R(5)-1\right)/4\right)$, but there is no mention of DEMR in the text; the occurrence has nothing to do with the geometric concept of DEMR but is only one of the myriad of occurrences in mathematics of these numbers.

n. Abu Bakr (?). Parts of the book on mensuration have been discussed in Busard [1968]. From the index it appears that the pentagon is not discussed. For the conjectures on the date of this author, see Busard [p. 68]; he mentions some other books on mensuration by Arabic authors [p. 69].

o. Al-Kashi (15th century). Youschkevitch says that al-Kashi's *On Measures* contains area calculations for polygons [1976, 112].

p. Yusuf al-Mu'taman ibn Hud (reign, 1081-1085). J. Hogendijk, who is editing parts of *Istikmal* (see Hogendijk [1985a]), has informed me that section 1 of species 2 of species 3, which deals with Book X of the *Elements*, contains the following result dealing with the classification of a line divided in EMR (cf. XIII,6; XIII,11; and Taisbak [1982]).

Proposition 23. If a line is divided in EMR, then
(a) if the smaller segment is "rational," the line is a "first binomial";
(b) if the greater segment is "rational," then the line is a "fifth binomial";
(c) if the line is "rational," then the smaller segment is a "fifth apotome."

These results go beyond those of Book XIII as far as the classification of line segments is concerned.

q. Al-Kuhi (fl. 970-1000); see Hogendijk [1985b*].

ii. Equations

Of particular interest to us in this section will be certain quadratic equations and their solutions. What follows immediately are some general remarks concerning the treatment of quadratic equations by Arabic, and indeed other, much later European mathematicians.

The Arabic mathematicians stated their problems in verbal form (I shall give several examples of this). However, when analyzing their treatment, I will resort to modern notation and, in particular, will denote the unknowns by letters and use exponential notation. Negative numbers are completely absent in the statement of the problems. Because of this, the equations are always stated in—or can be reduced to—one of the following three formats with all coefficients positive:

$$\text{A I:} \quad x^2 + ax = b$$

$$\text{A II:} \quad x^2 + b = ax$$

$$\text{A III:} \quad x^2 + ax = b.$$

The labelling follows Gandz [1938, 407], whom I also followed in Section 5 regarding the labelling of the Babylonian equations. This order is the same as that found in al-Khwarizmi's presentation.

Further, only real positive solutions of equations are recognized. Thus, in cases A I and A III only the positive solution is stated, whereas in case A II only the situation where there are two positive solutions is considered.

A. Al-Khwarizmi (before 800 to after 847)

i. Algebra

In addition to giving examples which illustrate the three types of quadratic equations, al-Khwarizmi presents further examples. Among these is the following problem [al-Khwarizmi–Rosen, 51]: "If some one says: 'I have divided ten into two parts; I have multiplied the one by ten and the other by itself, and the products were the same'; then the computation is this: you multiply thing by ten; it is ten things. Then multiply ten less thing by itself; it is a 100 and a square less twenty things, which is equal to ten things. Reduce this according to the rules, which I above have explained to you.''

If we denote one part by x so that the other part is $10-x$, then the condition implies that $10x = (10-x)^2$. In ratio form the solution of this equation corresponds to finding x so that $10:(10-x) = (10-x):x$; in other words, the unknown x is the length of the *smaller* segment when a line of length 10 is divided in EMR.

I note that this problem translates into the type A II equation

$$x^2 + 100 = 30x. \quad (1)$$

This equation has the two positive solutions: $15 + R(125)$ and $15 - R(125)$. Since the first one would give a value of x greater than 10 it does not correspond to a solution of the original problem.

The question of interest is whether or not al-Khwarizmi had DEMR in mind when he presented this problem. A negative answer is given by Gandz [1938, 531], who says that al-Khwarizmi "knows nothing of the special importance of this problem which is taught by Euclid II[11] . . . ." It should be noted that Gandz [for example, pp. 519, 524] is of the view that al-Khwarizmi was not influenced by Euclid in the development of his methods and that while "algebraically [al-Khwarizmi] is ahead of Euclid [by] 1000 years, geometrically he is behind Euclid [by] 1000 years." I presume that it is this general outlook on the work of al-Khwarizmi which causes Gandz to make the specific assertion concerning the problem above. The viewpoint of Gandz, as I noted in Section 5,G, is disputed by some authors.

In considering the answer to the question, the following points should be noted.

a. The problem as given does correspond exactly to the geometric description of the problem of DEMR as stated in II,11; for $10x$ is the area of the rectangle on the whole line with the height, the smaller segment, being x, and $(10-x)^2$ is the area of the square on the larger segment. Of course if we thought of the problem strictly in terms of the ratio definition of VI,def.3, it might seem more reasonable to us to let the unknown be the larger segment. If we did this we would obtain $10(10-x) = x^2$ or in type A I form:

$$x^2 + 10x = 100. \quad (2)$$

The solutions of this equation are $-5 + R(125)$ and $-5 - R(125)$ of which only the first corresponds to a solution of the original problem.

b. A work such as the *Algebra* could not have been written in mathematical isolation. Under Caliph al-Mamun (reign 813-833), al-Khwarizmi was a member of the House of Wisdom in Baghdad; al-Ḥajjaj wrote the second of his two translations of Euclid under the same caliph (see Toomer [*DSB* 5, 358], Kapp [1934], and Murdoch [*DSB* 1, 438]). Gandz [1936b, 266], however, sees in the *Algebra* a reaction against the introduction of Greek mathematics.

c. There are eight problems which involve the division of 10 into two parts and which lead to non-degenerate quadratic equations (note that these are all of type A II and that this is also true of three of the other six non-degenerate equations). Thus the problem that I have discussed can be thought of as one problem among eight. On the other hand, it stands out, for it is the only problem in the *Algebra* involving a non-degenerate quadratic equation for which the answer is irrational.

It seems to me, then, that we must leave unanswered the question as to whether al-Khwarizmi had DEMR in mind when he wrote the above problem.

ii. Predecessors of al-Khwarizmi

The questions of the sources of al-Khwarizmi and whether or not there were any earlier Arabic algebraic texts have been much discussed (see Gandz [1936b] and Toomer [*DSB* 5]. In this regard, however, there does exist a manuscript, entitled *Logical Necessities in Mixes Equations*, taken from an algebra by Ibn Turk [Ibn Turk–Sayali] which Sayali considers to be older than that of al-Khwarizmi. If this were the case then it would be especially interesting to know if the missing part of the work by Ibn Turk also contained the equation (1). It should be noted that the only common equation that we know of is $x^2 + 21 = 10x$ (type A II) [al-Khwarizmi–Rosen, 15; Ibn Turk–Sayali, 163].

Mention should also be made here of the work on mensuration by al-Khwarizmi [al-Khwarizmi–Rosen; Gandz, 1932], which speaks of the area of the pentagon [Gandz, 70] but only in the context of the statement that the area of any regular polygon is the product of half the perimeter with the radius of the inscribed. Thus the concept of DEMR is not involved. Gandz believes that al-Khwarizmi's work is taken from a Hebrew work, *Mishnat ha-Middot*, which he dates to the 2nd century. If this were the case, one would wish to know if the original had at one time a more detailed analysis of the area of the pentagon, perhaps of the type that we saw in Hero's work (Section 25,A). However, a recent study [Sarfatti, 1968, 58] rejects this and on the basis of terminology, which the author says is influenced by Arabic but which lacks certain words from the Spanish period, assigns it to the period between the 9th and 12th centuries. Neuenschwander [1982] has made a comparative study of the *Mishnat ha-Middot*, al-Khwarizmi, and Hero.

B. *Abu Kamil (c. 850 to 930)*

There are two works by Abu Kamil which are of interest. In the first we find detailed calculations involving various quantities related to the pentagon and decagon; in the second we again find equations related to DEMR.

i. On the Pentagon and Decagon

One may be surprised upon turning to this work, especially after having read Euclid and Hero. First of all the approach is different from Hero's, for now everything becomes exact—as opposed to the approximations of Hero—and the determination of quantities is reduced to the solution of equations. Secondly, the concept of DEMR is not brought in until problem 17 and even there it is only used to re-prove an earlier result.

The question of sources is very difficult. I mentioned in the last subsection that Euclid had already been translated into Arabic by the early 9th century and the same is true of Ptolemy (see Toomer [*DSB* 4, 202, and references]). Abu Kamil mentions Euclid explicitly (in problem 17) and seems to be acquainted with Ptolemy's work (in problem 1) but any connection with Hero's work on the pentagon is far from sure, a point already made by Levey [*DSB*, 31] with respect to the entire work of Abu Kamil.

My discussion is based in the first instance on the translation from the Arabic of Yadegari and Levey [Abu Kamil–Yadegari, Levey] but I shall bring in certain material from the Hebrew version of Mordechai Finzi (c. 1460) which has been translated into Italian [Abu Kamil–Sacredote] and from a Latin version which has been translated into German [Abu Kamil–Suter] as well as from the original Arabic manuscript [Abu Kamil–ms]. The lines on the diagrams are for the most part not visible. I have generally followed Abu Kamil–Yadegari and Levey, but I have made some changes based on the manuscript and the text.

In the first four problems we are given the diameter ($d = 10$) of a circle, and we are required to find the sides of the inscribed pentagon and decagon and then the circumscribed pentagon and decagon. Then in problems 5 and 7 (6 and 8) we are given the inscribed (circumscribed) pentagon ($a_5 = 10$) and we are required to find the diameter. In problems 9 and 10 we are given the inscribed and circumscribed decagon ($a_{10} = 10$) and we are required to find d. Problem 11 involves finding a_{15} (*Elements*, IV,16) when the diameter is given.

The surprising thing about these problems is that nowhere is the concept of DEMR brought in, even though there are many occasions where this would simplify matters (see, for example, the alternate treatment of these same problems by Fibonacci in Section 31,B). Consider, for example, problem 1 (Fig. VIII-1) where we are told that $d = 10$ and we are required to find a_5. Let us designate the unknown value of a_5 by x. First we are told that it is obvious that $HL = 1/10 \cdot x^2$ (this follows from consideration of the similar triangles HLD and $H\dot{D}H$; see VI,8,corollary). The Pythagorean Theorem gives $JD = 2DL = R\left(4x^2 - 2/5(1/10x^4)\right)$. We are told that it is also obvious that $AB \cdot JD + AB \cdot AB = JD \cdot JD$; this follows from Ptolemy's Theorem (*Almagest* I,10; see Pederson [1974, 58], for example, which states that in a quadrilateral the sum of the product of alternate sides is equal to the product of the diagonals). All this leads to the quadric equation $x^4 + 3125 = x^2$ which is solved for x^2 which in turn gives $x = a_5 = R\left(62\ 1/2 - R(781\ 1/4)\right)$.

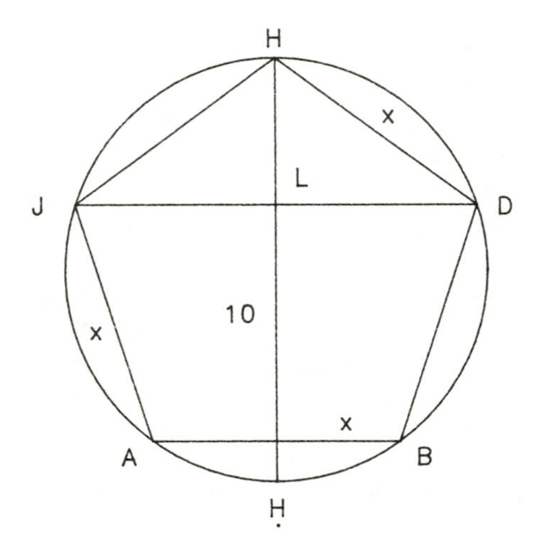

FIGURE VIII-1. Problem 1

Again in problem 2, Abu Kamil uses Ptolemy's Theorem to find a_{10} if $d = 10$, and this time a linear equation is obtained. Thus he has avoided using XIII,9 (used in problem 17) which would lead to $(5+x)(x) = 25$.

I also point out that in problem 5 (find d if $a_5 = 10$) Ptolemy's Theorem is used to show first that the diagonal satisfies $x^2 = 10x + 100$. This could have been obtained immediately from XIII,8, as was done, for example, by Francesca (Section 31,C, problem 2).

Problems 12 to 14 involve the equilateral triangle with the problem 15 involving an equilateral, but *not* the equiangular pentagon inscribed in a square (cf. Abu Kamil–Sacerdote [189] for Finzi's comment on the impossibility of circumscribing a square about a regular pentagon).

In problem 16 we are required to find the side of a pentagon whose area is 50. Abu Kamil uses problem 5 (see the discussion above) which relates the diameter to the side (on p. 28, line 4, read 100 instead of 200) and then expresses the area in terms of the side and radius to obtain an equation. Thus once again DEMR does not appear.

It is only when we arrive at problem 17 (Fig. VIII-2), where we are required to find the side $ZW(=x)$ of a decagon whose area is 100, that we encounter DEMR.

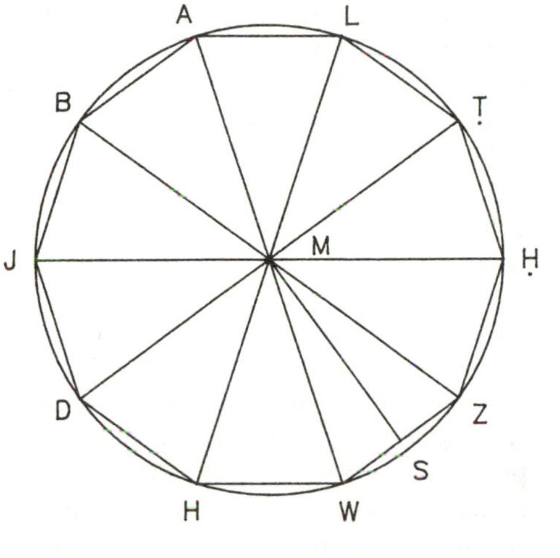

FIGURE VIII-2. Problem 17

Again the author starts off by relating the diameter and side, and for this purpose he states the generalized version of the result of problem 9—which, as I mentioned, was done without recourse to DEMR—to obtain

(3) $$ZM = 1/2x + R(1\ 1/4x^2).$$

All of a sudden and without stating that he is doing so, Abu Kamil re-proves relationship (3) using properties of DEMR. This development presents certain difficulties, so I will present it step by step with comments.

a. If a line is divided in EMR and the larger segment is added to the original line, then sum:larger segment = original:smaller segment.
Comment: XIII,5 says that the sum is divided in EMR, that is, sum:original = original:larger. Of course, since the original line was divided in EMR, this last ratio is equal to larger:smaller, and at first glance one would not think too much about the difference between XIII,5 and statement a. However, the Arabic proof of XIV,** [Herz-Fischler, 1985, section 3B] does not treat such matters lightly and makes use of the Ratio Lemma (Section 24,A). In view of this and the fact that Abu Kamil seems to be using XIV,** in step d, I do not think that we can simply take step a as being immediate from XIII,5. The translation given in Abu Kamil–Yadegari and Levey [p. 30] says only: "if a line is divided into 2 parts. . . ." However, the original Abu Kamil manuscript does indeed speak of DEMR (the text and translation are given in Appendix II,B,1). DEMR is also explicitly mentioned in the Hebrew and Latin texts. Furthermore, instead of giving the proportion, the Hebrew and Latin texts state that if the new line is divided in EMR, then its smaller segment is the original larger segment.

b. Euclid said: $a_6 + a_{10}$ is a line divided in EMR.
Comment: This is XIII,9.

c. Thus, if we add ZM $(=r = a_6)$ and $ZW = (a_{10})$, we obtain a line divided in EMR.

d. If we let $ZW = x$, then

(4) $$(x/2 + (ZM - x))^2 = 5(x/2)^2.$$

Comment: While c talked about adding a_6 and a_{10}, this is not what is involved in (4); for $x = a_{10}$ is the larger segment when $ZM = a_6$ is divided in EMR and $ZM - x$ is the smaller segment so that (4) corresponds to XIII,3. Thus in writing (4), Abu Kamil is implicitly using XIV,** and not XIII,9 which, as I shall show presently, would have led to an impasse. I do not know why Abu Kamil wrote down b and c. Since he was a contemporary of Thabit ibn Qurra and Qusṭa ibn Luqa, who were both connected with an Arabic edition of the *Elements* which contained XIV,** (see Herz-Fischler [1985, section 3B, appendix]), we must suppose that Abu Kamil was familiar with the result. But perhaps Abu Kamil was trying to justify the use of XIII,3 via a direct appeal to XIII,9 and that is why he has statement a; possibly he was influenced by the Arabic proof of XIV,**.

e. In (4) write $5(x/2)^2$ as $5/4x^2$, then take the square root of both sides. Finally, add $x/2$ to [again] obtain (3).
Comment: Both the Hebrew and Latin texts remark on the double proof of (3) because they say: "But it was already shown that . . . ."

Problem 17 is now finished by finding the altitude MS of triangle ZMW in terms of the radius and half side (Pythagorean Theorem) and then finding an expression for the square of the area of this triangle which from the given is equal to 100.

An obvious question for the modern reader is why Abu Kamil, if he felt it necessary to obtain a second derivation of the relationship between ZM and x, did not use the proportion given directly by XIII,9. If we proceed to do this we have $(ZM+x){:}ZM = ZM{:}x$ from which we obtain $ZM^2 = ZMx + x^2$. While we moderns would not even flinch at this equation and would simply treat x as a constant in solving for ZM in terms of x, Abu Kamil would have found himself at an impasse (perhaps theoretical and/or psychological), for this quadratic involves the *product* of the two unknowns. As far as I know this type of situation was never treated by the early Arabic mathematicians. Note that the equation obtained above from XIII,3 did not have to be solved as a quadratic equation, which too would have involved product terms; it only required the taking of square roots. In this regard note that if we tried to use XIII,1 applied to the line segment $ZM + x$ as given by XIII,9 we would obtain $(ZM + (ZM+x)/2)^2 = 5((ZM+x)/2)^2$. To solve this by first squaring the right-hand side, as Abu Kamil does in solving (4), would once again give us a cross product term.

I emphasize that we cannot be sure that Abu Kamil even tried to set up an equation using proportion theory. It is possible that the tradition that influenced him would direct him to the use of the results XIII,1-4 in the same way as Hero used XIII,1 (see Section 25,A). Again we cannot be sure if Abu Kamil first tried to use XIII,1 instead of XIII,3.

I also remark that in their footnote Yadegari and Levey [p. 50 fn. 45] obtain the same "illegal" equation $ZM^2 = xZM + x^2$ using Abu Kamil's result, namely, sum:larger = original:smaller. They derive this equation because of their statement, "obviously Abu Kamil has made use of a proportion to arrive at his conclusion [that is, (4)]." Note also that these authors [fn. 44] have assumed that (4) comes from XIII,1 instead of from XIII,3.

In problem 18 (Fig. VIII-3) we are given that the $72°-72°-36°$ triangle BHD has area 10 and we are required to find the side $DH(=x)$ of the pentagon.

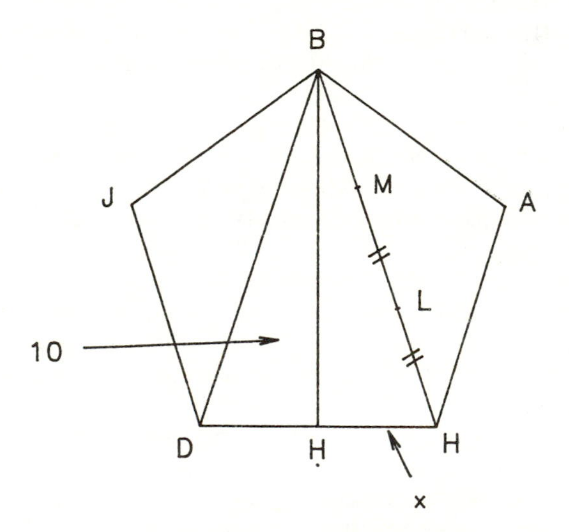

FIGURE VIII-3. Problem 18

Part of the text is garbled [Abu Kamil–Yadegari and Levey, 32, lines 1-4] but from what follows it is clear that Abu Kamil is referring to XIII,3 (both the Hebrew and Latin texts explicitly give the statement of XIII,3). Citing XIII,8 he divides BH into EMR at M and lets L be the midpoint of the larger segment MH. Since $MH = HD = x$ and $LH = x/2$, XIII,3 tells us that $BL^2 = (\text{smaller} + 1/2 \text{ larger})^2 = 5(x/2)^2$ and thus $BL = R(1\ 1/4x^2)$ and $BH = BL + LH = x/2 + (R(1\ 1/4x^2))$. This is exactly the same expression for the side of the isosceles triangle that was obtained in problem 17, and indeed the author repeats the same steps which, because in both cases the triangle has area 10, leads to the same numerical answer. However, it is only at the end of the repeat of the computation of problem 17 that we read: "Then it is clear that triangle BHD of this pentagon which we have described is similar to triangle ZMW of the decagon which we have described before . . . ." This last statement indicates that Abu Kamil was reluctant to simply state at the beginning of problem 18 that the solution reduces to the critical part of problem 17. This reluctance may have been due to the situation in the *Elements* where Euclid does not specifically relate the triangle of the pentagon and the central angle of the decagon; see my discussion in Section 2,A,i. Instead he apparently feels obliged to go through a computation using XIII,3. I also note that once again, as in problem 17, the use of a proportion, in this case via XIII,8—which gives $BH:x = x:(BH-x)$—would lead to a quadratic involving the product of the unknowns. But also once again we cannot be sure if Abu Kamil even tried to set up an equation using proportion theory.

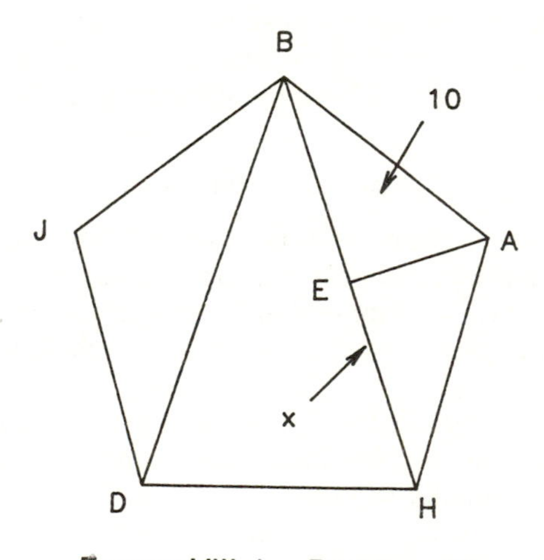

FIGURE VIII-4. Problem 19

In problem 19, it is now triangle ABH (Fig. VIII-4) which has area 10, but instead of the side it is the diagonal $BH(=x)$ that we are required to find. Abu Kamil starts off by citing XIII,1. After stating that $5(BH/2)^2 = 1\ 1/4x^2$ he writes: "It is clear from what we described that line AH is the square root of $1\ 1/4\ x^2 - 1/2\ x$." Note that Abu Kamil has given the formula for dividing a line of length x in EMR, although he does not state this fact explicitly (we will come across this formula on several occasions, for example, in Subsection D in connection with Ibn Yunus, where it is also presented without justification). I suspect that when Abu Kamil says, "what we described," he is referring to the use in problem 18 of XIII,18. This latter result is what is needed, in conjunction with XIII,1 to obtain

$(AH + x/2)^2 = 1\ 1/4x^2$. The conclusion of Abu Kamil's statement follows from this latter equality upon taking square roots. Note once again how Abu Kamil goes back to Euclid instead of using what has already been done, for in the last two problems he has shown that diagonal $= 1/2$ side $+ R(1\ 1/4(\text{side})^2)$. I note as well that in problem 5 Abu Kamil had equated x^2 with $R(x^4)$ (see also his *Algebra* [Abu Kamil–Levey]) so that presumably he would see nothing wrong with pulling the $(\text{side})^2$ term out or the radical and then solving for the side in terms of the diagonal (see the technique in problem 5). Again the technique of squaring both sides would lead to a quadratic equation with the product of unknowns. The problem is finished off in the usual manner—by dropping the perpendicular AE and then finding an expression for the area.

In problem 20, the last one (Fig. VIII-5), we are given a regular decagon and told that the area of one of the 18°–18°–144° triangles formed by three adjacent vertices K, T, H has area 10 and we are to find the side KH of the pentagon.

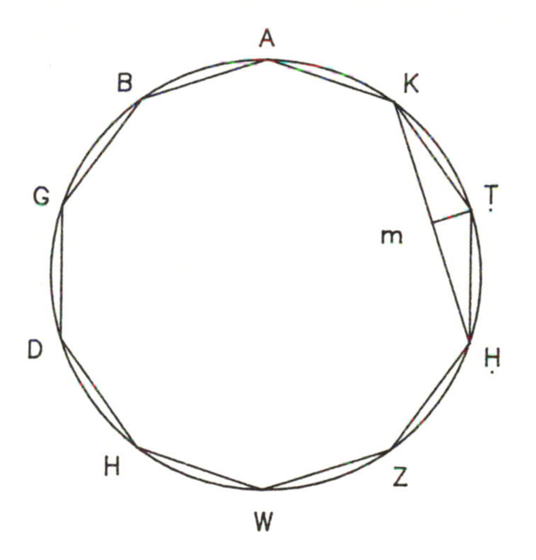

Source: Abu Kamil manuscript

FIGURE VIII-5. Problem 20

In problem 5 Abu Kamil had found the diameter of the circle circumscribed about a pentagon of side 10. Abstracting from the particular case, he announces the general relationship $d^2 = 2(a_5)^2 + R(4/5(a_5)^4)$, from which one can find the radius in terms of the side of the pentagon. Now the author relates the radius and the sides $HK(a_5)$ and $KT(a_{10})$ via XIII,10. The solution then proceeds as usual by finding the altitude TM, etc.

As I have indicated, the text presents us with several problems and we see versatility and a command of Euclid combined with apparent oversight. One difficulty is caused by our lack of knowledge of the sources—aside from Euclid—of Abu Kamil. This, of course, is a general problem in early Arabic mathematics.

One thing to be remarked about the text is that it does not contain, at least in the form of problems, what one might consider the basic calculations related to the pentagon, namely, given the diameter or side, find the diagonal and area. We did see, however, that the problem of finding the diagonal given the side is done as part of problem 5 and that problem 16 contains a calculation of the area in terms of the unknown side. Thus what we have is a more advanced text whose existence suggests that at some point, before Abu Kamil's *On the Pentagon and Decagon*, there was an Arabic work, probably a book on plane mensuration, that contained these basic computations as problems in their own right. Unfortunately, to my knowledge there is no such treatise now existing. If we look back to Hero (Section 25,A,i) we find that, just as in Abu Kamil's problem 16, the calculation of the area is done by subdivision into five triangles and via the Pythagorean Theorem; but aside from this, the basic spirit and approach of the two authors are quite different. It is possible that early Arabic approaches to the calculation of the area of a pentagon are reflected in the work of Fibonacci (see Section 31,B).

It should also be noted that in his preface Abu Kamil talks about "Other subjects . . . which were difficult for the mathematicians of our time." This sentence appears after all the pentagon and decagon topics are mentioned in a general way. Thus it would seem that the difficult problems were 12 to 15. Does this mean that the pentagon and decagon problems were not difficult and perhaps done by others? Or is this just a way of speaking, to which we should not attach too much significance?

ii. Algebra

This work has been translated from a Hebrew version by Finzi (in Abu Kamil–Weinberg and Abu Kamil–Levey; see also Karpinski [1914], Gandz [1936a], and Levey [*DSB*]. As in the work by al-Khwarizmi, we find various problems in which 10 is divided into parts which satisfy certain conditions. Two of these are of interest to us.

Problem 42 [Levey, 150; Weinberg, 95; the problem numbering is Levey's] reads: "If one says that 10 is divided into two parts, each one is divided by the other, then the sum of the quotients equals the root of 5."

Abu Kamil shows how to solve the problem in three different ways. In the first method, one part of the line is called x and the statement leads to the equation:

(5) $$(10-x)/x + x/(10-x) = R(5).$$

This is reduced to a quadratic equation whose solutions are $5 \pm R(225 - R(50000))$. In the second method, the ratio $(10-x)/x$ is first considered as an unknown x'

which satisfies $x'(R(5)-x')=1$. The quantity x' is solved for and this in turn leads to equation (2) for x so that the parts are $R(125)-5$, $15-R(125)$. In the third method, Abu Kamil proceeds by taking $(10-x)/x$ as 1 unit. He ends up with the solution of method 1 and then reduces it to the form of method 2 (see Levey [25, 78]).

Problem 43 [Levey, 154; Weinberg, 99] reads: "If one says that 10 is divided into two parts, each is divided by the other, and the products of each quotients by itself, then when added gives 3 as a sum . . . ."

This statement is equivalent to the equation

$$((10-x)/x)^2 + (x/(10-x))^2 = 3. \tag{6}$$

Abu Kamil remarks that, since the product of the terms which are squared is 1, we will obtain the square of the sum of the two terms if we add 2 to both sides. Thus this problem becomes problem 42 in the first form.

The question once again is whether or not Abu Kamil had DEMR in mind when he composed these problems. First we notice that al-Khwarizmi's problem does not appear in Abu Kamil's work. If we look at problem 43, the number 3 reminds us of XIII,4 which states that $(\text{whole})^2 + (\text{short})^2 = 3(\text{longer})^2$. Algebraically this becomes $10^2 + (10-x)^2 = 3x^2$ or $(10/x)^2 + ((10-x)/x)^2 = 3$. To convert this equation to the equation of problem 43, note that $10/x = x/(10-x)$ because of the definition of DEMR. Despite this possible derivation, the stated solution of the problem suggests that problem 43 was simply obtained from problem 42 by squaring.

Thus, if we turn to problem 42, the $R(5)$ on the right side reminds us of XIII,1,3. However, application of these results does not seem to lead fairly directly to the equation and indeed the simplest way seems to be via XIII,4 in the manner illustrated above and in Abu Kamil's solution to problem 43.

If we consider the numbers $R(125)-5$ and $15-R(125)$ which appear as answers to problems 42 and 43, then we notice that they never appear in *On the Pentagram and Decagon*. The only numbers similar to these are $5 + R(125)$ and $10 + R(500)$ which appear in problems 5 and 9, but which are not there specifically associated with DEMR.

Again, as in the case of al-Khwarizmi, we must remember that problems 42 and 43 are but two of forty-one problems—out of a total of sixty-nine, some degenerate—which deal with the division of 10 into two parts. However, there are several which have irrational solutions. Nevertheless the term "$R(5)$" is certainly intriguing and again, as in the case of al-Khwarizmi, I must leave unanswered the question as to whether or not Abu Kamil had DEMR in mind when he composed problems 42 and 43. Fibonacci (Section 21,B,iv), however, explicitly states the relationship of his version of problem 42 to DEMR.

C. *Abu'l-Wafa' (940-998)*

The *Book of Geometric Constructions* has been summarized by Woepcke [1855] and by Suter [Abu'l-Wafa'—Suter]. The chapter and problem numbers differ in the two manuscripts and I have used those given by Woepcke (see Suter [p. 96] for the correspondences between the two manuscripts).

In Chapter II, problem 3, a pentagon whose side is a given line AB is to be constructed [Woepcke, 328] (see Fig. VIII-6). One takes $BC = AB$ and then with D, the midpoint of AB as the centre, and DC as the radius, the point E on the extension of AB is obtained. Abu'l-Wafa' has constructed line AE which is the extension of AB in the sense that if we divide AE in EMR then we obtain the original line. This is the result that I have called II,11′ in Section 2,C; Abu'l-Wafa' is using the construction of Ptolemy that was discussed in Section 26,A. Next one constructs the triangle AZB with $AZ = ZB = AE$. This gives the "triangle of the pentagon of which one has great need in many constructions." Then about A,Z, and B we swing arcs of radius AB to obtain points H and T, the last two vertices. This construction by Abu'l-Wafa' may be compared with the construction, which also uses II,11′, that I gave in Section 2,C.

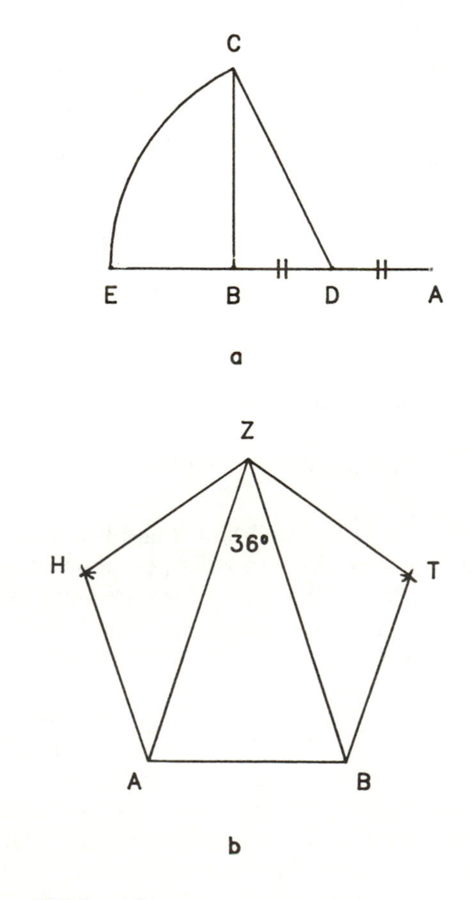

FIGURE VIII-6. Chapter II, Problem 3 (not in text)

In problem 4 [Woepcke, 233; Hallerberg, 1959, fig. 6] one is to perform the same construction as in problem 3, but this time with the compass having a *fixed* opening equal to the given side of the pentagon. The recent discovery of the Arabic translation of a part of Pappus' *Collection* shows that Abu'l-Wafa' was inspired, in his work with the fixed compass, by a Greek text (see Toomer [1984, 37]). For a brief history of construction with a fixed compass opening, see Hallerberg [1959] and Sections 31,A,6; 32,C.

The very ingenious construction of Abu'l-Wafa' can best be understood with reference to Figure VIII-6 and the drawing of XIII,8. In problem 3 the compass could be adjusted which meant that the arc of radius DC in Figure VIII-6 could be swung down to E. In problem 4, point E is obtained by constructing a triangle similar to right triangle DBC with the hypotenuse DC lying along line AB and the leg of length $1/2 \cdot AB$ along CD. Consider the diagram accompanying XIII,8 (Fig. I-29). Line EH is the side of the pentagon, and the extension point that has just been determined corresponds to B. Since the fixed opening of the compass equals the side of the pentagon, an arc can be swung about E and B to find A. If we now extend line AH by the length $HC = a_5$, the new extended line will once again equal the diagonal so that we once more have constructed the triangle of the pentagon.

In problem 10 [Woepcke, 330], the decagon with a given side is constructed. This uses the $72°-72°-36°$ triangle of problem 3 but now Z is the centre of the circle (cf. the comments on XIII,9 in Section 2,A). In problem 11 the same is done with a fixed opening, using problem 4.

In Chapter III, problem 9 [Woepcke, 331; Suter, 102], a pentagon is inscribed in a circle as in Ptolemy (Section 26,A) and then in problem 10 this is done—using problem 4 above—with a fixed opening equal to the radius of the circle. In problem 16 [p. 333; p. 105], a decagon is to be inscribed in a circle. According to Suter's indications this is done constructing the pentagon and halving the arc and also by means of XIV,**. In Chapter IV, problem 4, a circle is to be circumscribed about a pentagon [p. 333, p. 106].

Chapter VI involves inscribing certain polygons in others. In problems 16 (17) [Woepcke, 337], we are to inscribe (circumscribe) an equilateral triangle about a regular pentagon, and in problems 18 and 19 the triangle is replaced by a square. In problem 20 the summit of the pentagon is required to fall on the diagonal of the square. A solution (perhaps approximate or false) is only given for 18.

In Chapter XII the surface of a sphere is divided into spherical triangles. Although the number corresponds to the number of faces of the regular polyhedra (there are also some exact and approximate constructions corresponding to semi-regular polyhedra), Abu'l-Wafa' does not seem to talk specifically about the polyhedra themselves. In problem 9 [Woepcke, 354], the surface of the sphere is to be divided into twenty triangles. DEMR only enters the picture indirectly, for a great circle must be divided into ten equal arcs and these correspond to the sides of the decagon. The construction is completed by using these division points to swing arcs to form the polar triangles. The construction in 10 is, according to Woepcke, similar to that of Euclid's construction of the icosahedron. The division of the surface into twelve pentagons is done in problem 11 by joining the centres of the triangles of problem 9. In problem 12, one constructs the side of the cube (XIII,15) and then divides it in EMR which gives (XIII,17[a],corollary) the edge of the dodecahedron; this is then used for the construction. In problem 13 one joins the centres of the twelve triangles to once again solve problem 9, in the "dual" manner of problem 11.

D. *Ibn Yunus (died c. 1009)*

The tenth chapter of *Al-Zij al-Hakimi al-Kabir* (*The Large Astronomical Handbook Dedicated to the Caliph al-Hakim*) of Ibn Yunus has been translated by Schoy [1923] and contains material of interest.

The first point of interest is that the author does not work with chords (Section 26) but uses the Indian Sine Function (Section 29) which satisfies the relationship (Fig. VIII-7) $\text{Sine}(\theta) = 1/2 \cdot \text{chd}(2\theta)$. The Indian Sine Function, whose value depends on the radius of the circle, should not be confused with the modern trigonometric function "$\sin(\theta)$"; the relationship between the two functions is given by $\text{Sine}(\theta) = r \cdot \sin(\theta)$.

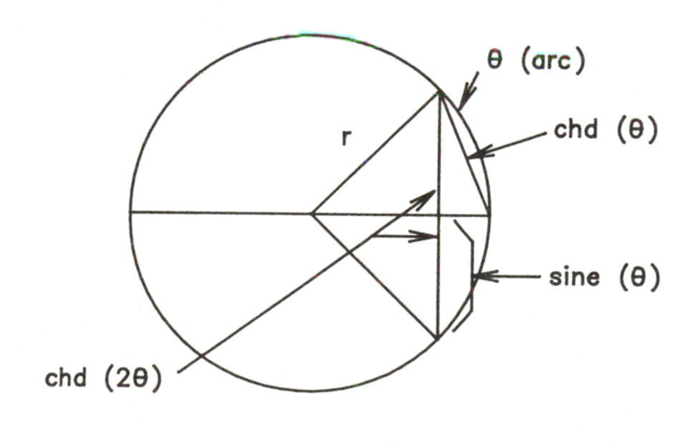

FIGURE VIII-7. Indian Sine Function

Ibn Yunus divides the circumference of the circle that he is working with into 360 "degrees" and then, just as Ptolemy did, he divides the radius of the circle into 60 "parts." Thus $\text{Sine}(30°) = a_6/2 = r/2 = 30$. $\text{Sine}(18°) = a_{10}/2$ and $\text{Sine}(36°) = a_5/2$.

In his text [p. 374] Ibn Yunus first cites XIII,9 and says this carries over so that when Sine(30°) and Sine(18°) are joined together along a line then the new line is also divided in EMR. Then he states that if Sine(30°) is divided in EMR we obtain Sine(18°) as the larger segment. This is simply XIV,** with a factor of 2. To actually divide Sine(30°) in EMR we are told to calculate $R\left((30)^2 + (15)^2\right) - 15$. We recognize this calculation as saying that, to find the larger segment when we divide a line of length x in EMR, we compute $R\left(x^2 + (x/2)^2\right) - x/2$. No reason is given, but recall the discussion of problem 18 and 19 of Abu Kamil. Was this result considered well known? Furthermore, why is XIII,9 itself brought in when it is not used? Is it mentioned as some sort of justification for XIV,** (see problem 17 of Abu Kamil). Looking ahead, I note that the use of the above formula without justification occurs in the works of al-Biruni (Section 28,E,i) and Fibonacci (Section 31,B,ii).

Returning to the calculation of Ibn Yunus we find the value, in sexagesimal, Sine(18°) = 18;32,27,40,15. Now Sine(36°) is obtained by invoking XIII,10 in the form $\text{Sine}^2(18°) + \text{Sine}^2(30°) = \text{Sine}^2(36°)$. This leads to Sine(36°) = 35;16,1,36,52. These values may be compared, after multiplication by 2, with the values obtained by Ptolemy (Section 26,A) for chd(36°) and chd(72°).

E. Al-Biruni (973 to after 1050)

There are two works by al-Biruni which are of interest to us. In the first, *The Book on the Determination of Chords in a Circle*, the principal object is to show that the lengths of the chords are "determined." Almost as an afterthought we are told how to perform the calculation. On the other hand, in *Canon masuidius* we find the main emphasis on the practical computation but with the proofs of *The Book on the Determination of Chords* repeated. For the dating of these works I have followed Kennedy [*DSB*, 151].

i. The Book on the Determination of Chords in a Circle *(completed by 1027)*

This work has been published by Suter [al-Biruni–Suter]; see also articles by Saidan and Saud in Said [1979, 681-90, 691-705]. In the first part of the book al-Biruni discusses two results dealing with "broken chords" in a circle, and he gives many different proofs of these theorems which had been proposed by various authors. It is the second broken chord theorem which is of interest to us because it is used in the second part of al-Biruni's book in connection with the decagon.

The second broken chord theorem (Fig. VIII-8) states that, if A and G are two points on a circle and if D is the midpoint of arc AG, then for any point B on the arc ADG we have $AD^2 = AB \cdot BG + BD^2$. Since I could only consult the German translation, I cannot tell if al-Biruni stated the results in terms of multiplication as Suter does or if he used the area terminology of the Greeks. This in itself might not mean too much; I have been informed by Dr. J. Hogendijk [1983] that Arabic translations of Greek works sometimes used the word "multiplication" where the Greek used "rectangle." Al-Biruni gives nine proofs of this result, including four of his own. Of all the proofs the simplest, which uses Ptolemy's Theorem, is due to al-Biruni himself; see al-Biruni–Suter [29, proof 6].

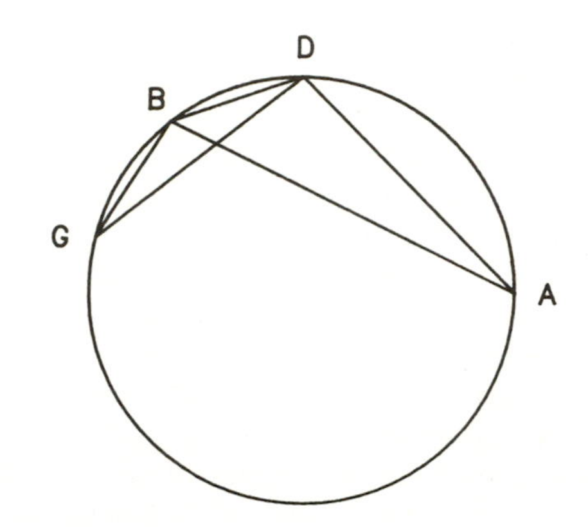

Source: Al-Biruni–Suter [26, fig. 20]

FIGURE VIII-8. The Second Broken Chord Theorem

In section 29 of *Chords* [al-Biruni–Suter, 62], al-Biruni gives what we would presently call an existence proof of the fact that the side of the decagon is "known" (i.e., determined) if the radius of the circle is given. The proof is as follows (Fig. VIII-9):

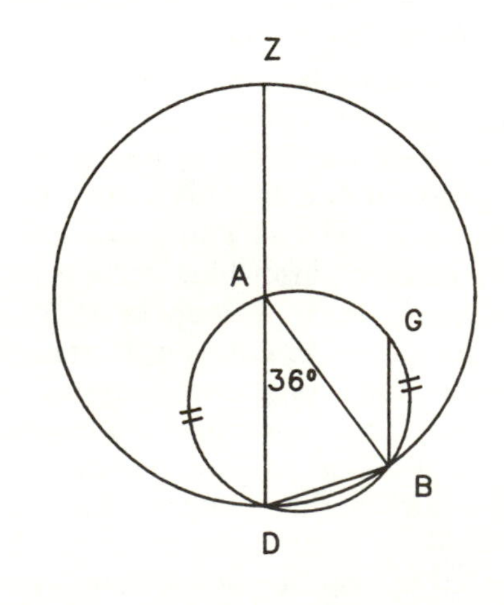

Source: Al-Biruni–Suter [63, fig. 59]

FIGURE VIII-9

In a circle of diameter ZAD let DB be the side of the decagon and consider the $36°-72°-72°$ triangle. We now circumscribe the small circle about the triangle and, in anticipation of the chord theorem, point G is picked so that arc AD = arc DBG. Since arc DB is 1/5 of the small circle, arcs AD and AGB—cut off by equal chords AD and AB—are both 2/5. Together these statements imply that arc DBG is 2/5 and arc BG = arc DB = 1/5 so that chords BG and DB are equal. From the second broken chord theorem $AD^2 = AB \cdot BG + BD^2 = AD \cdot DB + DB^2 = (AD+BD)BD$. Thus, says al-Biruni, "line ADB is proportioned just like a straight line which is divided at point D in e.m.r.; the larger segment is AD and is known because it is the radius of the given circle. The small segment is BD, the side of the decagon, which is thus also known."

After this proof we are told, without being given any details (see the discussion concerning Ibn Yunis in Subsection D), how to perform the actual calculation: take 1/4 of the diameter from the square root of 5/4 of the radius squared. For the pentagon al-Biruni simply invokes the result that, if the chord corresponding to an angle is known, then the chord corresponding to the double angle is also known. No formula is given, however.

Notice that the theoretical statement above concerning DEMR is nothing but a paraphrase of XIII,9. Thus the object of this proof appears to be to reveal that a proof showing that the side of the decagon is determined can be done by means of the broken chord proposition alone. This, together with the great number of proofs given for the two broken chord propositions and the statement concerning the pentagon, further indicates that this treatise is more concerned with virtuosity and perhaps philosophy (cf. discussion of Euclid's *Data* in Section 5,J and Berggren [1985, 23]) than with computation.

ii. Canon Masuidius

This is an astronomical treatise named after the Indian ruler to whom it was dedicated. It is the third part dealing with trigonometry and the determination of chords which is of interest to us; this portion has been translated by Schoy [al-Biruni–Schoy].

Al-Biruni starts out by giving the formulae for the sides of regular polygons [p. 2]. For the side of the pentagon we are given the following (Schoy has given the formulae in modern notation so I do not know what the original format was; I have corrected the term "r/4" given by Schoy to "[r/2]"):

$$a_5 = R\big((R(2r \cdot 2r \cdot 5/16) - [r/2]^2 + r^2\big).$$

No formula is given for the side of the decagon and we are simply told that it appears in the determination of the side of the pentagon.

In effect, if we look at the expression for a_5 we see that the first term is equal to $R(1\ 1/4r^2) - r/2$; it is just the algebraic formula for obtaining a_{10} in terms of r that we saw in our discussion of Ibn Yunus—it is the algebraic formulation of XIV,**. Since $r = a_6$ the formula for a_5 is just the algebraic version of XIII,10. Indeed, when we arrive at the derivation of the formulae, we find that al-Biruni has followed the ideas of the proofs of XIII,9,10; again because of the translation it is difficult to make a precise comparison.

Following the proofs of the formulae al-Biruni gives the same proofs as in *The Book on the Determination of Chords* to show theoretically that these chords are determined. This only increases my suspicion about the philosophical nature of the book.

Section 29. India

DEMR appears in mathematical texts from India, mainly in connection with computation of the trigonometric functions. Recall the discussion of Section 28,D (see Fig. VIII-7) where it was pointed out that the Sine function, which depends on the radius, was used by Indian mathematicians. The different values of R that were used are discussed by Gupta [1978]. The sides of the pentagon and decagon can be expressed in terms of the Sine function as $a_{10} = 2 \cdot \text{Sine}(18°)$ and $a_5 = 2 \cdot \text{Sine}(36°)$.

For the bibliographical references to the original material, the reader is referred to the articles cited. For biographical references, Bose [1971; *DSB*] may be consulted. A general history of geometry in India may be found in Amma [1979]. Although al-Biruni (Section 28,E) did at least some of the work of interest to us in India, I have placed him in the section on Arabic mathematics. His work predates that of the mathematicians mentioned here.

a. The earliest Indian works dealing with trigonometry do not seem to have considered 18° or 36° [Pingree, *DSB*; Gupta, 1976, 2] with the principal angles being multiples of 7 1/2°.

b. Bhaskara II (b. 1115), in his geometrical work *Jyotpatti*, states without proof the following result [Gupta, 1976]:

$$\text{(7)} \qquad \text{Sine}(18°) = \big(R(5r^2) - r\big)/4$$

This is just the relationship Sine(18°) = $a_{10}/2$ stated above.

In Bhaskara's *Lilavati*, we read that to find the side of the pentagon inscribed in a circle we are to multiply the diameter by 70534/12000 (see Bhaskara–Colebrooke [p. 91], Gupta [1975], and Amma [1979, 192]. Again no reasons are given by Bhaskara.

c. Gupta [1976] has published the following "proof" which is due to Laksminatha Misra (c. 1500). I include it here for the reader's pleasure even though it is false:

By definition Sine(90°) = $r = (R(25r^2) - r)/4$. Since 18° can be written as 18° = (1/5) · 90°, we can (assumes the author) find Sine(18°) by replacing the 25 in the formula for Sine(90°) by (1/5) · 25 = 5. This indeed gives precisely formula (7).

d. Following is an incomplete proof of (7) from the *Marici* of Munisvara (1638) [Gupta, 1976, 5] (see Fig. VIII-10). In a circle of radius $r = OX = OY$, let arc YM = 36°. Draw semicircle OX about the midpoint C of OX and draw the arc MD about Y. Assume that the two arcs meet at the single point T on line YC. Then Sine(18°) = $YM/2 = YT/2 = YC/2 - TC/2 = (R(r^2 + (r/2)^2) - r/2)/2$ which is equivalent to (7). Regarding the question of why the arcs just happen to meet at the one point T, the text simply says that the proof is dealt with in detail in a foreign work.

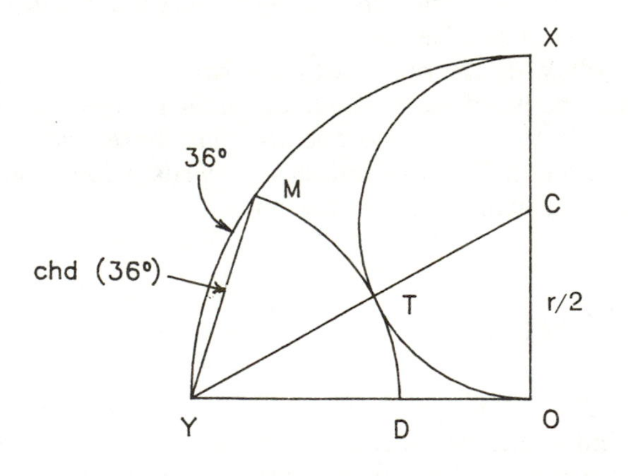

Source: Gupta [1976, fig. 1]

FIGURE VIII-10

The above construction and the missing detail can be explained in the following way. Think of Y and C as given points and draw the arc OTX of radius $r/2$. Draw arc MTD *of radius* YT. The circles are tangent at precisely the point T on the line YTC connecting the centres (III,12).

Now we fix our attention on the triangle YOC and arcs DT and OT to the exclusion of the rest of the construction. What we have is precisely Hero's construction (Section 25,B) for dividing OY in EMR at D. In other words, $YM = YD$ is precisely the greater segment when OY is divided in EMR and by XIV,** (since $OY = r = a_6$) $YM = YD = a_{10}$. I suspect that this is what the "foreign" (perhaps Arabic) source had in mind in the original proof.

e. Another proof of (7) by Munisvara appears in his *Siddhantasarvabhauma* (c. 1645) [Gupta, 1976, 6]. The essence of the proof is the lemma

$$(a_{10})^2 + (r \cdot a_{10}) = r^2. \tag{8}$$

This result, which reminds us of XIII,10, is proved by trigonometric rather than geometric means. The proof depends on the fact that 18° and 72° are complementary angles. Since trigonometric functions of each of these angles can be expressed in terms of functions of the angle 36°, by double angle formulae we can obtain an equation for Sine(18°).

Now (8) can be written in equation form as

$$(2x)^2 + (2x)r = r^2 \tag{9}$$

where x = Sine(18°). If we solve equation (9) we obtain formula (7).

f. The following geometric proof by Jagannatha (b. 1652) is given by Gupta [1976, 7] (see Fig. VIII-11).

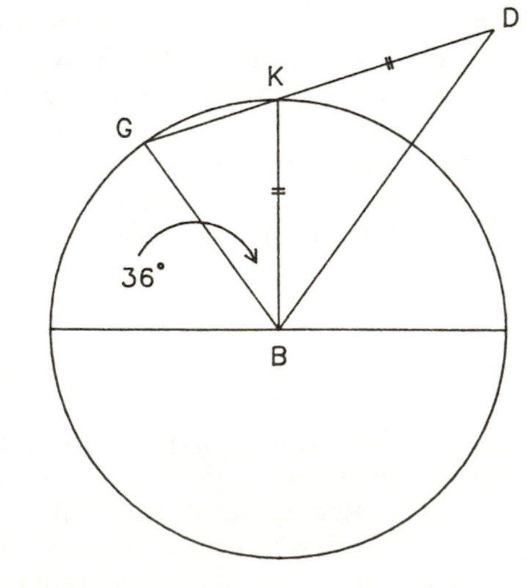

Source: Gupta [1976, 8]

FIGURE VIII-11

Let $KG = y$ be the side of the decagon and extend KG by an amount equal to the radius. Then we are essentially in the set-up of the proof of XIII,9 and thus have that triangles KBG and BGD are similar. Taking the radius to be 60, proportionality gives us $y{:}60 = 60{:}(60+y)$ or $y^2 + 60y = 3600$. The solution of the equation is given to be $y = 37;4,55,20$.

A similar proportion apparently based on Ptolemy's construction is also mentioned by Gupta [p. 9].

g. Pingree [*DSB*] in his study of mathematical astronomy in India speaks of the following three trigonometric tables that contain values for Sine(18°) (details of how the tables were constructed are given in the article):

(i) Mahendra (late 14th century) in his *Yantrarja*, which is based on Islamic sources, takes $r = 3600$ and has [p. 626, table XI,1] Sine(18°) = 1112;28.

(ii) In his *Siddhantasarvabhauma*, Munisvara takes $r = 191$ and has Sine(18°) = 59;1,20,6 and Sine(36°) = 112;16,1,8 [p. 615; table VIII,25].

(iii) Kamalakara in his *Siddhantatattvaviveka* (completed 1658) takes $r = 60$ and has Sine(18°) = 18;32,27,40,15 and Sine(36°) = 35;16,1,36,52 [p. 617, table VIII,29]. Note that these are the same values given by Ibn Yunis (Section 28,C).

Section 30. China

Benfey [1974] has pointed out the existence of two spherical incense burners with a pierced dodecahedral pattern from China which were deposited in the Japanese Imperial Treasure House in the 8th century. The article discusses other later examples and possible sources and influences.

In general there seems to have been relatively little interest in Euclidean type geometry in early China. The Pythagorean Theorem was known, although the date when the result first appeared is disputed, and there were various formulae for the volumes of pyramids and spheres. For details see Mikami [1913], Needham [1951, 103], Struik [1963], Nakayoma and Siven [1973, 96], Swetz and Kao [1977], Wagner [1978, 1979, 1985], and Swetz and Tsian-se [1984].

CHAPTER IX

EUROPE: FROM THE MIDDLE AGES THROUGH THE EIGHTEENTH CENTURY

This final chapter deals with the transmission and the development of DEMR in Europe from the medieval period through the 18th century. I have split the chapter in half with Kepler at the beginning of the century marking the division point.

Section 31. Europe Through the 16th Century

A. *Authors Consulted*

The earliest works examined here represent for the most part the beginnings of the transmission of knowledge from the Arabs. It is not until we reach Fibonacci in the early part of the 13th century that we find anything involving DEMR, and here we find that even the material borrowed from Abu Kamil is presented at a more sophisticated level. Then comes a period of more than 250 years when there is no sign of any development; in fact the texts represent a regression after Fibonacci. This is followed by the quite interesting work of Francesca, Cardano, and Bombelli during the Italian Renaissance, as well as by a few minor contributions by other European mathematicians.

I also discuss two numerical examples and approximations to the pentagon. I have checked all the planar calculations, which can be taken as correct unless otherwise stated. For the three-dimensional problems I have only checked the problems of the form: given D or e_{12} (e_{20}) find V_{12}(V_{20}). For modern day formulae suitable for checking other quantities, see Beard [1973, 53] and Williams [1972, 66, 67].

As was the case with the Arabic period, which in fact overlaps with the period discussed here, all the authors I have consulted are listed. For the authors whose works contain nothing or little of interest a few words in this introduction will suffice. I have split the authors into groups instead of following a strictly chronological order.

i. The Middle Ages

The Middle Ages or medieval period may be considered as extending from the 5th century to sometime in the 15th century. This period overlaps with the Renaissance especially in Italy, which began in the early 15th century. (For a general survey of mathematics in the Middle Ages, see Mahoney [1978] and Youschkevitch [1961]).

There are many medieval handbooks of geometry that are not included in this list either because I was unaware of their availability in critical editions or because they exist only in manuscript form. Some of these are summarized, with respect to their Archimedean content, by Clagett [1978, especially part II, chap. 3]. The ones that I have seen indicate that it is unlikely that

DEMR makes an appearance in any of them; the same is true for the works listed in Evans [1976].

a. "Boethius" (11th century)

In *Geometria II* [Folkerts, 1970], whose relationship to the 5th century Roman writer is not clear, we find the statement of II,11 [p. 128], but of the five diagrams from various manuscripts [p. 237] only one (*m*) is anywhere near being correct, a sign perhaps that this result was not understood. Similarly the diagrams [p. 238] that are associated with IV,11 [p. 131] show no understanding of the statement; I note that IV,10 is missing (for the dating, see p. 105).

b. Abraham bar Ḥiyya (Savasorda) (fl. before 1136)

In "On the Measurement of Figures with More than Four Sides," the fifth part of his *Book of Geometry* [Abraham bar Ḥiyya–Vallicrosa, par. 127, p. 91], there is a general discussion of the determination of the area of plane figures by dividing them into triangles. We find [p. 90, fig. 85] a regular pentagon divided into five triangles and this leads to the result, which also appeared in al-Khwarizmi's work on mensuration (Section 28,A,ii), that the area is the "semi-diameter" multiplied by half of the perimeter (see Fig. IX-1). However, we are never told how to find the side in terms of the diameter. The area of an arbitrary plane figure [p. 91, fig. 86], in particular a pentagon, is to be obtained by dividing it into triangles (see Fig. IX-2). Figure IX-1 also appears in the Latin translation (1145) of the *Book of Geometry* by Plato of Tivoli [Abraham bar Hiyya–Curtze, 119]. I suspect that the pentagon appears here and elsewhere simply because the pentagon comes after the triangle and the square in the ordering of plane figures by the number of sides. (The development of this portion of the text by Fibonacci is discussed in Subsection B,i.)

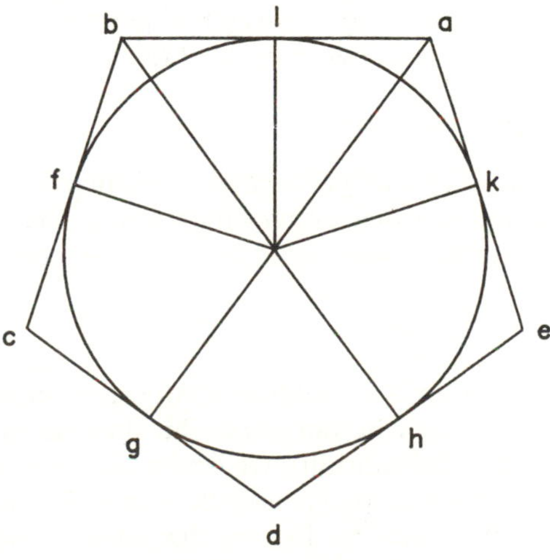

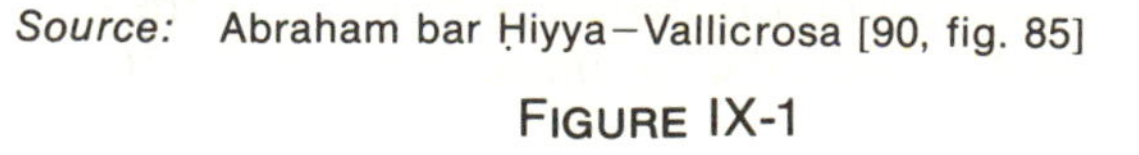

Source: Abraham bar Ḥiyya–Vallicrosa [90, fig. 85]

FIGURE IX-1

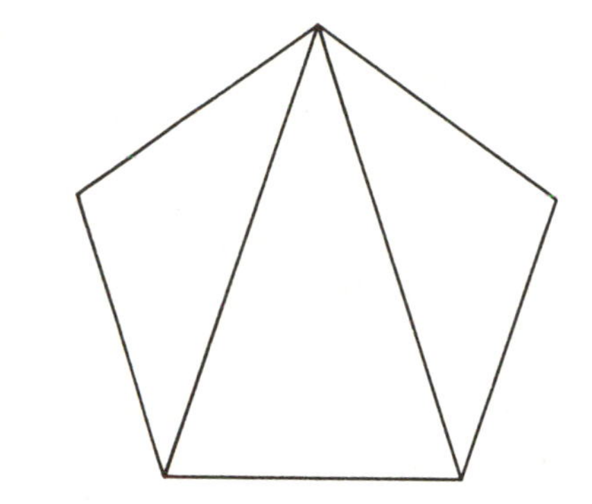

Source: Abraham bar Ḥiyya–Vallicrosa [91, fig. 86]

FIGURE IX-2

For discussions of the passage of Arabic mathematical knowledge into 12th century Spain, and in particular to bar Ḥiyya and Plato of Tivoli, see Haskins [1924, 14] and Glick [1979, 257].

c. *Practical Geometry* (13th century?)

Victor [1979] has edited two medieval texts: *Artis cuiuslibet consummatio* (13th century? [p. 105]) and *Pratike de geometrie* (13th century [p. 101]). In problem I,15 of the former work [p. 159], one finds the area of an equilateral pentagon. However, the pentagon is not regular, rather it consists of an equilateral triangle built upon a square which makes for an easier solution, compared to a regular pentagon. In the second work, it appears as if the author is dealing with a regular pentagon but we are simply told to divide it into five triangles (see the footnote in the text and the method of Abraham bar Hiyya above).

d. Jordanus de Nemore (fl. c. 1220)

From the discussion of his work in Grant [*DSB*] and Mahoney [1978], it appears that this author did not leave any work of interest to us.

e. Bradwardine (c. 1295-1349)

From the description in Molland [1978], Bradwardine's *Geometria speculativa* is more philosophical than technical and contains nothing of interest.

f. Johannes de Muris (fl. 1317-1345)

Parts of different manuscript versions of *De arte mensurandi* have been discussed by Clagett [1978, part I, chaps. 3,4; part IV, Appendix II]. I do not know if this work—which seems from what Clagett has published to be at a higher level than other handbooks—contains anything of interest (see also Poulle [*DSB*].

g. Dominicus de Clavasio (fl. mid-14th century)

The *Pratica geometriae* of Dominicus de Clavasio was written in Paris in 1362 and a Latin version has been published by Busard [Dominicus–Busard]. Construction 14 of Book II [p. 563] reads: "To find the area of any polygonal, equilateral and equiangular figure. Triangles and squares were discussed above. Now we must discuss the others, namely, the pentagon, hexagon, heptagon, octagon, monagon, decagon etc. Wherein it should be known that the pentagon is said to be a plane figure contained by 5 sides, and the hexagon by 6, the heptagon by 7, the octagon by 8, etc."

Following this is a discussion of how one finds the area of a regular hexagon by dividing it into six equilateral triangles. Then we read: "And if the figure were a pentagon, it is divided into 5 similar triangles, and if a heptagon . . . ."

In construction 16 [p. 565] we are told that to find the area of any polygonal figure, in particular a non-equilateral pentagon, we divide it into triangles.

ii. Versions of the Elements *and Scholia*

I have not made a detailed study of the variations and additions that appear as part of the various editions; those I read consisted of minor variations. The scholia of the critical edition [Euclid–Heiberg, V] are without interest except for the one discussed in Section 14. For a scholium proving XIV,**, which appears in manuscripts of Gerard of Cremona's version of the *Elements*, see Busard [1974] and Herz-Fischler [1985, Section 4,F].

a. Hermann of Carinthia (?) (fl. 1140)

The first six books of a 13th-century manuscript presumably authored by Hermann of Carinthia have been published by Busard [Euclid–Hermann]. The definition of DEMR is missing from the beginning of Book VI [p. 113]. In the diagram of IV,11 [p. 92], we find the pentagram completed, unlike in the critical edition.

b. Campanus of Novara (d. 1296)

In connection with IV,10, Campanus has diagrams covering various possible cases [Euclid–Campanus]. For the proofs that he gave of XIV,**, see Herz-Fischler [1985, Section 4,E]. See Appendix II,1 for his admiration of DEMR.

I have seen several indications in the literature, unfortunately without precise references, that Campanus proved the incommensurability of the segments of a line divided in EMR (see Itard [1961, 38] and Smith [1973, I, 218]). I have checked the obvious locations for such statements and unfortunately have been unable to find the commentary. See Murdoch [1968] for a discussion of the general nature of Campanus' comments.

iii. Italy from Fibonacci through the Renaissance

For comparison purposes I have grouped the various Italian mathematicians together, but this grouping is not meant to imply any lack of contact with the rest of Europe. (On mathematicians in Renaissance Italy, see Rose [1975].) Most striking is the great difference in level between Fibonacci and his successors, a situation that persists until the work of Francesca more than 250 years later. Fibonacci's influence is discussed by Vogel [*DSB*, 612]. Additional references which may be of interest are given in Francesca–Arrighi [p. 13].

Since not all the dates of the works are known, I have placed the dated works first, works whose dates are not known second, and then works which I was unable to consult.

a. Fibonacci (ms. 1220)

Discussed in Subsection B.

b. Gherardi (ms. 1327)

The table of contents of Gherardi's treatise [Gherardi–Arrighi, 66] indicates that it has a few sections on the "rules of measurement," but since only a few folios are devoted to them it is unlikely that the development is advanced.

c. Giorgio Martini (ms. 1436)

La practicha di gieometria [Giorgio–Arrighi] contains some applied geometry and surveying but nothing of interest to us.

d. Gherardo (ms. 1442)

This version of Fibonacci's *Practica geometriae* [Gherardo–Arrighi] has a whole series of problems and figures [p. 88] treating non-regular pentagons but no theoretical material involving DEMR.

e. Mainardi (ms. 1448)

Mainardi's *Practica geometriae* [Mainardi–Curtze, 378] contains various generalities about the area of polygons and in particular of an irregular pentagon.

f. Alberti (ms. 1485)

If we look at Leoni's 1726 edition [Alberti–Leoni, plate XIII; it is not clear whether or not the illustration comes from Bartoli's 1568 edition], we find a construction of a regular decagon inscribed in a circle where the side is obtained from the radius by dividing the latter in EMR. The method used is essentially that of Hero but with the last arc swung out to meet the circumscribing circle. However, I was unable to find any reference to polygons in the text and a check of the illustrations in the critical edition [Alberti–Orlandi, vol. II] showed that this illustration was not in the original. On what seems to be known about Alberti's mathematical level, see Gille [*DSB*, 96].

g. *Codice ottoboniano latino 3307* (c. 1400-1460?)

Despite the promising title, this Florentine manuscript [Arrighi, 1968] turns out to deal only with arithmetic and algebra.

h. *Renaissance Mathematics in Siena*

In Gazzaia (1433)–Nanni [1982], the area of a pentagon of side 8 is calculated, using the formula $A = 3a^2 - a$, to be 184, whereas the correct value is 110.1; see Section 25,A,iv. See also Franci–Rigatelli [1981*].

i. Francesca (ms. 1470s?)

Discussed in Subsection C.

j. Da Vinci (c. 1500)

See vii,b below.

k. Paccioli (printed edition, 1509)

Discussed in Subsection D.

l. Cardano (printed edition, 1545)

Discussed in Subsection E.

m. Bombelli (printed edition, 1572)

Discussed in Subsection F.

iv. 16th Century Non-Italian Authors

a. Candalla (1566)

Discussed in Subsection G.

b. Ramus (printed edition, 1599)

Discussed in Subsection H.

c. Stevin (1585)

Discussed in Subsection I.

d. Magirus (c. 1597)

Discussed in Section 32,A,i.

v. Pre-1600 Numerical Approximations to DEMR

a. Unknown Annotator to Paccioli's *Euclid* (16th Century)

Discussed in Section J,i.

b. Holtzmann (1562)

Discussed in Section J,ii.

c. Mästlin (1597)

Discussed in Section J,iii.

vi. Fixed Compass and Straight-Edge Constructions

In Section 28,C we saw that Abu'l-Wafa' gave several constructions involving a fixed opening of a compass and a straight edge. In his history of fixed compass constructions, Hallerberg [1959, 237] briefly discusses the work of several Italian mathematicians in this domain. In particular, Ferrari gave a treatment of the first six books of the *Elements* and Tartaglia gave solutions for most of the construction problems in Euclid. I do not know if the treatment of DEMR and the related construction problems involve anything new as far as DEMR is concerned. Ramus (Subsection H) was familiar with the fixed compass constructions of Benedetti (see Hallerberg [1960, 129], and it may be that his construction of the pentagon, which I will present, is based on the work of Benedetti. The remarks by Ramus were in turn the inspiration for the work of Mohr (Section 32,C).

vii. Approximate Constructions of the Pentagon (Subsection K)

(a) *Geometrica Deutsch* (1484).
(b) Leonardo da Vinci (c. 1500).
(c) Dürer (1525).
(d) Benedetti (1585).

B. Fibonacci (c. 1170 to after 1240)

For general surveys of Fibonacci's work, one may consult Vogel [*DSB* 3] and Youschkevitch [1961, 371]. The Arabic sources of his works are discussed by Bortolotti [1929]. All references are to Fibonacci–Boncompagni.

i. Planar Calculations

In chapter III, part 3, "On the Dimensions of Planar Figures Having More than Four Sides," of his *Practica geometriae*, Fibonacci shows how to calculate the area of polygons, in particular the regular pentagon, by dividing the surface into right triangles. This part of the text is based in part on the *Book of Geometry* by Abraham bar Ḥiyya (see Section A,i,b and Abraham bar

Ḥiyya–Curtze [p. 5, and fn. 119ff.]). There is no mention of DEMR, but we are told [vol. II, p. 85] that if the diameter of a circle is 8 then the side of the inscribed pentagon will be $R(40-R(320))$ and the diagonal will be $R(40+R(320))$ "as will be demonstrated in its proper place." (The calculations are presented in part 4 [p. 105].)

Ptolemy's name is not mentioned—although the *Almagest* is referred to but it is his construction and proof for the side of the pentagon that we are given (see Section 26,A). The terminology is "algebraic" (cf. the discussion in subsection iii of "Geometric Subtleties" where this construction is repeated but with the diagram upside down). The calculations are done for $d = 8$. Approximations are made, and Fibonacci obtains $(a_5)^2 = 22\ 1/9$ which gives $a_5 = 4 + 4$ pedes $+ 3$ uncie $+ 17$ puncta. We learn [p. 23] that there are 6 pedes (feet) in 1 unit and 18 uncie (inches) in 1 pedes. I could not find that part of the text which states how many puncta (points) there are in 1 uncie, but in the table [p. 96] the maximum value in the puncta column is 17, which suggests that the uncie was divided into 18 puncta. This would give $a_5 \doteq 4 + 4/6 + 3/6 \times 18 + 17/6 \times 18 \times 18 = 4.7032$, as opposed to the exact value of 4.7023. To find a_5 from a circle of diameter d we are told to calculate $(d/8) \cdot (4 + 4$ pedes . . .).

Next Fibonacci shows in two ways how to calculate the diagonal *be* (Fig. IX-1) of the pentagon. Note that XIII,8 is not used; indeed DEMR is not mentioned or used, except of course in Ptolemy's construction, in this section.

In the first method we draw the diameter bz and the side (a_{10}) of the decagon ze thus forming the right triangle bez. Since $bz = 8$ and $(a_{10})^2$ was calculated in the course of Ptolemy's method as $24 - R(320)$ we have $(be)^2 = 40 + R(320)$.

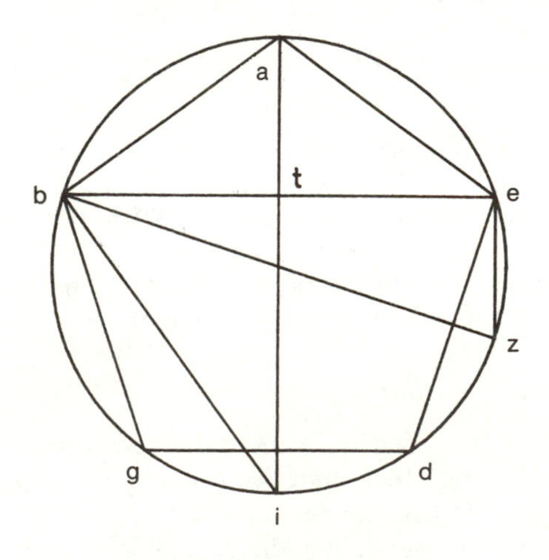

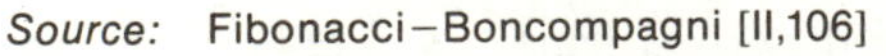
Source: Fibonacci–Boncompagni [II,106]

FIGURE IX-3

In the second method we consider the right triangle abi and the altitude bt. Then $ia{:}ab = ab{:}at$ (VI,8). Since $ab^2 = a_5^2 = 40 + R(320)$ and $ia = 8$ we obtain $at = 5 - R(5)$ and then $ti = ai - at = 3 + R(5)$. By III,35 $(be)^2 = 4(bt)^2 = 4(ta)(ti) = 40 + R(320)$. To find the area of the pentagon, Fibonacci uses the preliminary result to the second proof of XIV,6 (see Section 24,A) which states in numerical terms that $A_5 = (3/4d) \cdot (5/6\ d_5)$. The area of the decagon is said to be given by the formula $5/4 \cdot d \cdot a_5$ (reason?). Approximations are given for these areas as well as for other figures inscribed in the circle.

ii. *Volume Computations of the Dodecahedron and Icosahedron*

Although Hero presented the numerical answers to volume computations for the dodecahedron and icosahedron (see Section 25,C), he gave no indication of the details of the computation. Nor was there any evidence of any subsequent numerical computations in the later Greek and Arabic periods, although Pappus compares the volumes of the dodecahedron and icosahedron (see Section 27). In the *Practica geometriae* of Fibonacci, however, we shall find several approaches to the computations which show a complete mastery of Books XIII and XIV (on a manuscript source for Book XIV, see Herz-Fischler [1985, Section 5,C]).

In Chapter 6, "On the Measurement of Bodies," of the *Practica geometriae* [II, 195], Fibonacci presents the problem of finding the volume (V_{12}) of the dodecahedron inscribed in a sphere whose diameter is 6. I note that this problem precedes that of finding the volume of the icosahedron, whereas in the critical edition the construction of the icosahedron (XIII,16) precedes that of the dodecahedron (XIII,17). The same was true of Hero and we will find the same order in all the works of the Italian Renaissance that I shall discuss. However, in the listing of the various results of Book XIII [p. 162], the construction of the icosahedron is listed before that of the dodecahedron. The construction of the edges of the five solids (XIII,18), however, seems to be associated with Book XIV [p. 162].

The diagram associated with the problem is of independent interest, for Fibonacci is actually combining three diagrams in one and mixing two- and three-dimensional representations. The pentagon *abcde* represents one of the faces of dodecahedron inscribed in a circle with centre at f and diameter ag. The point h represents the centre of the sphere and the lines connecting h with the five vertices of the pentagon represent the edges of one of the twelve pyramids into which the dodecahedron can be decomposed. The right triangle *akil* has been constructed by taking ak as one-third of ai and then drawing the altitude lk. (This is just the diagram of XIII,15 reversed.)

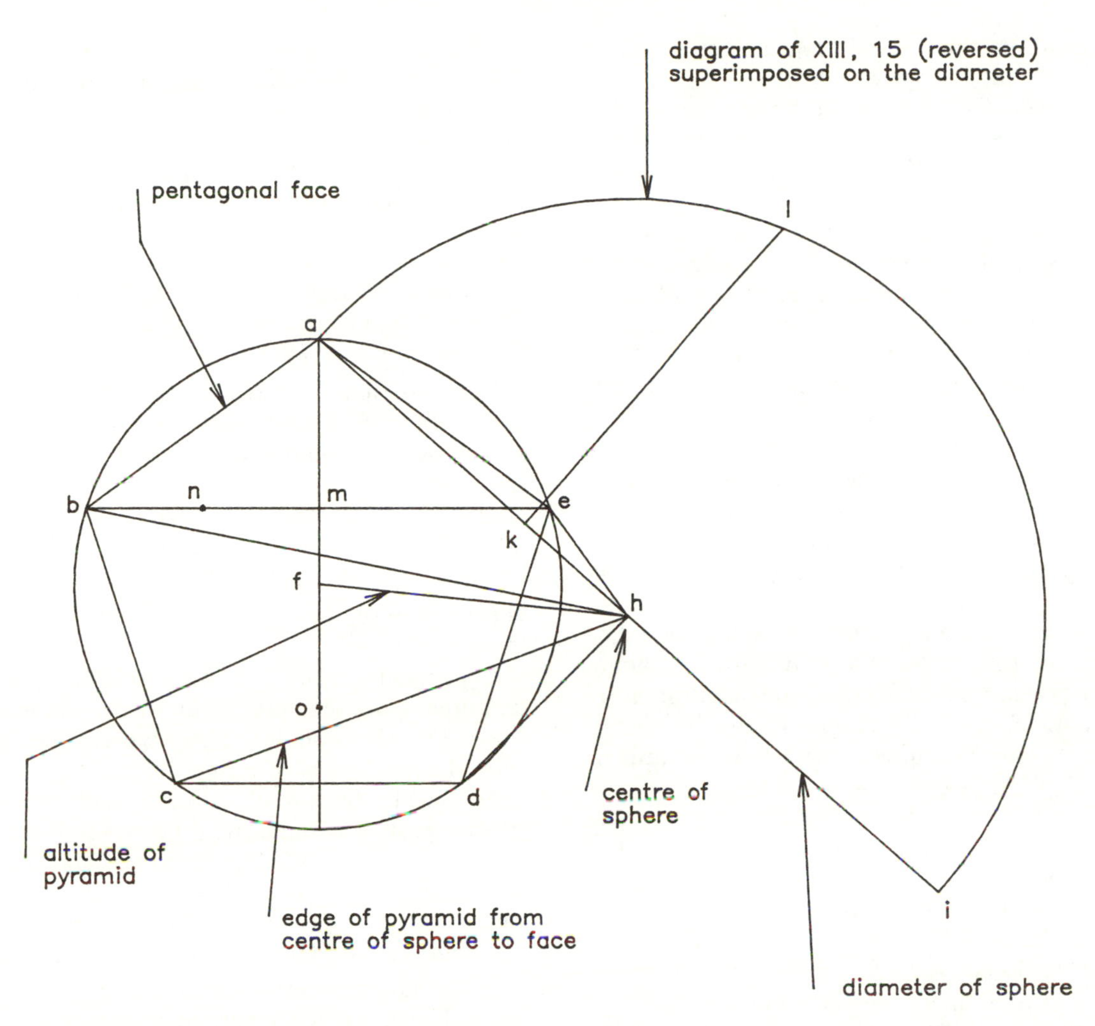

Source: Fibonacci–Boncompagni [II,196]

FIGURE IX-4. Dodecahedron

After stating that V_{12} will be twelve times the volume of one of the pyramids, Fibonacci uses similar triangles to obtain $ia:al = al:ak$ and then invokes—without reference—XIII,13′(i) in the proportion form of XIII,18 to obtain $ai^2:al^2 = ia:ak = 3$, from which $al^2 = 12$ (presumably the possibility of simply putting $ia = 12$ and $ak = 2$ in the first proportion was unacceptable or unnoticed). Invoking Euclid XIII,15,8,17[a],corollary, Fibonacci states that al is the side of the cube inscribed in the sphere of diameter ai and thus also the diagonal be of the pentagon and that if we now divide be in EMR we will have the side ae of the pentagon.

Now there is a new heading in the text, "The Method of Dividing a Line in Extreme and Mean Proportion." Despite this, we are not given any theoretical details. We are simply told to add one-quarter of be^2 to be^2, then take the root and from this subtract one-half of be which will give us $R(15) - R(13)$ for this particular numerical case.

The question is how did Fibonacci obtain this result or justify it for himself. We have already seen in Subsection i how he obtained, without using DEMR, the diagonal of the pentagon given the side, but in fact the problem here is to find the side given the diagonal by means of DEMR. However, in problem 19 of the next chapter of the *Practica geometriae* (subsection iii), we find that in Fibonacci's rendition of Abu Kamil's work he explicitly ties in DEMR—via the statement of XIII,8—when obtaining the above formula (there it is developed from XIII,1 and not proportion theory). We also find an explicit statement of how to divide 10 into EMR (via II,11 or VI,def.3) in the *Liber abaci* (see Subsection iv).

Next he invokes the lemma to XIV,2 (Section 24,A) which states that, with respect to a pentagon inscribed in any circle, $S(d_5) + S(a_5) = 5S(a_6)$ which gives $(a_6)^2 = 6 - R(7\ 1/5)$. Following this, an application of the Pythagorean Theorem applied to the triangle ahf (hf

being the altitude of the pyramid and ha the radius of the sphere) gives $(hf)^2 = 3 + R(7\ 1/5)$.

It now remains to find the area of the pentagon and to compute it, as in Subsection i, using the preliminary result to Proof 2 of XIV,6 that states $A_5 = (3/4d) \cdot (5/6d_5)$.

In the actual computation it is $(V_{12})^2$ which is first computed; this involves a series of manipulations and combining of terms and leads to $3240 + R(5832000)$ for the square of the volume. Then invoking Euclid X as a theoretical support, Fibonacci gives V_{12} as $R(2700) + R(540)$ and this in turn is approximated from above by 71 1/5 (true value, 75.19994 . . .).

Turning to the problem of finding the volume (V_{20}) of the icosahedron inscribed in the same sphere of diameter 6, Fibonacci once more uses Book XIV, this time in the form of the key result—namely, Theorem 2—in which the circle circumscribing the triangle of the icosahedron is one and the same as the circle circumscribing the pentagon of the dodecahedron. Since the radius of the circle, and thus also the altitude of the pyramid, has already been computed for the dodecahedron, it remains to find the area of the triangle. To find the side of the triangle, we have recourse to XIII,12 which tells us that $S(a_3) = 3 \cdot S(r)$, giving $a_3 = 18 - 3R(7\ 1/5)$.

At this point we might expect to see the Pythagorean Theorem used to find the altitude of the triangle. But rather than doing this, Fibonacci uses some similar triangle arguments to obtain the altitude $ae = 3/4\ ad$ and $ae{:}af$ is "sexqualtera" (i.e., 3:2; the same relationship was shown to hold between ao and af in the dodecahedron problem). Then follows a series of computations which amount to computing V_{20} as 20 (volume of the pyramid from the centre g of the sphere) and in turn the volume of the pyramid as 1/3 · base · altitude. The final result for $(V_{20})^2$ is $3240 + R(5832000)$. To obtain a numerical value Leonardo approximates $R(5832000)$ from above by 1449 (true value, 1448.92) and so $(V_{20})^2$ is less than $3240 + 1449 = 4689$. V_{20} is the square root of this latter which is approximated from above by 68 1/2 (true value, 68.48).

In the above calculation the radius of the circumscribing circle of the triangle of the icosahedron was obtained from XIV,2 as the radius of the circle of the pentagon in the case of the dodecahedron, but now in a new set of calculations [p. 200] we are shown how we could compute this radius directly.

In the diagram (Fig. IX-5) ah is the diameter of the sphere and g is the centre. Point k is picked so that $kh = 4 \cdot ak$ and then altitude ik and right triangle aih are drawn (this is just the diagram of XIII,16[a],corollary(i) reversed). Invoking—without references—XIII,13′(i), Fibonacci states $(ah)^2{:}(ai)^2 = ah{:}ak = 5$, and since $(ah)^2 = 36$, we obtain $(ai)^2 = 7\ 1/5$. Using Euclid XIII,16[a],corollary(i), he obtains that ai is the radius of the defining pentagon of the icosahedron. Now to find the side of the defining pentagon, and thus the side of the triangle of the icosahedron, Fibonacci uses XIII,10, that is, $S(a_{10}) + S(a_6) = S(a_5)$. To find a_{10} Fibonacci uses XIV,** (Section 24) which states that a_{10} is the radius divided in EMR. This latter calculation is in turn done by evoking, again without commentary, the same formula that we saw in connection with the dodecahedron giving $a_{10} = 3 - R(1\ 4/5)$ and $a_3 = 18 - R(64\ 4/5)$. The use of XIII,12 now shows once again that $(af)^2 = 6 - R(7\ 1/5)$.

Not satisfied with these computations, Fibonacci continues [p. 201] to display his mastery of Book XIV and numerical manipulation. First he uses XIV,8 in the form $d_5{:}a_3 = e_6{:}e_{20} = V_{12}{:}V_{20}$. Earlier we saw that the diagonal $= R(12)$ and that $(a_3)^2 = 18 - R(64\ 4/5)$, and now this latter is approximated by $10 - 1/20$. Then working with squares and using the proportion and the earlier approximation 75 1/5 for V_{12} he obtains 4689 as an upper bound for $(V_{20})^2$.

Once again [p. 201] Fibonacci relates V_{12} and V_{20} but this time using the ratio given in the Summary (iv) of Book XIV (Section 24). This involves the ratio of the sum of the squares of a line divided in EMR and the larger part to the sum of the squares of the line and the smaller part. For theoretical purposes Fibonacci considers a line ab divided in EMR at g with gb being the larger part. At point g the vertical line dg, equal in length to ab, is erected. Then $bd^2 = dg^2 + gb^2$ and $dg^2 + ga^2$ are the terms whose ratio is desired.

For computational purposes, the length ab is taken to be 10. This number appears in connection with various algebraic and geometric problems involving DEMR (see Subsection iii).

To find the larger segment when we divide ab in EMR, we are told to take $R(5/4\ ab^2) - ab/2$ (the text actually says "mean"; there are several corruptions in this part of the text) instead of the "add 1/4" formula that we saw earlier and will see in later writers. To find the smaller segment one would expect the text either to say nothing or to simply subtract the larger segment from the line. However the text actually tells us to calculate $3/2\ ab - R(5/4\ ab^2)$, which of course is equivalent. One obtains $db^2 = 250 - R(12500)$ and $ad^2 = 450 - R(112500)$. These are approximated (from above) by 138 1/5 and 114 3/5 so that $(V_{12})^2{:}(V_{20})^2 = 138\ 1/5{:}114\ 3/5$ (= 1.2059; the true answer is 1.2060).

The section closes [p. 202] with the case where the diameter is 7. We are told that the volume varies with the cube of the diameter, so that using the approximate value 75 1/5 for the dodecahedron in the case when the diameter is 6 we multiply by 7^3 and divide by 6^3 to obtain 119 5/12. Fibonacci also states that the same sort of thing can be done for the icosahedron and other solids.

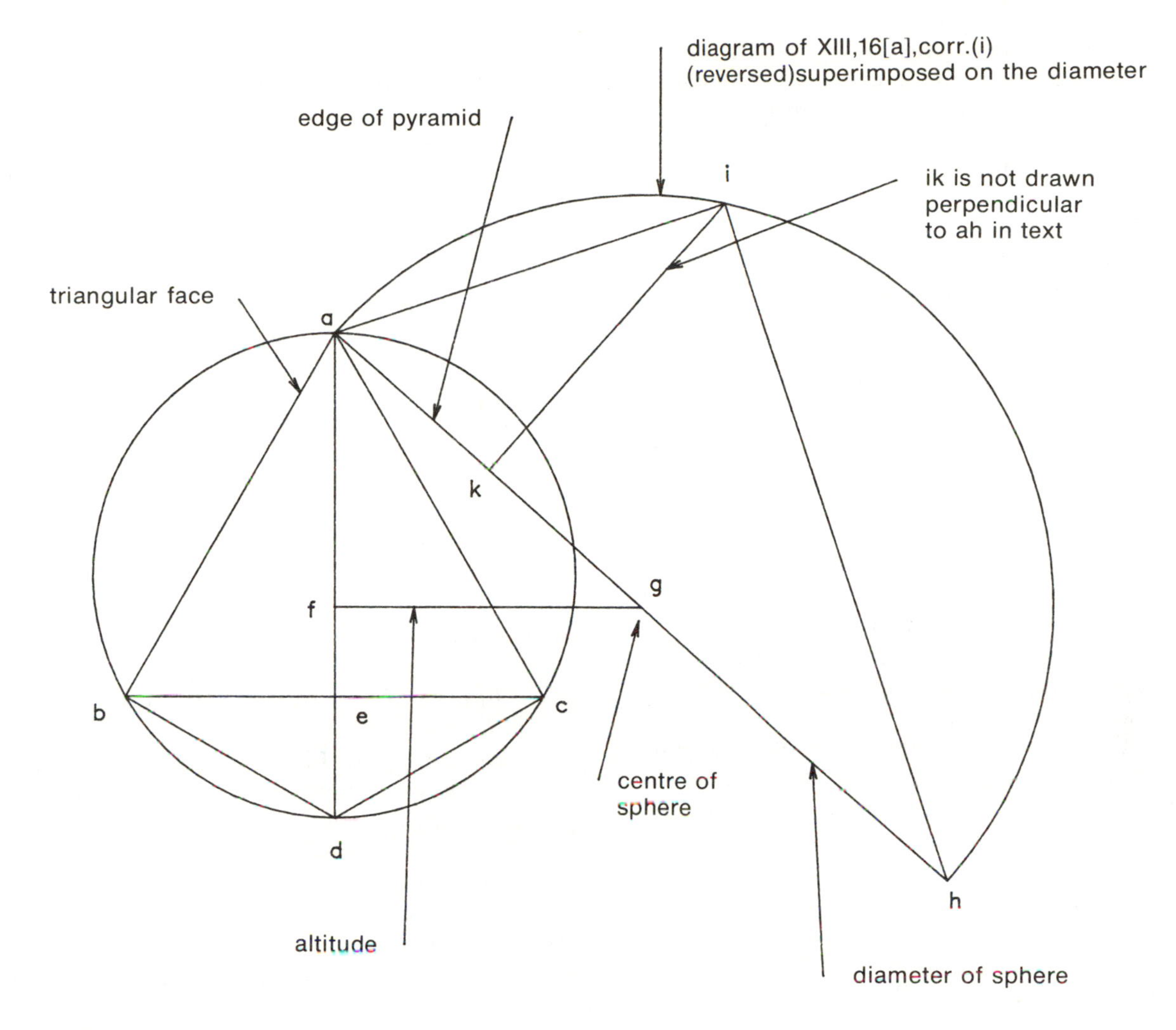

Source: Fibonacci–Boncompagni [II,200]

FIGURE IX-5. Icosahedron

iii. Fibonacci and Abu Kamil

In the eighth chapter of his *Practice of Geometry* [Fibonacci–Boncompagni, II, 207], Fibonacci has taken most of the problems from Abu Kamil's *On the Pentagon and Decagon*, referring to them as "certain geometrical subtleties (Section 28,B,i). Note that this places these two-dimensional problems after Fibonacci's treatment of the three-dimensional problems of his Chapter 6. While much of Abu Kamil's material is more or less taken over as is, we also find Fibonacci presenting other approaches. In particular we shall find that he sometimes directly applies results involving DEMR in problems where Abu Kamil did not. The relationship between the work of Fibonacci and Abu Kamil was first pointed out by Sacerdote [Abu Kamil–Sacerdote, 175]. A further discussion was given by Suter [Abu Kamil–Suter, 38].

It should be noted that Fibonacci decided to rearrange problems 5-10 in the same order as 1-4 (i.e., inscribed a_5, a_{10} then circumscribed a_5, a_{10}) and to group the solutions of 5 and 7 which are really the same problem (problem 8 which corresponds to 6 is not carried over from Abu Kamil). Thus we find Abu Kamil's problems carried over in the order 1,2,3,4,5,7,9,6,10. For ease of comparison, I have used the same numbers and order as in the case of Abu Kamil. Only those problems involving DEMR will be discussed.

Problem 2 [p. 209] involves finding the side of the decagon inscribed in a circle whose diameter is 10. The first solution is essentially that of Abu Kamil. But when we look at the second solution we find that just as in his section dealing with planar calculations (Subsection i), Fibonacci has followed Ptolemy's construction proof and method of performing the calculations but with the diagram upside down. He computes a_{10} as $R(31\ 1/4) - 2\ 1/2$ and $(a_5)^2$ as $62\ 1/2 - R(781\ 1/4)$ and points out that this latter answer was already obtained in problem 1 [p. 207] which involved precisely the finding of a_5 given the diameter. However, there Fibonacci had followed Abu Kamil's method that I explained in Section 28,B,i. In his version of problem 1 Fibonacci obtains the approximation 5 7/8. Again as in the case of Abu Kamil I point out that the use of XIII,9 and VI,def.3 would have led directly to the simple equation $(5+x)(x) = 25$.

The solutions to problems 5 and 7 of Abu Kamil, namely, how to find the diameter of the circle if the side of the inscribed pentagon is given to be 10 (see Section 28,B,i for a description of problem 5), are essentially taken over by Fibonacci [p. 210]. Then for a third method of solution [p. 211, line 12], he invokes the lemma to XIV,2, that is, $S(d_5) + S(a_5) = 5 \cdot S(r)$. We are given that $a_5 = 10$ and in the first solution d_5 was shown—via Ptolemy's theorem, not via XIII,8—to be equal to $5 + R(125)$. Thus r^2 and then d^2 can be calculated, and so d can be calculated as $R(200 + R(8000))$ (see also problem 9 below).

After presenting Abu Kamil's solution to problem 9 [p. 211]—finding the diameter of a circle whose inscribed decagon has side 10—Fibonacci displays his prowess in two new solutions.

We let the diameter of the circle be $2x$ (Fig. IX-6) and draw the chords ac and $df\ (= a_5)$ and the diagonals ad and cf. By Ptolemy's theorem, we have $af \cdot cd + ac \cdot fd = ad \cdot cf = (cf)^2$. In this last formula $fa \cdot cd = (2x)(10)$ and by XIII,10 $ac \cdot fd = (a_5)^2 = (a_{10})^2 + (r)^2 = 10^2 + x^2$. But if we look at right triangle acf, we have $cf^2 = (af)^2 - (ac)^2 = 4x^2 - (a_5)^2 = 4x^2 - (10^2 + x^2)$. This leads to the equation $2x^2 + 20x + 200 = 4x^2$, and solving it we obtain $r = x = 5 + R(125)$ or $d = 10 + R(500)$.

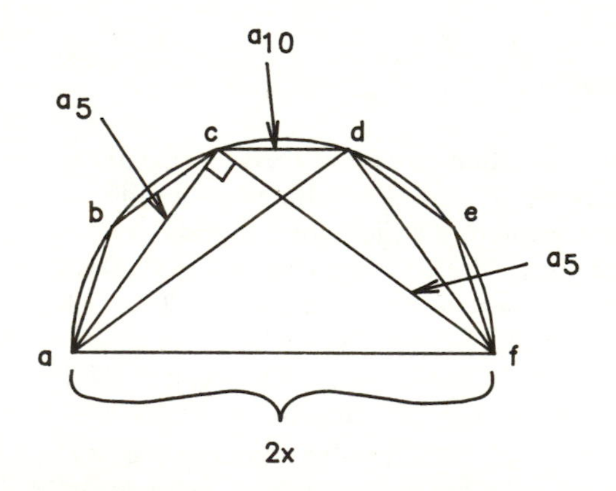

Source: Fibonacci–Boncompagni [II,211]

FIGURE IX-6. Problem 9, Solution 1

The second new solution (Fig. IX-7) involves Ptolemy's construction that Fibonacci had already used in his second solution of problem 2. If we take $r = gh = x$, then $(ik)^2 = (gi)^2 = (gh)^2 + (hi)^2 = 1\ 1/4x^2$. Therefore $10 = a_{10} = hk = ki - hi = R(1\ 1/4x^2) - 1/2x$. If we transfer the $(1/2)x$ and square both sides, we obtain $100 + 10x = x^2$. Note that this equation could have been obtained from XIII,9 and VI,def.3. Solving this gives $d = 2x = 10 + R(500)$ and Fibonacci then gives the general formula $d = a_{10} + R(5(a_{10})^2)$ as the relationship between the side of any decagon and the circumscribing circle which in turn is illustrated for the case $a_{10} = 6$.

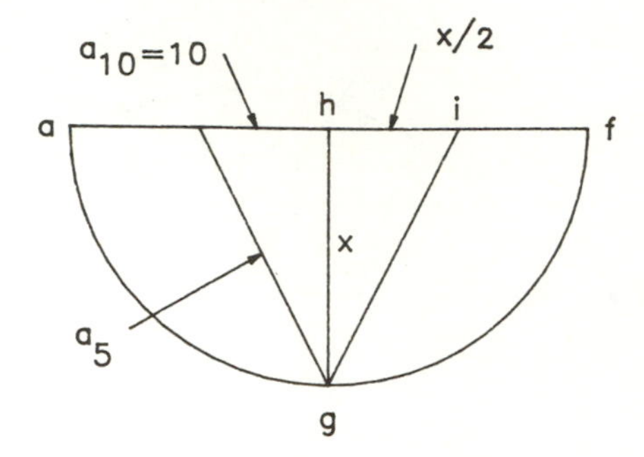

Source: Fibonacci–Boncompagni [II,212]

FIGURE IX-7. Problem 9, Solution 2

Instead of stopping, Fibonacci continues on and uses the same figure to show how to compute d in terms of $a_5(=10)$, as in problems 5 and 7 above. One has $100 = (a_5)^2 = (gk)^2 = (gh)^2 + (hk)^2 = r^2 + (a_{10})^2 = x^2 + (R(1\ 1/4x^2) - 1/2x)^2 = 2\ 1/2x^2 - R(1\ 1/4x^4) = (2\ 1/2 - R(1\ 1/4))x^2$ which gives $x^2 = (1/2 + R(1/20))(100) = 50 + R(500)$ so that once again $d^2 = 200 + R(8000)$. Finally the general formula $r^2 = (1/2 + R(1/20))(a_5)^2$ is given.

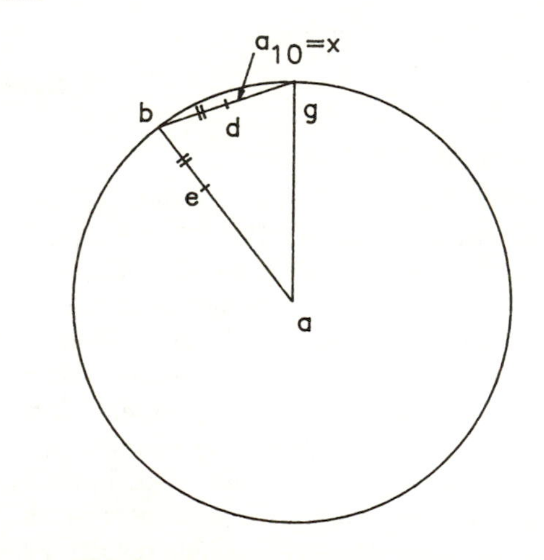

Source: Fibonacci–Boncompagni [II,215; diagram and text]

FIGURE IX-8. Problem 17

With regard to problem 17 [p. 214]—finding the side of a decagon whose area is 100—I mentioned in Section 28,B,i that Abu Kamil obtained the desired equation twice. The first method used problem 9, which in Abu Kamil's case did not involve DEMR. The second method involved DEMR—indeed this was the first problem for which Abu Kamil used DEMR—but the treatment was hesitant or perhaps confused and not clearly separated from the first method.

In Fibonacci, however, both treatments are precise and clearly separated although abbreviated (read 1600 for 2600 on p. 215, line 15). Indeed Fibonacci explicitly quotes XIV,** of Section 24 [p. 215, line 17; see Herz-Fischler, 1985, section 5,C] which is the first result on his list [p. 162] of theorems in Book XIV.

In the text, point e is taken on ab with $eb = bd$ where d is the midpoint of the side bg $(=x)$ of the decagon. Then we are simply told that according to Euclid $(ae)^2 = 5(eb)^2$. This is just XIII,3: S(smaller + 1/2 larger) $= 5 \cdot S$(1/2 larger) because by XIV,** 1/2 larger segment of $ab = 1/2 a_{10} = bd = eb$ and further smaller + 1/2 larger = total − 1/2 greater = $ba - be = ae$. This gives $(ae)^2 = 1\ 1/4x^2$ from which one has again $ab = bc + ae = 1/2x + R(1\ 1/4x^2)$ (the text has 1/2 40).

In his treatment of problem 18—finding a_5 given that the area of the 36°−72°−72° triangle agd (Fig. IX-9) is 10—we saw how it was only at the very end that Abu Kamil states that the triangle is similar to the one in problem 17. Fibonacci does not go through the whole procedure involving finding the altitude at and setting up an equation via the area [p. 215; in Fig. IX-9, z divides ag in EMR and i is the midpoint of ag], but rather having obtained the relationship $ag = R(1\ 1/4x^2) + x/2$, he then states that the triangle is similar to that of problem 17 and that we thus obtain the same answer. Again we may ask why this was not stated right away (see the comment on Abu Kamil's version). I also note that Fibonacci explicitly gives the proportion $ag{:}gz = gz{:}za$ (that is, XIII,8) whereas Abu Kamil avoids giving the actual proportion.

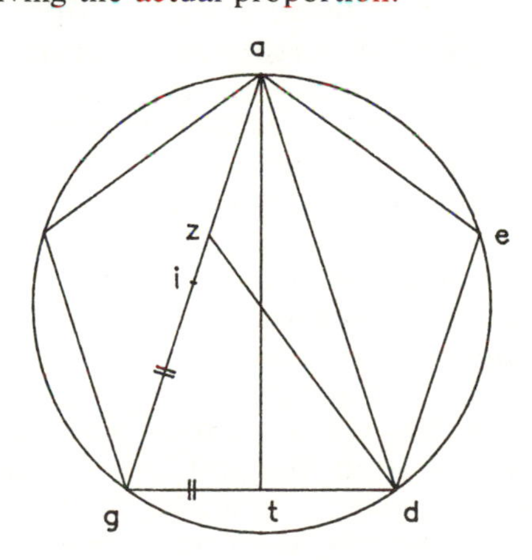

Source: Fibonacci–Boncompagni [II,215]

FIGURE IX-9. Problem 18

Again in problem 19—finding the diagonal of the pentagon if the area of the triangle aed in Figure IX-9 is equal to 10—Fibonacci is more precise than Abu Kamil, for he explicitly gives the result of XIII,8 in terms of DEMR and gives the intermediate statement $(ab + bz)^2 = 5(ze)^2$ derived from XIII,1.

Fibonacci's solution to problem 20 is the same as that of Abu Kamil. I note, however, that whereas Abu Kamil has abstracted from problem 5 ("we have already shown . . .") Fibonacci states the general result without commentary.

iv. Equations from Abu Kamil's Algebra

In Section 28,B,ii I considered Abu Kamil's treatment of two problems and discussed the question of whether these problems had their origin in the concept of DEMR or whether the appearance of the numbers related to DEMR was an accident.

The first problem is repeated in Fibonacci's *Liber abaci* (which translates as Book of Computations) [Fibonacci–Boncompagni, I, 434] as "Ten is to be divided into two parts and the greater is divided by the smaller and the smaller by the larger and I put together those things which come out of the division and they were the root of 5 'denarii'."

We find Fibonacci taking up Abu Kamil's solutions but with added details [p. 435, line −6 corresponds to Abu Kamil's third solution whereas p. 436, line 26 corresponds to the second]. For example, we find him giving the theoretical justification for the technique used in Abu Kamil's second method [p. 436].

What really interests us, however, is the following statement [p. 438]:

"So we can also proceed in these other [text: "his aliis aliis"] ways but what we have said should be enough and you know according to this division [text: "et scis secundam hanc divissionem"] that 10 is divided in mean and extreme proportion; as 10 is to the greater part so is the greater to the smaller: wherefore 10 being multiplied by the smaller part i.e. by 15 − R(125) makes it equal to the greater part [i.e., $R(125) - 5$] multiplied by itself.

"If you want to divide 10 in this proportion [i.e., in EMR] take the greater part as x, the smaller part is 10 − x. Multiplying this by 10 gives 100 − 10x and multiplying x by itself gives x^2 which is equal to 100 − 10x. Therefore add 10x to both sides and we will have $x^2 + 10x = 100$ 'denarii' and solve this [lit. "do it in these terms"] using algebra [text: "alzebra"] etc."

There are two possible interpretations of this text. (i) Fibonacci is saying that he has shown us three ways to solve the given problem, but that he recognizes the numbers have to do with EMR and that if we do not believe him we can go ahead and check by multiplying.

Further, if somehow we knew ahead of time that we are dealing with EMR then we could have found the solution by solving the equation $x^2 + 10x = 100$. (ii) Somehow we should have seen ahead of time that we are dealing with EMR so that we can reduce the solution of the problem to the solution of the equation.

Although Davis [1977, 33] appears to favour the second interpretation, it seems to me that the first is to be favoured for several reasons. (1) Since the Latin "hoc"—used in the ablative "his"—usually looks backwards in the sentence, Fibonacci appears to be comparing the three solutions already obtained ("these other ways") with the upcoming DEMR. (2) The statement "according to this division" seems to refer to the actual numbers that appear in the solution of the given problem. (3) Whereas Abu Kamil's next problem—number 43; essentially $x + y = 10$; $(x/y)^2 + (y/x)^2 = 3$—is solved by reducing it to the one just discussed, Fibonacci replaces the 10 by 12 and the 3 by 4. Because of this, the solutions now have no relation to those numbers obtained when the line—of length 12 in this case—is divided in EMR. If Fibonacci had had DEMR at the centre of his thoughts in dealing with the previous problem, we would have expected him, in view of Abu Kamil's problem number 43, to continue on with it in the following question.

v. The Rabbit Problem, Fibonacci Numbers

In his *Liber abaci* [I, 283; a translation appears in Struik, 1969, 2], Fibonacci presents the problem of determining how many pairs of rabbits there will be in each generation if the first pair breeds one pair in the first and each succeeding month and if it is assumed that this pair and every succeeding pair gives birth to a new pair in the second month after their birth. Thus at the end of the second month we will have the original pair plus their offspring (i.e., two pairs). Then at the end of the third month we have the two from the original month plus a new pair giving $2 + 1 = 3$, while at the end of the fourth month we have $3 + 2 = 5$, etc. Fibonacci shows that the numbers of pairs will be 2,3,5,8,13,21,34,55,89, 144,233,377 in the first through twelfth months respectively "and in this way you can do it for the case of [an] infinite number of months." As I will discuss in more detail in Section 32,B, the ratios of the successive terms of the Fibonacci sequence are approximations to $(1 + R(5))/2$, the numerical value associated with DEMR. At this point I merely state that there is no indication whatsoever that Fibonacci realized that any connection existed.

vi. Summary

If we compare Fibonacci's computational approaches with those that I have discussed previously we find much that appears to be new.

As far as the diagonal of the pentagon is concerned, we saw that Abu Kamil (problem 5) performed the calculation by using Ptolemy's theorem to obtain a quadratic equation. This technique is quite different from the two methods presented by Fibonacci. I note that the use of a right triangle (VI,8) to obtain a proportion as in Fibonacci's second solution occurs in Abu Kamil's problem 1, but we cannot draw any conclusions from the use of such a simple result, which in fact appears several times in Fibonacci's work.

When we consider the area of the pentagon, Fibonacci's technique of using the preliminary result contained in XIV,6 is different than that of Hero (Section 25,A), Abu Kamil (problem 16), and Abraham bar Hiyya (introduction to this chapter), where the basic idea is to find the altitude going from the centre of the circle to the side and then multiply this by the side of the pentagon to obtain the area of one of the five triangles into which the pentagon has been decomposed.

The volume computations of Fibonacci are the first ones for which the details are given. The various manipulations using the results of Book XIV indicate a high level of knowledge and ability.

Again when we turn to Fibonacci's presentation of the problems from Abu Kamil's *On the Pentagon and Decagon*, we find new methods of solution that again display a deep understanding and ability.

We must conclude then that either there were several, now lost, works from which Fibonacci obtained his material or he was responsible for a significant raising in the level of the applications of the properties of DEMR to various computational problems.

C. *Francesca (c. 1415 to 1492)*

While much has been written about Pierro de la Francesca the painter (see, for example, Hendy [1968] and Clark [1969]), relatively little has been written about his mathematical work or his sources and training. For the little that seems to be known, one may consult Clagett [1978, 383], Davis [1977], and the introductions in Francesca–Mancini and Francesca–Arrighi.

The two texts of interest to us are *Trattato d'abaco* [Francesca–Arrighi] and *De quinque corporibus regularibus* [Francesca–Mancini]. Clagett [p. 384, 390] suggests that *Trattato* was written before 1477-78 and that *De quinque corporibus* was written in the late 1480s.

I shall thus start with *Trattato* as the basic text and indicate how, if at all, the problems are treated differently in *De quinque corporibus*. As we shall see, most problems were more or less taken over from one text to the other. This situation should be compared with the treatment of the semi-regular polyhedra, which as Clagett [p. 398] notes underwent major additions. I shall also discuss the few problems in *De quinque cor-*

poribus which are not in *Trattato*. There is some use of DEMR in the discussion of the semi-regular polyhedra in Part IV of *De quinque corporibus* but I shall not discuss these.

Whereas the problems in *De quinque corporibus* are numbered, there is no numbering in *Trattato*. For ease of reference and because some of the problems are solved in two ways or contain sub-problems of interest to us, I have given numbers to all of the problems and sub-problems, while at the same time I have indicated the page and line in the critical edition of Arrighi. Davis has labelled [1977, table II; see also Jayawardene, 1976, appendix] the problems of *Trattato* by folio and number, but neglects multiple solutions and sub-problems. For clarity I have indicated the problems using symbolic notation. A reference to *De quinque corporibus* without any additional statements in the discussion means that the problem has been more or less taken over as such. It is to be understood that all pentagons, dodecahedra, and icosahedra are inscribed in the circle or sphere in question.

We shall see in examining Francesca's work that it is far removed from the Fibonacci text. Some techniques are the same, for example, in problem 7 Francesca uses XIII,12, 17,corollary to calculate e_{12} given the diameter (this, however, is not the only way to do this; see Bombelli's method, Section F) just as Fibonacci did (Section B,ii). But even here there is a difference, for Fibonacci quotes a formula and Francesca sets up an equation. One basic difference between the authors is that in the volume calculations it is the diameter that is fundamental for Fibonacci, whereas for Francesca it is the edge that is fundamental (as it was for Hero, see Section 25,C; recall also that Pappus compared volumes for equal surface areas, see Section 27,B). Francesca has more problems than Fibonacci had and his investigation of the icosahedron is quite different and contains a new result (problem 12). On the other hand, Fibonacci displayed, as we have seen, a great virtuosity in his manipulation of the results of Book XIV as well as in his planar results, including his new techniques for solving Abu Kamil's problems (none of which appear in Francesca's work). The numbers in the texts are not the same and, despite the level of the problems, the texts of Francesca display much less finish and even a certain unsureness—in particular, regarding Ptolemy's construction (see problem 5.1).

PROBLEM 1 [p. 189, line 1; *De quinque*, XXIX, p. 506]. Given $d_5 = 12$ find a_5.

We are told that when the diagonal is divided in EMR we obtain the side (XIII,8); thus letting x be the larger segment we obtain $x^2 = 12(12-x)$ and so $a_5 = x = R(180) - 6$. Note that in problem 7 x is taken to be the smaller segment.

PROBLEM 2 [p. 189, line 16; *De quinque*, XXX, p. 507]. Given $a_5 = 4$ find d_5.

Again we are given the statement of XIII,8. The chord is then $x + 4$ with x being the smaller segment. This leads to $x(x+4) = 16$ and $x = R(20) - 2$; $d_5 = R(20) + 2$.

PROBLEM 3 [p. 190, line 7; *De quinque*, XXXIV, p. 509]. Given $a_5 = 4$ find the altitude [''cateto''] AF (Fig. IX-10).

From problem 2 we know $AD = d_5 = R(20) + 2$; also $DF = 2$. Thus using the Pythagorean Theorem $AF = R(R(320) + 20)$.

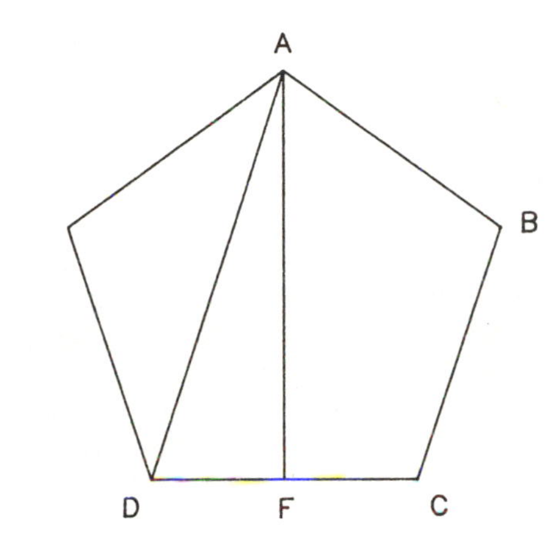

FIGURE IX-10. Problem 3 (not in text)

PROBLEM 4 [p. 190, line 23; *De quinque*, XXXV, p. 509]. Given $a_5 = 4$ find the altitude AH constructed upon the diagonal BE.

From problem 2 we know that $BE = d_5 = R(20) + 2$. Thus using the Pythagorean Theorem (Fig. IX-11) with $BE/2$ and AE we obtain $AH = R(10 - R(20))$.

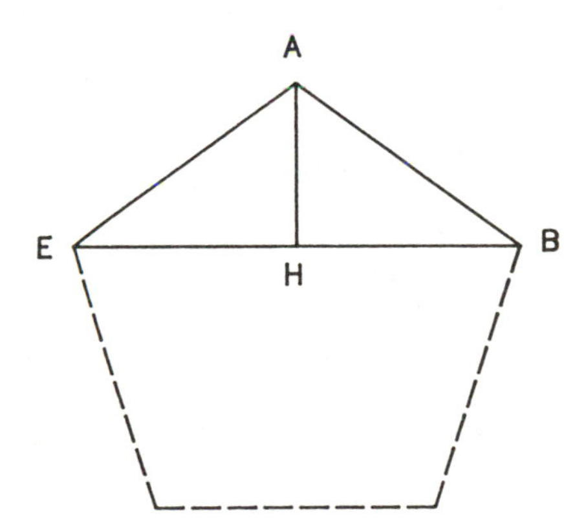

FIGURE IX-11. Problem 4 (not in text)

PROBLEM 5 [p. 190, line -8]. Given $a_5 = 4$ find A_5.

Problem 5, method 1 [p. 190, line -5]. This method involves solving sub-problem 5.1 below.

Problem 5.1 [p. 190, line 5; *De quinque*, XXVII, p. 505]. Given $a_5 = 4$ find d.

The text states that $d{:}4 = 4{:}R(10) - R(20)$; actually the portion "$d$ is to 4 as . . ." is missing, but that this is what Francesca had in mind is clear from what follows. From this one obtains $d = R\big(R(204\ 4/5) + 32\big)$.

There is no mention at this point in *Trattato* of how the proportion was obtained, but later on in the text—see problem 21—we read: "And on this we will cite an authoritative statement of Ptolemy who says that if the diameter of the circumscribed circle is 4 then the side of the pentagon will be [the root of] 10 − R(20)." Approximately the same statement occurs in connection with problem 9. Since these two problems appear in other parts of the text, it is possible that when problem 5.1 was written the source of the statement was not known. It is perhaps somewhat strange that Ptolemy's construction is not given and in fact one could interpret, although this is far from certain, the wording "we will cite an authoritative statement of Ptolemy . . ." to mean that Francesca in fact only knew the result and not the construction. Recall that Fibonacci (Section B,i) gave Ptolemy's construction and proof and evaluated a_5 when $d = 8$.

We find the same sort of situation occurring in *De quinque corporibus* where in the opening paragraph of the section on the pentagon [p. 505] we read: "For any regular [text: "equilateral"] pentagon the square of the diameter of the circle in which it has been inscribed is to the square of the side just as 16 is to 10 less the root of 20," and then this statement is referred to in the first problem, number XXVII, of the section which is essentially the same as sub-problem 5.1. Again in problem XXIX of part two, just as in the corresponding problem 9 of *Trattato*, we find the reference to Ptolemy.

Problem 5, method 1, *continued*. Having found the diameter of the circumscribing circle, Francesca calculates the area of the pentagon from the formula $A_5 = 3/4\,D \cdot 5/6\,d_5$. This result, contained in the second proof of XIV,6, had already been used by Fibonacci; see Section B,i. The value of d_5 is in fact given in the statement but was calculated in problem 2. I remark that, because of all the radicals appearing, Francesca works at first with the squares of the quantities involved; this technique is often used but I will not mention it again. The final answer is $A_5 = R\big(400 + R(128000)\big)$.

Problem 5, method 2. This solution is best described as "slick." Instead of dividing the pentagon into five triangles, see problem 6, method 1, Francesca uses a division into four triangles as in Figure IX-12. The three triangles ABE, EBK, and ECD all have the same base d_5 (problem 2). The fourth triangle, CKB, has base $a_5 = 4$ (by XIII,8). The altitudes of all four triangles are the same and are equal to AH (problem 4). The area of a triangle formula now gives the result.

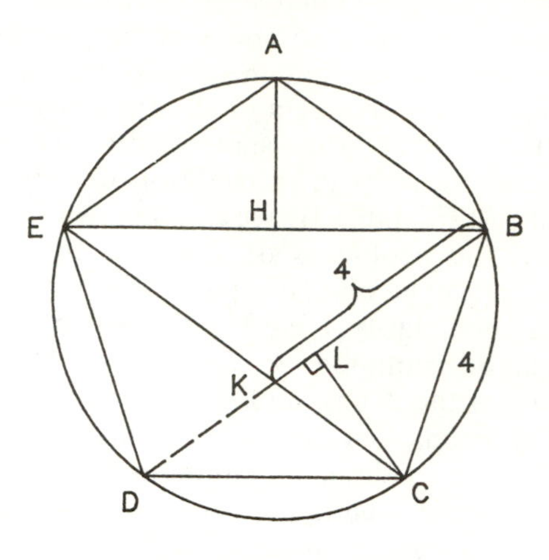

Figure IX-12. Problem 5, Method 2 (figure constructed from text)

Problem 6.1 [p. 192, line 6; *De quinque*, XXVIII, p. 505]. Given $d = 12$ find a_5.

This problem and 6.2 (find the area) are combined as one problem in *Trattato*, but they are separated in *De quinque corporibus*.

Francesca states [p. 192, line 8; in *De quinque*, XXVIII, Euclid XIII,8 is not mentioned] XIII,8,9,10 but only uses the latter two. First he lets $x = a_{10}$ and then uses XIII,9 with $a_6 = 6$ to obtain the equation $(x+6)x = 6^2$ giving $a_{10} = x - R(45) - 3$. Then XIII,10—$S(a_5) = S(a_6) + S(a_{10})$—shows that $a_5 = R\big(90 - R(1620)\big)$. Note how this problem could have been solved more simply by using the proportion discussed in problem 5.1.

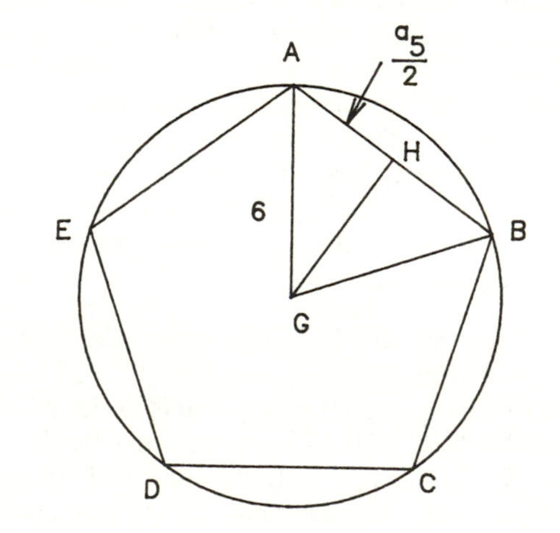

Figure IX-13. Problem 6.2, Method 1 (not in text)

PROBLEM 6.2 [p. 192, line 6; *De quinque*, XXXVI, p. 510]. Given $d = 12$ find A_5.

PROBLEM 6.2, METHOD 1 [p. 192, line -5]. One divides the pentagon into five triangles (Fig. IX-13), then considers triangle AGB whose side is a_5 and uses the Pythagorean Theorem to find the altitude. Multiplying the area of this triangle by five gives $A_5 = R\big(R(32035130\ 5/16) + 7593\ 3/4 - R(11533007\ 13/16) - R(6407226\ 9/16)\big)$!

PROBLEM 6.2, METHOD 2 [p. 193, line -8; *De quinque*, XXXVI, p. 510]. In *De quinque corporibus* Francesca uses the preliminary result of the second proof of XIV,6 just as he did in problem 5, method 1 of *Trattato*. However in *Trattato* (Fig. IX-14), he more or less goes through the derivation of the result. More specifically he divides the pentagon into five triangles, and then considers triangle ABG. Then $BE \cdot AG$ = area of four triangles since 1/2 BE is the altitude and AG is the base. But, he asks, if $AG = 6$ as a base gives four triangles, what base will give five triangles? The answer is $AK = 5/4 \cdot 6 = 7\ 1/2$. Thus the area of the pentagon is $7\ 1/2 \cdot BE$.

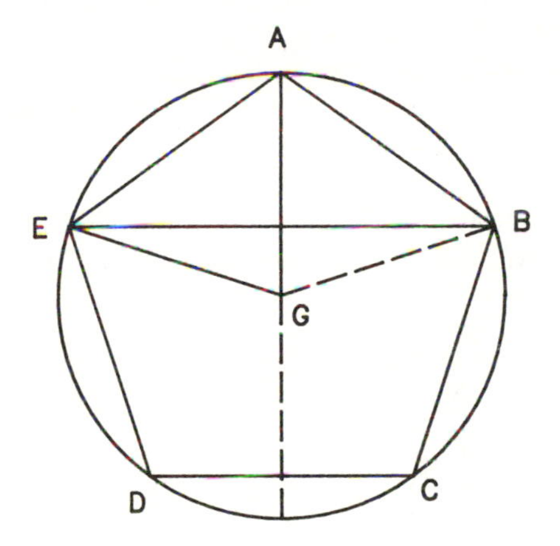

FIGURE IX-14. Problem 6.2, Method 2 (dotted lines not visible on reproduction in text)

In *Trattato*, a_5 had been found in 6.1 and here in 6.2 we are simply told that $BE = d_5 = R\big(R(1620) + 90\big)$. Presumably this was to be done by the use of XIII,8 which had been quoted, but not used, in 6.1. In *De quinque corporibus*, XXXVI, Euclid XIII,8, is quoted but no details are given.

The answer obtained by method 2 is $R\big(R(512781\ 1/4) + 5062\ 1/2\big)$ in the *Trattato* (correct) and $R\big(R(5125681\ 1/4) + 5042\ 1/2\big)$ in *De quinque corporibus* (Paccioli, Section E, has *6* and *6*).

Next we come to the problems involving the inscription of the dodecahedron and icosahedron in a sphere. In *De quinque corporibus* these are in part II where the numbering starts all over.

PROBLEM 7 [p. 235, line 1; *De quinque*, XXVIII, p. 532; summary, p. 539]. If $D = R(48)$ or if $D = 12$ find e_{12}.

Let $D = R(48)$ [p. 532]. By XIII,15 $D^2 = 3(e_6)^2$ giving $e_6 = 4$ (this is why D was taken to be such a strange number!). Next we know from XIII,17[a],corollary that when we divide e_6 in EMR we obtain e_{12} as the larger segment. This leads to the equation $(4-x) \cdot (4-x) = 4 \cdot x$ so that $e_{12} = 4 - x = R(20) - 2$. I shall describe the method of *De quinque corporibus*, XXVIII in connection with problem 8. The case $D = 12$ is discussed in the summary [p. 539].

PROBLEM 8 [p. 235, line 23; *De quinque*, XXVII, p. 532]. If $e_{12} = 4$, find D.

In *Trattato* this is done by working with proportions in connection with the result of problem 7. In *De quinque corporibus* this problem is done first by using XIII,17[a],corollary and XIII,15 as in problem 7 of *Trattato* but in reverse order. This gives $e_6 = R(20) + 2$ and $D^2 = 72 + R(2880)$. Since problems 7 (XXVIII) and 8 (XXVII) have been interchanged in *De quinque corporibus*, we might expect that in XXVIII Francesca would simply use proportions applied to the result of XXVII. But instead he solves this problem exactly as he solved it in *Trattato*. Then he says that if $(e_6)^2$ turns out not to have an exact square root (understood: "if your diameter just happens not to be equal to $R(48)$"!) then one works with proportions, for example, if $D = R(51)$ then $e_6 = R(17)$, and the use of proportions, using the sides of the cubes not the diameters, gives $e_{12} = R\big(25\ 1/2 - R(361\ 1/4)\big)$.

PROBLEM 9 [p. 236, line 4; *De quinque*, XXIX, p. 533]. If $e_{12} = 4$ find S_{12}.

To find S_{12} one simply finds the area A_5 of one of the pentagonal faces and multiplies by 12. The area of a pentagon whose side is 4 was already found in problem 5, and yet here we find Francesca going through more or less the same calculations, that is, he finds d as in problem 5.1 and d_5 as in problem 2 (the value for d_5 is assumed in the statement of 5.1, presumably from problem 2). Here he works directly with $5/8\, d \cdot d_5$ instead of $3/4\, d \cdot 5/6\, d_4$. As mentioned in connection with problem 5.1, there is a reference here to Ptolemy, whereas in problem 5.1 there was no indication of where the result concerning the diameter came from. This redoing of the calculations, coupled with evidence of some revision, is all the more puzzling in view of the next problem (9) where the text specifically mentions "fifth result of [the section dealing with] the pentagon" (i.e., with problem 5.1).

The answer is $S_{12} = R(57600 + R(2{,}654{,}000))$; note that the final answer in *De quinque corporibus* has 576000 although the correct figure is given in the text two lines above.

PROBLEM 10 [p. 237, line 21; *De quinque*, XXX, p. 534]. If $e_{12} = 4$ find V_{12}.

As was the case with Fibonacci, Francesca uses the formula $12(1/3 \cdot A_5 \cdot$ altitude of pyramid), which in fact was probably the only method available. However $12A_5 = S_{12}$ and this, as is pointed out in the text, was computed in problem 9. The altitude is found, as in Fibonacci—but again this was probably the only way available—by using the Pythagorean Theorem applied to the diameter of the sphere and the radius of the circle. Despite the fact that the diameter of the sphere had just been found in problem 8, Francesca once again repeats a calculation. Now comes the difference with Fibonacci's technique. While the latter used the preliminary result to XIV,2, Francesca refers to problem 5.1 where this calculation was done—see the comment on problem 9—presumably via Ptolemy's construction.

The final answer given is $R(R(3{,}964{,}928{,}000) + R(3{,}276{,}800{,}000) + R(3{,}171{,}942{,}400) + 64000)$ (= 490.44—correct! *De quinque corporibus* has 6400, apparently due to copying the wrong term).

PROBLEM 11 [p. 238, line 16; *De quinque*, XXXI, p. 535, also summary, p. 539]. If $D = 12$ find e_{20}.

We saw in problems 7 and 8 how Francesca used XIII,17[a],corollary to find e_{12}, and I remarked that Fibonacci had also done the same thing. Now the techniques diverge, for Francesca uses XIII,18 to find e_{20} instead of the preliminary result of XIV,2 and XIII,12 as Fibonacci did in the course of his calculation of the volume of the isocahedron (Section 31,B,ii).

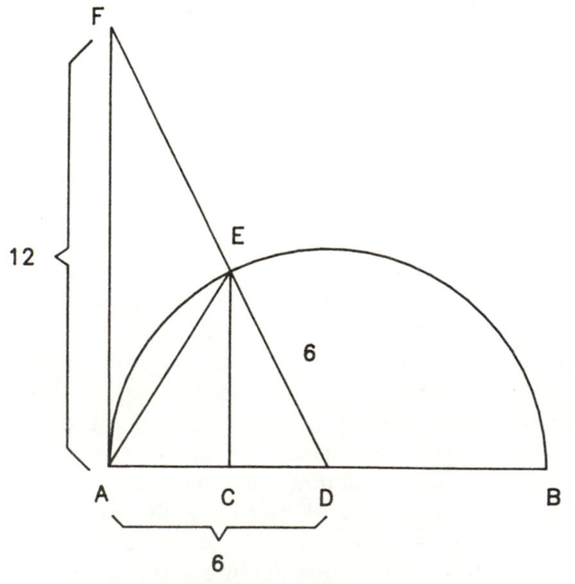

FIGURE IX-15. Problem 11 (based on the text and XIII,18)

We recognize in Figure IX-15 the diagram of XIII,18, explicitly mentioned in the text, except that in the *Elements* it is the line MB, symmetric with AE, which is shown to be e_{20}.

The actual calculations, some of which duplicate steps in the proof of XIII,18, proceed as follows: $FD^2 = AF^2 + AD^2 = 144 + 36 = 180 = 5 \cdot AD^2$. Since $\Delta ECD \approx \Delta FAD$, we have $6^2 = ED^2 = 5 \cdot CD^2$ so that $CD = R(7\ 1/5)$ and $AC = 6 - R(7\ 1/5)$. A double use of the Pythagorean Theorem gives $AE^2 = AC^2 + CE^2 = (43\ 1/5 - R(1036\ 4/5)) + 28\ 4/5 =$ (*De quinque* only) $72 - R(1036\ 4/5)$.

PROBLEM 12 [p. 239, line 11; *De quinque*, XXXII, p. 536]. If $e_{20} = 4$ find D.

Instead of using problem 11 and proportions, Francesca performs the calculations directly employing the following result which is stated without proof.

Result: "[If in the diagram of XIII,18] . . . AE and EB are joined together in a straight line they will form a line divided at point E in EMR with the larger segment being EB and smaller segment being AE, equal to 4, is the side of the icosahedron."

I can think of two possible proofs of this: one uses XIII,9, the other XIV,**.

(i) From the proof of XIII,18 (see Fig. IX-16), $AC = a_6 = EC$; $CB = a_6 + a_{10}$. The similarity of triangles ECB and AEB gives $AE{:}EB = a_6{:}(a_6 + a_{10})$, and the result follows from XIII,9 converse (not proven in the *Elements*; it was employed by Ptolemy, see Section 26,A).

(ii) If we use the similarity of triangles ECA and AEB, we have $AE{:}EB = a_6{:}a_{10}$ and the result follows from XIV,**.

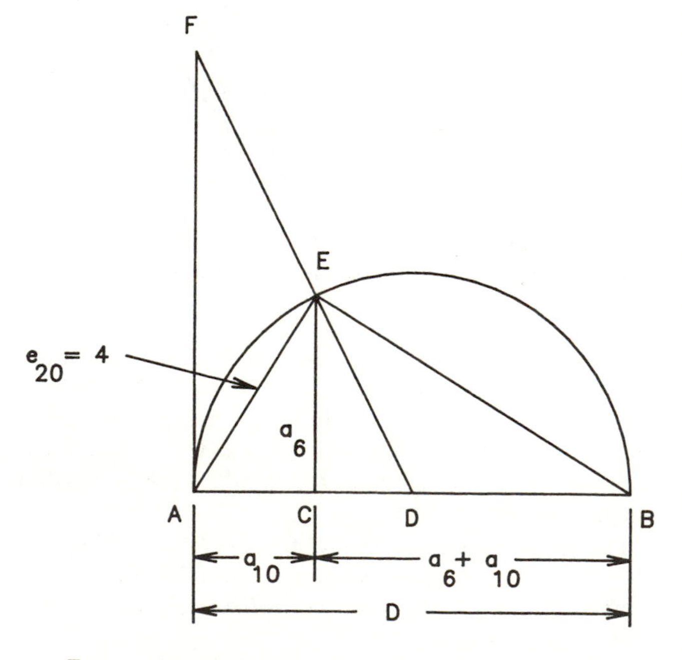

FIGURE IX-16 (based on the text and XIII,18)

Using the result and letting $EB = x$, Francesca obtains the equation $4(4+x) = x^2$ or $EB = x = 2 + R(20)$; see problem 2. Then from the right triangle AEB, we have $AB^2 = AE^2 + EB^2$, which gives $D = AB = R(40 + R(320))$.

Problem 13 [p. 240, line 19; *De quinque*, XXXIII, p. 536]. If $e_{20} = 4$ find S_{20}.

The area of one of the triangles is calculated, after finding the altitude (details are not given until problem 16), and then $S_{20} = 20 \cdot \text{area} = R(19200)$.

Problem 14 [p. 240, line -9; *De quinque*, XXXIII, p. 537]. If $S_{20} = 200$ find e_{20}.

One uses problem 12 (for one triangle) and proportions giving $e_{20} = R(533\ 1/3)$.

Problem 15 [p. 241, line 5; *De quinque*, XXXV, p. 537]. If $S_{20} = 200$ find D.

We know e_{20} from problem 14 and then problem 12 gives $D = R(R(R(429188\ 26068/32489)) + R(3333\ 1/3))$.

Problem 16 [p. 241, line 24; *De quinque*, XXXVI, p. 538]. If $e_{20} = 4$ find V_{12}.

The radius of the sphere is known from problem 12 and the radius of the circle circumscribed about the triangle can be computed from $2/3 \cdot$ (altitude). The altitude of the pyramid upon the triangular base is computed via the Pythagorean Theorem; see problem 12. The final answer is $V_{20} = R(9955\ 5/9 + R(91{,}022{,}222\ 2/9))$ (= 139.63—correct).

The following problems are only found in *De quinque corporibus*.

Problem 17 [*De quinque*, XXXVII, p. 538]. If $V_{20} = 400$ find e_{20}.

One uses problem 16 and proportions giving $e_{20} = R((806400)^{1/3} - R(597{,}196{,}800{,}000))$.

Problems 18,19,20 [*De quinque*, XXXI, XXXII, XXIII, p. 507]. Find a_5, d_5, d for a certain pentagon given $a_5^2 + d_5^2 = 21$ ($a_5^2 + d_5^2 - d^2 = 20$; $a_5^2 + d_5^2 + d^2 = 40$).

If $d = 4$ then $(a_5)^2 + (d_5)^2 = 20$ (no details are given, but see problem 5.1 and 6.2, method 2). Using proportions, one obtains that a sum of 21 corresponds to $a_5, d_5 = R(10\ 1/2 \mp R(22\ 1/20))$ and $d = R(16\ 4/5)$. Problems 19 and 20 are solved in the same way.

The following problem is only found in *Trattato*.

Problem 21 [p. 208, line -9]. If 5 circles are inscribed in a circle of diameter 12 as shown (Fig. IX-17), find the diameter of the smaller circles.

There are several corruptions and/or mistakes in the text and the letters on the diagram are illegible, but the following represents the idea of the corrected solution. If we call the diameter of the circle that passes through the centres of the five smaller circles x, then the distance between centres will be $12-x$. However, this distance between centres is also the side of the pentagon inscribed in the inner circle. By an "authoritative statement of Ptolemy" (see problem 5.1), we know that if $d = 4$ then $a_5^2 = 10 - R(20)$ and using proportions we have $5/8x^2 - (R(20)/16)x^4 = a_5^2 = (12-x)^2$. This quadratic equation in x^2 is solved with the answer given being $R(7200 - R(50{,}181{,}120)) - (60 - R(2880))$.

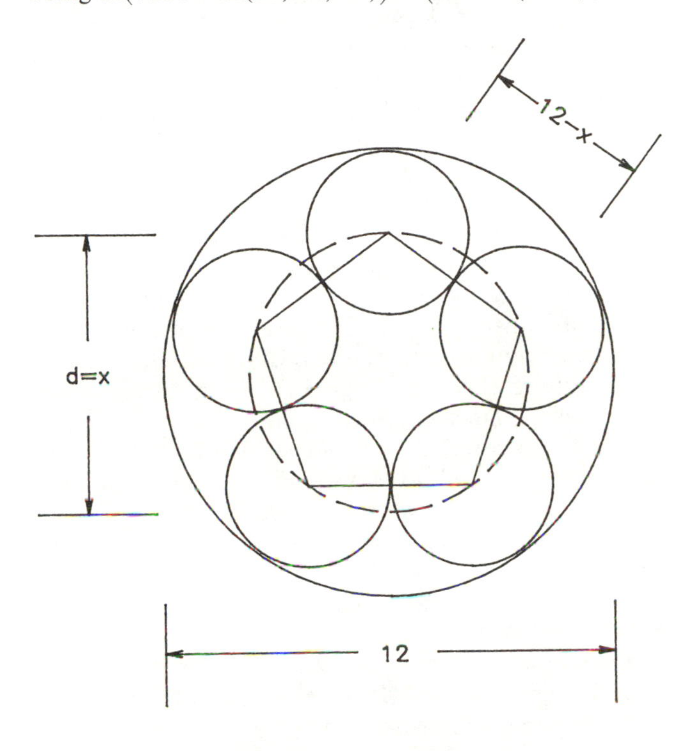

Figure IX-17. Problem 21

We now see that Francesca has presented a fairly complete range of what we might call basic problems—those involving sides, diameters, areas, and volumes. There are many more than in Fibonacci and some of the techniques (see, for example, problem 5, method 2, problems 11,12) are quite interesting and different from what we have seen before. Furthermore, we now find the "algebraic" problems 18, 19, and 20, as well as the very interesting geometrical problem 21, which are certainly not "basic" problems. Whether all or some of these problems represent a collection of results built up since the time of Fibonacci or whether they are mainly due to Francesca I cannot say.

D. *Paccioli (c. 1445-1515)*

Perhaps no mathematician has plagiarized as much and with so little change in details and has stirred so much controversy and such vehement reactions as Luca Paccioli (I have followed the British Library and Paris' Bibliothèque Nationale spelling, although this seems to be a cause of controversy). For details of his life and references to his works, see Clagett [1978, 416], Rose [1975, chap. 6], and Jayawardene [*DSB*].

As far as plagiarizing and reaction is concerned, consider what Mancini [Francesca–Mancini, 416] has to say: "The text of [Paccioli's *Libellus in tres partiales tractatus divisus quinque corporum regularium et dependentium active perscrutationis D. Petro Soderino principi perpetuo populi florentini a M. Luca Paciolo . . .* (*A Little Book on the Five Regular and related Bodies, divided into three partial tracts, specially dedicated to D. Petrus Soderinus, perpetual leader of the Florentine People by M. Luca Paccioli . . .*); this was bound in with the *Divina proportione*, see below] placed under meticulous line by line comparison with the [manuscript of Francesca's *De quinque corporibus*] is, aside from very rare variances, the literal Italian version of the treatise of Piero de la Francesca composed in Latin. Publishing the translation, Paccioli did his best to give an original appearance to his work and with foxy malice [Italian: *volpina malizia*] he places his own name beside that of Soderino [Gonfalonier for Life of Florence] . . . although he had already dedicated the entire volume [i.e., *Divina proportione* which is also the name applied to the entire bound volume] to him. Then omitting to specify if he was the author or the translator he took advantage of the faith of the reader."

Thus we see that there is no need for us to consider the appearance of DEMR in Paccioli's *Libellus*. The only question of interest to scholars concerning this book seems to be whether Francesca wrote it in Italian or Latin and whether Paccioli translated from the Latin or has copied, at least in part and perhaps using *Trattato*, the original Italian. On this question, see Clagett [1978, 390, 391 fn. 23] and Davis [1977, 108].

According to Davis [1977, appendix 1, table 1], Paccioli also took problems on the dodecahedron and icosahedron from Pierro's *Trattato* and included them in his 1494 book, *Summa de arithmetica, geometrica, geometria, proportioni et proportionalita*. Clagett [p. 421] speaks of a 1478 manuscript *Trattato di arithmetica e d'algebra* "which has the same character" as Francesca's *Trattato*. I do not know if there are problems on DEMR in this former work and if so how close they are to Francesca's problems.

As a final comment on Paccioli's "literary borrowings," I mention Cardano's 16th-century comments [Cardano–Witmer, 7 fn. 2] on Paccioli's use in his *Summa* of a 1202 manuscript (Fibonacci?) dealing with algebra and Agostini [1925] who, after discussing Paccioli's word for word inclusion in his *Summa* of an 1481 mercantile text, says that one should not accuse Paccioli of plagiarism for having used in his book what is available to everybody and is not personal intellectual property.

Our interest in Paccioli therefore lies in his 1509 book *Divina proportione*. Actually, the bound version consists of three distinct works. The first part is what I am referring to when I speak of *Divina proportione*, and it is its contents that I shall discuss hereafter. The second part is entitled *Tracto de l'architectura* (this part is reprinted in Paccioli–Bruschi [1978]; see also Paccioli–Winterberg [p. 126-53, 285-337]). *L'architectura* is essentially a discussion of proportions and its relationship to architecture, lettering, and the structure of humans. The theory is based on the work of the Roman architect Vitruvius and the various Renaissance interpretations of it. Because of the various false statements in the literature, the earliest apparently being that of Montucla [1799, vol. 1, 551; this is not in the original edition of Montucla, 1758, vol. 1, 455], I insist upon the fact that Paccioli does not recommend the use of DEMR in determining the proportions of works of art and architecture. The third part is the *Libellus* which, as I mentioned above, is just an Italian version of Francesca's *De quinque corporibus*.

There are two aspects to *Divina proportione* proper: the strictly mathematical and the "philosophical." In Paccioli's text (on the clarity of which see Rose [1975, 263]) these aspects are intertwined, but here I shall only discuss the strictly mathematical content of the work. A discussion of the other aspect, including Paccioli's use of the term "divine" in connection with DEMR, will be found in Appendix II. The references given are, respectively, the chapters in the original and the page of the Italian and German texts in Paccioli–Winterberg.

In Chapter VII [p. 44, 195] we find the statement of XIII,1 followed by an explanation of the use of the term "mean and two extremes" in connection with DEMR. This is tied in with VI,17 and the idea of a "continuous" proportion, for which the example $9:6 = 6:4$ is given. Without a reference (see XIII,6), we are told that the "divine proportion" cannot be rational.

In Chapter VIII [p. 47, 198] we are told, without details that if 10 is divided in EMR then the segments are $R(125) - 5$ and $15 - R(125)$. Recall that these numbers appear in Section 31,A,iv in connection with Fibonacci's solution of an equation taken from Abu Kamil's *Algebra*. These numbers are used in Chapter X [p. 49, 201] to illustrate XIII,1 (given in Chapter VII). These same numbers are also used in Chapters XI and XXI in connection with the statements of XIII,1,2,3,5,4,9,9 again (stated in an equivalent way in XVII in a form that Paccioli calls the converse), 8; XIV,** (see Section 24,A; here the division of 12 is compared to the division of 10), XIV,7. In Chapter XX we are told that one cannot construct the pentagon without DEMR. In none of the above is there any real discussion or analysis; there is simply a statement of the result and the throwing out of some numbers.

In Chapters XXVI to XXXI we find the constructions, with details now, of the five regular polyhedra as well as the construction of the edges (XIII,18). Among the contents of the rest of the book are inscription results from Book XV and the discussions of various other non-regular polyhedra.

Thus from a mathematical viewpoint, *Divina proportione* contains nothing new of interest and therefore stands in marked contrast to Francesca's *De quinque corporibus* which is bound in the same volume. Our sole interest in this book lies in the use of the term "divina proportion" in connection with DEMR and to the "philosophical" statements concerning DEMR. It is to these aspects that we shall return in Appendices I and II.

E. Cardano (1501-1576)

For a discussion of the life of Cardano, his contributions to the solution of the general cubic equation, and the controversy surrounding some of his publications, see the foreword and preface to *The Great Art* (1545) [Cardano–Witmer].

We find DEMR explicitly mentioned in three problems; in addition there are six other equations whose solutions are numerically related to DEMR, but where there is no mention of DEMR.

In Chapter XXXV, problem 17 [p. 213] reads: "Find two numbers the square of the second of which is equal to the product of the first and the sum [of the two], and the sum of the squares of which is 10. You see clearly that if x is postulated as the sum of them it will be divided according to the proportion having a mean and two extremes and the parts will be $R(5/4x^2) - 1/2x$ and $1/2 - R(5/4x^2)$."

The first condition of the problem is just the numerical version of the statement of II,11 and Cardano is here using, without any further comment, the formula for the larger and smaller segments.

The solution now continues. "The [sum of the squares] of these will therefore be $5x^2 - R(20x^4)$ and this will equal 10. Therefore from the rule for adding plusses and minuses, $5x^2 - 10 = R(20x^4)$. Hence the parts will be $R(2\ 1/2 + R(5)) + R(2\ 1/2 - R(5))$ and $R(2\ 1/2 + R(5)) - R(2\ 1/2 - R(5))$."

The next problem, problem 18 [p. 214], is: "Find three proportional numbers the first and second of which equal the third and the sum of the squares of the first and second of which is 10. Divide x into two parts, the sum of the squares of which is 10. These will be $1/2\,x + R(5 - 1/4\,x^2)$ and $1/2\,x - R(5 - 1/4\,x^2)$. Multiply x by the smaller and the product is the square of the greater. Otherwise divide x according to a proportion having a mean and two extremes, then square the parts, and the sum of the squares will be 10. Hence the parts will be $R(22\ 1/2 + R(405) - R(12\ 1/2 + R(125))$, $R(12\ 1/2 + R(125)) - R(2\ 1/2 + R(5))$, $R(10 + R(80))$."

From the penultimate sentence it seems that the first sentence should read "the *sum* of the first and second of which equal the third . . . ." Thus problem 18 appears to be the same as problem 17 and in fact, despite the difference in appearance of the answers, a numerical check indicates that both sets of answers are the same. The "solution" of 18 is not entirely clear to me (on this, see p. xxiii].

In the last problem of the book [Chapter XXXIX, problem 13, p. 252], the problem is: "Find a number the fourth power of which plus twice its cube is one more than the number," which leads to the equation $x^4 + 2x^3 = x + 1$. Then following, at least for me, a cryptic discussion which leads back to the same equation, we read: "We already know the ratios of the quantities, since the whole times the first equals the square of the sum of the second and the third. Hence this sum must be divided according to a proportion having a mean and two extremes. The smaller part of this is 1, so the remainder (which is the greater part) is $R(1\ 1/4) + 1/2$ . . . ."

In addition to these three problems, we have the following ones. In Chapter I ("On Double Solutions"), Cardano illustrates [p. 20] the possibility of having "true" (i.e., positive) and "false" (i.e., negative) solutions by the equations $x^3 + 4 = 3x^2 + 5x$ and $x^3 + 3x^2 = 5x + 4$. The first has the solutions $R(1\ 1/4) - 1/2$, 4 and $-(R(1\ 1/4) + 1/2)$, while the solutions of the second equation are precisely the negatives of the first. The number $a = R(1\ 1/4) + 1/2$ is the numerical value associated with DEMR, and the solutions $-a$, $1/a$ of the first equation satisfy the quadratic $x^2 + x = 1$, whereas the roots a, $-1/a$ of the second equation correspond to the quadratic $x^2 = x + 1$.

Problem 5 [p. 43] of Chapter V reads: "Find the number which, when its cube is subtracted from it and when the square root of the remainder is added to the remainder, yields the first number."

Cardano solves this by letting x^2 be the remainder after the cube root is subtracted from the number. Since when we add x we obtain the original number once again, it must be that what we have added (i.e., x) was equal to what we subtracted (i.e., the cube root of the original number), in other words, the original number was x^3. Thus we have the equation $x^2 + x = x^3$, which Cardano reduces to $x^2 = x + 1$ and then says: "Hence $x = R(1\ 1/4) + 1/2$."

In the last chapter (XXXIX) problem 1 [p. 235] reads: "Find three proportional quantities of which the square of the first is equal to the [sum of the] second and the third, and the square of the third is equal to the [sum of the] squares of the first and second. Since the square of the third is equal to the squares of the second and first, let $x^4 = x^2 + 1$. Therefore x, or the ratio is $R(R(1\ 1/4) + 1/2)$."

Problem 2 [p. 236] reads: "Find three proportional numbers the third of which is equal to [the sum of] the second and first, and the square of the first of which is equal to the sum of the second and third. Let the first be 1, the second x and the third x^2. Since the third is equal to the sum of the second and first $x^2 = x + 1$ and the ratio accordingly is $R(1\ 1/4) + 1/2$."

In Problem 10 [p. 250] the equation $x^4 = x^2 + 1$ results from the following statement: "Find three proportional numbers the sum of which is 8 and the square of the third of which is equal to the sum of the squares of the first and second."

From the above examples we obtain the impression that Cardano was well acquainted with the numerical quantity $R(1\ 1/4) + 1/2$ and its relationship to DEMR and to the equations $x^2 + x = 1$ and $x^2 = x + 1$ and assumed that his readers too would have the properties at their command. I suspect that some of the problems were made up by starting with these quadratics and then obtaining a more complicated equation and verbal problem.

F. *Bombelli (1526-1572)*

Rafael Bombelli's *Algebra* [Bombelli–Bortolotti] was published in its entirety for the first time only in 1966. The first three books were published in 1572; the fourth and fifth books were published in 1929 by Bortolotti and based on the manuscript he had found. The chapter headings and most of the drawings of the fourth and fifth books, which I shall use, are due to Bortolotti [p. xliv]. For details of Bombelli's life, see Rose [1976, 146].

While some of the material involving DEMR as such contains something "new," there is nothing that will surprise us. The same is certainly not true for the computations involving the dodecahedron and icosahedron. The basic approach, which is radically different from anything that we have seen, is characterized by its minimal use of the various results involving EMR that were employed by his predecessors. The edges of the solids are 4 (recall that Fibonacci took the diameter as the basic length), and I assume that Bombelli took these from Paccioli's version of Francesca's *De quinque corporibus*.

(1) BOOK III, PROBLEM 80 [p. 355]. There are really two separate questions contained here. The problem itself states that one is to divide 10 such that the larger segment multiplied by itself will equal the smaller segment multiplied by 10. This is the same problem, except for the difference in the way it is expressed, that we saw in Section B,iv regarding Fibonacci's solution of Abu Kamil's equation (see also Section D on Paccioli); the resulting equation and solution are the same. Then we are told more generally: "If a quantity is divided according to the proportion having a mean and two extremes it will be in two parts such that the square of one will equal the product of the other [part] by the [given] quantity." Then without details we are given the formula in verbal form, again the one stated by Fibonacci (Section B,ii): larger segment of $a = R(a^2 + 1/4\ a^2) - a/2$.

The second question of problem 80 requires us to divide $10 + 2x$ in EMR, and this is solved by the use of the formula just provided, giving $R(125) - 5 + R(5)x - x$ (one must be careful in reading Bombelli; the radical operation is only attached to the first symbol unless L-shaped brackets are used). At first glance this division of an unknown in EMR would appear to be new, but we have already seen a similar operation in Abu Kamil's problem 20 in *On the Pentagon and Decagon*.

(2) BOOK 3, PROBLEM 38 [p. 360]. In this problem for the first time we find DEMR entering into the statement of the problem itself. We are to divide 12 into two parts such that, when the first part is divided in EMR, four times the resulting larger segment will equal the resulting smaller segment multiplied by the second of the two parts of 12.

Letting the larger segment of the first part be x, Bombelli uses problem 80 above to state that the smaller segment will be $R(1\ 1/4)x - x/2$. Note that he is implicitly using the fact that, when the larger segment of a line divided in EMR is itself divided in EMR, we obtain the smaller segment. Adding together the two segments, we obtain $R(1\ 1/4)x + 1/2x$ which is the first part resulting from the division of 12. The second part is thus $12 - R(1\ 1/4)x - 1/2x$. Using the statement of the problem to set up an equation, Bombelli obtains that $x = R(180) - 10$ and that the divisions of 12 are $10 - R(20)$ and $2 + R(20)$.

(3) BOOK 4, PROBLEM 93 [p. 580]. Instead of the numerical formulation of problem 80 above, Bombelli uses the geometric formulation of II,11 (the word "parallelogram" is used instead of "rectangle") to state his problem; no name is given to the proportion involved (i.e., DEMR).

Without proof or source, Bombelli gives what we recognize as Hero's construction (see Section 25,B, Fig. VII-8a) with the entire square on ab being completed. The various quantities in the construction are computed for the case $ab = 12$ giving $ag = R(180) - 6$ (we saw these numbers in Paccioli, Chapter XIX). The answer is checked by multiplying, to which Bombelli adds that "this demonstration derives from algebra." He concludes with a reference to problem 80 above.

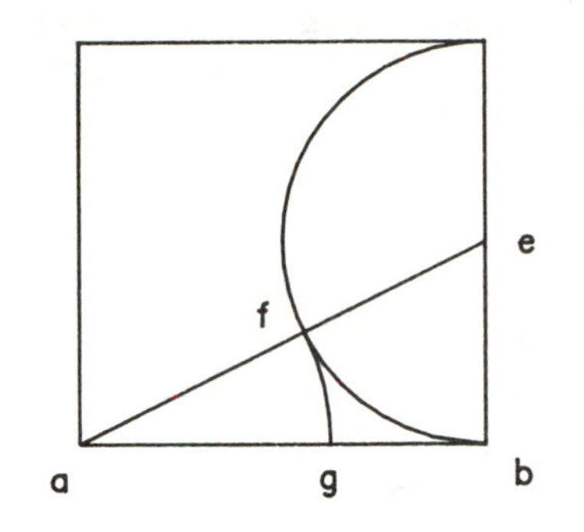

Source: Bombelli–Bortolloti [p. 580]

FIGURE IX-18

(4) BOOK V, SECTION 136 [p. 641]. Bombelli gives, without proof or source, Ptolemy's construction for the side of the pentagon. The computation is carried out for $d = 12$. Recall that this calculation was also done by Francesca in problem 5, but that Francesca went back to XIII,9,10. On the other hand, I indicated in connection with problem 5.1 that Francesca probably used a calculation based on Ptolemy's construction. Ptolemy's construction was presumably also available in a manuscript of Fibonacci's, or Ptolemy's work, or in Commandino's 1562 edition of Ptolemy.

(5) *Dodecahedron*, BOOK V, SECTIONS 153,154 [p. 657]. We saw as regards XIII,17[a] that Euclid viewed the dodecahedron as sitting on an edge. On the other hand, Pappus viewed the dodecahedron as sitting flat on one of the faces (see Section 27, Proposition 58). We now come to a third way of viewing the dodecahedron; namely, we can think of balancing the dodecahedron on a vertex. In addition to the diagram (Fig. IX-19a) given by Bortolotti, which may not have been in the manuscript, I have further illustrated the situation by using the top and front view of an orthographic projection (see Fig. IX-19b).

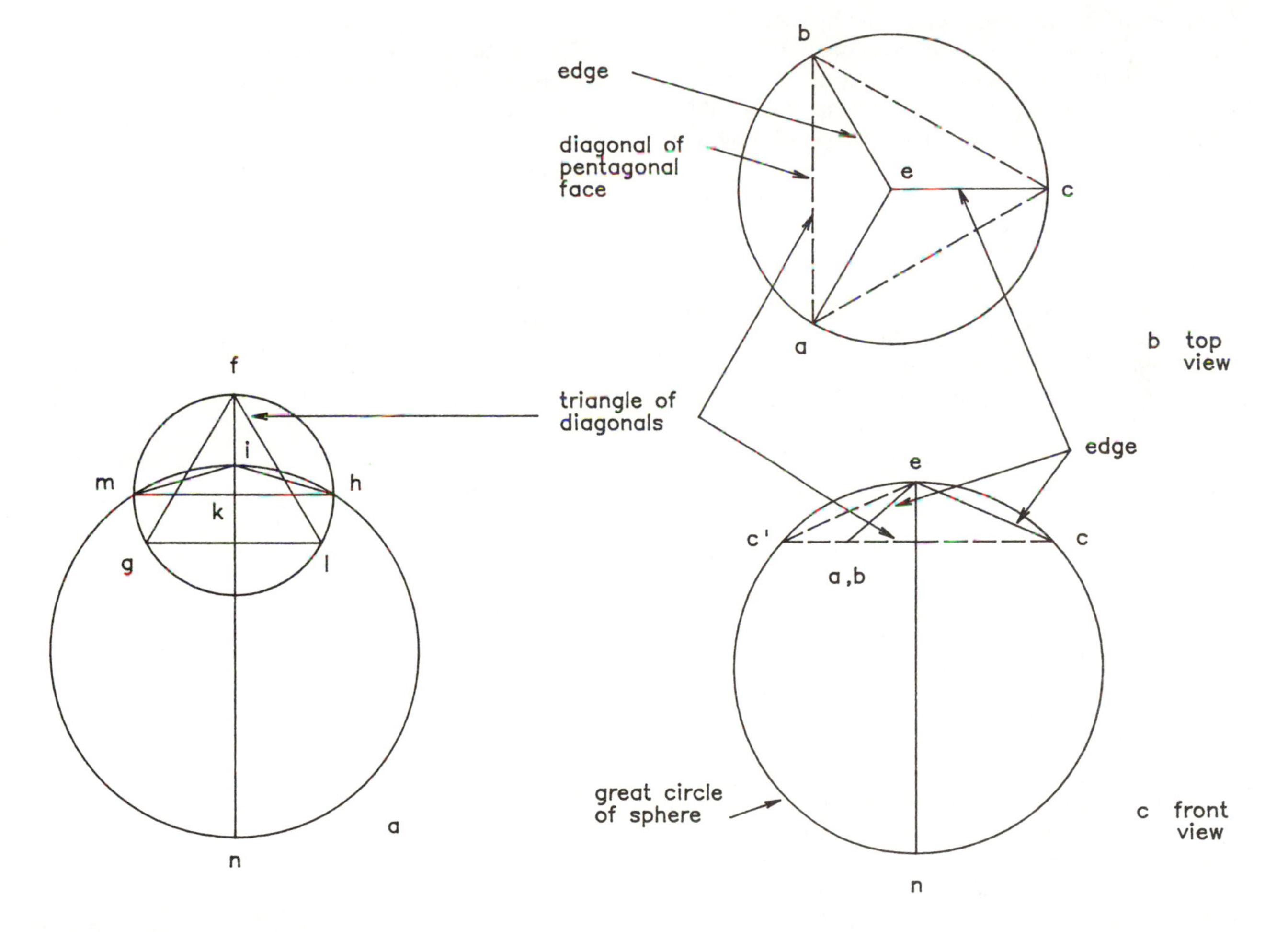

Source: Bombelli–Bortolloti [p. 658]

FIGURE IX-19. Dodecahedron (dimensions in top and front views exaggerated for clarity)

In Figure IX-19b when we look down at the dodecahedron from the top we will see the vertex *e* and the edges *ea*, *eb*, and *ec* of the adjacent pentagons. The lines *ab*, *bc*, and *ca* are precisely the horizontal diagonals of these pentagons. Thus the equilateral triangle *abc* formed by these diagonals will lie in a horizontal plane and this is represented in the front view by the horizontal line *cc'*. With the dodecahedron oriented as in the top view, the front view will show edge *ec* in true length (edges *ea* and *eb* will appear foreshortened). The sphere, in which the dodecahedron is inscribed, will appear as a circle which passes through the points *e*,*c*,*c'*.

Reversing what I have just said, Bombelli draws the equilateral triangle *fgl* with sides equal to the diagonal *bc* and circumscribes a circle about it (Fig. IX-19a). Let *k* be the centre and *mh*, which corresponds to *cc'* above, be the diameter. Then one draws the triangle *mih* with sides *mi* and *ih* equal to the edge *ce* of the dodecahedron; this corresponds to triangle *c'ec* above. Finally one circumscribes a circle about triangle *mih*. Its diameter *ni* is the diameter of the sphere.

Notice how Bombelli has avoided using the XIII,17[a], corollary employed by Fibonacci and Francesca (problems 7 and 8), which involves equating the diagonal with the edge of the cube, which in turn can be related to the diameter.

Turning to the actual computations [section 154] Bombelli assumes, as did Francesca, that $e_{12} = 4$. The section contains the following three problems.

(5.1) *Find the surface area.* Presumably using Ptolemy's construction as in problem 4 above, Bombelli states that if the diameter of the circumscribing circle of one of the pentagonal faces is 8, then $a_5 = R(40 - R(320))$. Thus using proportions, one obtains that if $e_{12} = a_5 = 4$ then $d = R(32 + R(204\ 4/5))$. This is just as in Francesca's problem 5.1 except there the basic proportion involved a diameter of 4 instead of 8 as is the case here. The area is now computed by splitting up the pentagon into five triangles just as in Francesca's problem 6, method 1 (although there d was given; the case where a_5 is given was solved by two other methods in problem 5). The final answer is the same as in Francesca's problem 9, although the method of computing the area was different there.

(5.2) *Find the diameter of the circumscribing sphere.* Here Bombelli uses the method of section 153 discussed above. To find the diagonal of the pentagonal face or, equivalently, the sides of the equilateral triangle *fgl*, he invokes XIII,8. Presumably using the method of problems 2 or 3 above, he says that if the diagonal is 6 then the side of the pentagon will be $R(45) - 3$, so that using proportions we have that if $a_5 = 4$ then $R(20) + 2$ (Francesca, problem 2, did this by solving an equation). This is the only place where the DEMR theorems are used.

The radius *mk* is found using XIII,12, and then the Pythagorean Theorem gives *ik*. Finally by VI,8, corollary—using the right triangle *mni* with altitude *mk*—we have $D = R(72 + R(2880))$ as in Francesca's problem 8.

(5.3) *Find the volume.* This is done essentially as in Francesca, problem 10. The answer is the same except that the first number ends with 800 instead of with the correct 8000.

(6) *Icosahedron*, BOOK V, SECTIONS 155,156 [p. 660]. If we balance the icosahedron on a vertex and look down from the top (Fig. IX-20), we see the defining pentagon *abcde* of the icosahedron as explained in connection with XIII,16[a]; Bombelli is viewing the icosahedron as Euclid did. With the icosahedron oriented as in the top view, the front view will show edge *ga* in true length (the other edges will appear foreshortened). The sphere in which the icosahedron is inscribed will appear as a circle which passes through the points *g*,*a*,*a'*.

Reversing what I have just said, Bombelli draws the pentagon (Fig. IX-20) *hipqk* with sides equal to the edge of the icosahedron and circumscribes a circle about it. Let *n* be the centre and *mo*—which corresponds to *a'a* above—be the diameter. Then one draws the triangle *mlo* with sides *ml* and *lo* equal to the edge *ga* of the icosahedron; this corresponds to the triangle *a'ga* above. Finally one circumscribes a circle about triangle *mlo*. Its diameter *rl* is the diameter of the sphere.

Turning to the actual computations [section 156] Bombelli assumes, as did Francesca, that $e_{12} = 4$.

(6.1) *Find the surface area.* As in Francesca, problem 13.

(6.2) *Find the diameter of the circumscribing sphere.* Recall that the pentagon *hipqk* is the defining pentagon of the icosahedron and thus its sides are the edges of the icosahedron (i.e., 4). But in computing the surface area of the dodecahedron, Bombelli had already computed the diameter of the circumscribing circle of a pentagon with side 4 (i.e., for a dodecahedron and icosahedron with the same *edge* the pentagonal face of the first has the same area as the defining pentagon of the second). Thus *mo* is known and the Pythagorean Theorem gives *nl*. By VI,8,corollary—using the right triangle *mlr* with altitude *mn*—we have $D = R(40 + R(320))$. Note that DEMR entered only in the original area computation of 5.1. This method may be compared with that of Francesca's problem 14.

(6.3) *Find the volume.* This is essentially done as in Francesca's problem 16; however, the answer obtained is now the equivalent $80 + R(3555\ 5/9)$.

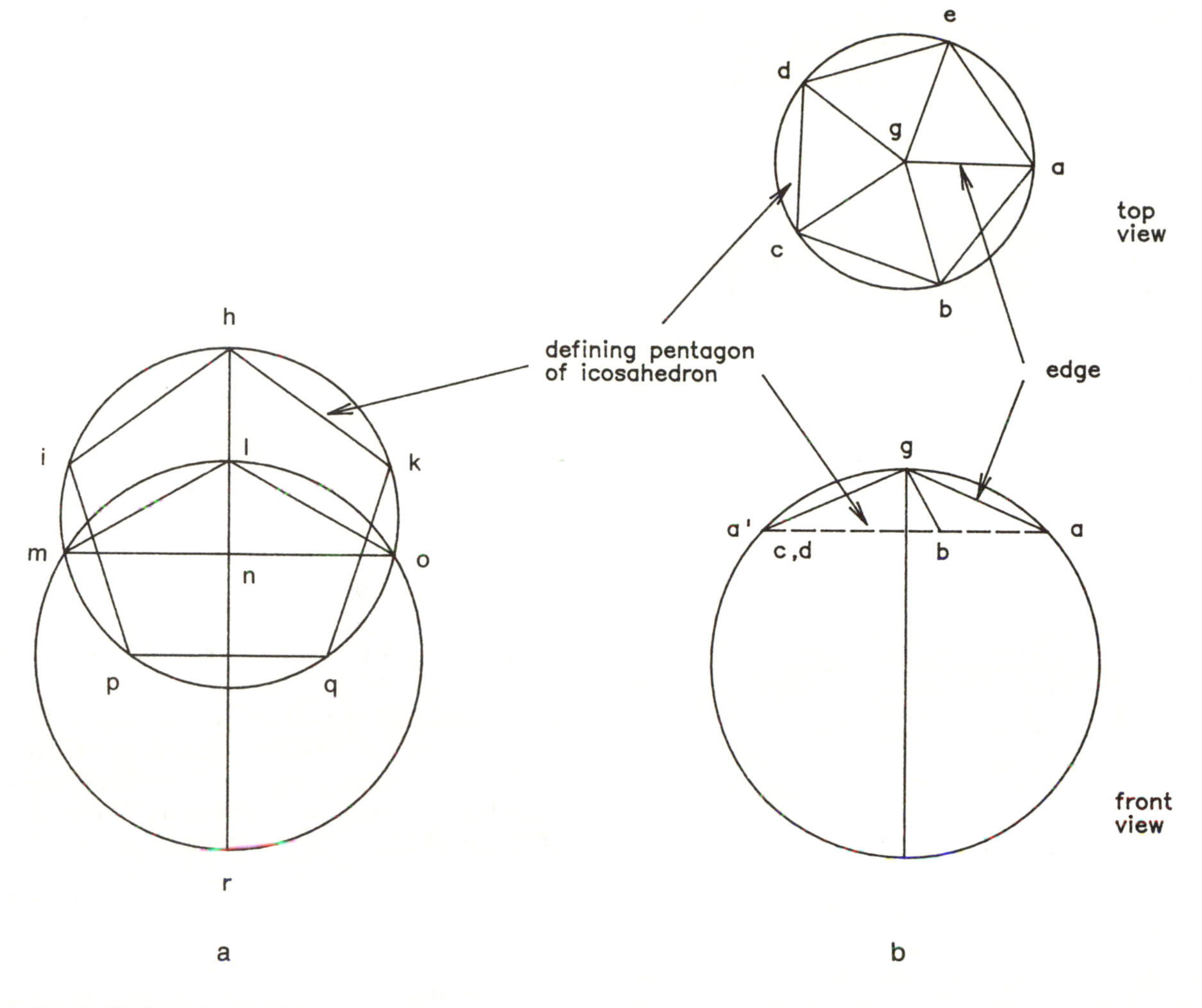

Source: Bombelli–Bortolloti [p. 661]

FIGURE IX-20. Icosahedron (dimensions in top and front views exaggerated for clarity)

Bombelli, it would seem from the unpublished manuscript that he left behind, was an independent thinker who took the dodecahedron and icosahedron problems of Francesca and looked at them in an original and quite striking way, which avoided using the results of Books XIII and XIV as much as possible.

G. Candalla (1502-1594)

At the end of his 1566 edition [Euclid–Candalla] of the *Elements* Candalla (= Flussas = Francois de Foix, comte de Candale) added on another "book" which dealt with various inscription problems involving polyhedra. In this sense it is a continuation of Book XV, but at a higher level. In particular, we find various relationships involving DEMR which occur in these inscription problems. I have used the version given in Euclid–Billingsley [Book XVI, fol. 445v.].

Proposition 3 states that if in a pentagon we drop the perpendicular AG, then it is divided in EMR by the diagonal BF. The proof employs XIII,8 which tells us that AD is divided in EMR at I. The use of similar triangles together with the Ratio Lemma of XIV gives the result.

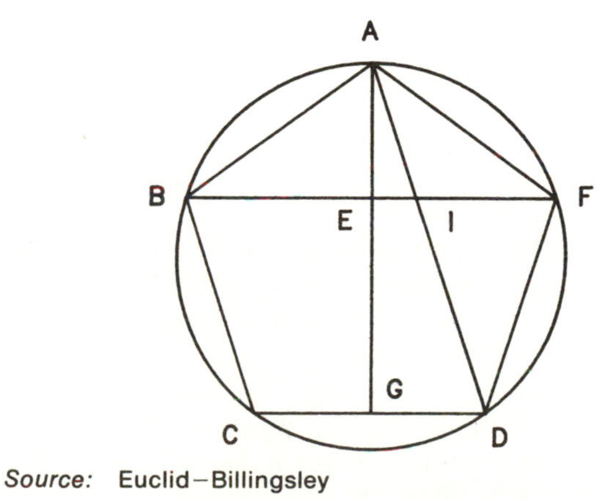

Source: Euclid–Billingsley

FIGURE IX-21. Proposition 3

In Proposition 4 (Fig. IX-22), we learn that if we cut the sides of one of the triangles ABG of a tetrahedron in EMR at I, M, and L, and if we draw the various lines to obtain triangle CDE, then this is the triangle of the icosahedron inscribed in the tetrahedron. This is based on Candalla's development of Book XV which also involves DEMR and is much more developed than Book XV of the critical edition. Proposition 5 states that with respect to the icosahedron inscribed in a tetrahedron we have S(smaller segment of e_4 divided in EMR) $= 2 \cdot S(e_{20})$.

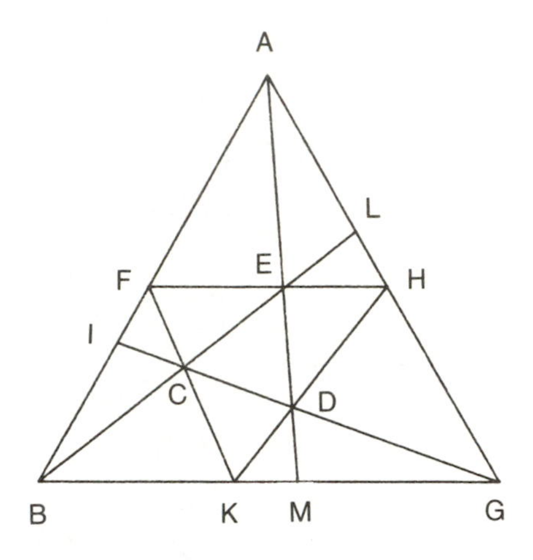

Source: Euclid–Billingsley

FIGURE IX-22. Proposition 4

The "converse proportion" of EMR is shown to hold between a dodecahedron and the cube inscribed in it in proposition 18. Other propositions in Candalla's addition involve EMR implicitly.

I note that Kepler used Candalla's *Euclid* in connection with his polyhedra theory of planetary motion; see Kepler [1597, 46].

H. Ramus (1515-1572)

An analysis of Ramus' *Geometry* [1599] is given by Verdonk [1966, section 58]. Hookyas [1958, 58] and Mahoney [*DSB* 2, 288] comment briefly on Ramus' concept of geometry and how he felt it should be developed and presented.

In Book XVIII Ramus discusses the inscription of the pentagon, and then in connection with XIII,8 [p. 126] he presents, without proof, the following construction of a pentagon with a given side.

One takes the given line AB and adds the larger segment BH to both sides giving points F and G. From these latter two points we swing arcs of length AB. These arcs meet the arcs, also of length AB, from A and B at points C and E. Point D is then obtained by swinging arcs from C and E.

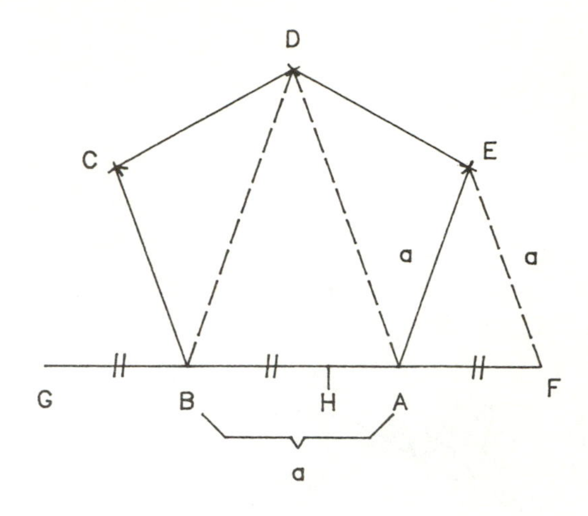

Source: Ramus [1599, 126]

FIGURE IX-23 (letters and dotted lines added)

That this method works may be seen as follows. Start with the pentagon $ABCDE$ and draw the triangle ABD and then the similar triangle FAE. By XIII,8 AB is equal to the larger segment when BD is divided in EMR. Thus AF is equal to the larger segment when $AB = AE$ is divided in EMR.

In Book XIX, "On the Measurement of Regular Polygons and Circles" [p. 130], the first result is that, if we divide a polygon into congruent triangles radiating from the centre, then the area of the polygon is (altitude of the triangle) · (1/2 perimeter). We already saw this result in connection with al-Khwarizmi (Section 28,A,ii) and Abraham bar Hiyya (Subsection A,1,a).

Now Ramus illustrates this with a regular pentagon (Fig. IX-24), but instead of assuming that just the side or just the radius of the circumscribed circle is given he assumes that both are given, the side being 12 and the radius being 10! The Pythagorean Theorem is then used to obtain that the altitude is 8, which in turn gives an area of 240.

Somehow Ramus knows that the right triangle formed by the half-side, altitude, and radius is approximately a 3–4–5 triangle. We have seen this previously in connection with Hero in Section 25,A. I note that Hero is mentioned in result 5 of Book XIII [p. 155].

The statement of result 11 of Book XXV [p. 165] is that of XIII,16[a],corollary. In connection with this Ramus appears to be calculating the volume of an icosahedron whose edge is 6. An answer of 519 11/37 is obtained whereas the correct answer is 471.25.

In connection with result 15 of Book XXV [p. 168], which is just XIII,17[a],corollary, Ramus calculates the volume of a dodecahedron whose side is 7 3/5 (!) and obtains a volume of 760, whereas the true volume is 3363.93.

For Ramus' comments on XIV,2 lemma and XIV,** and criticisms, see Herz-Fischler [1985, section 2, fn. 3].

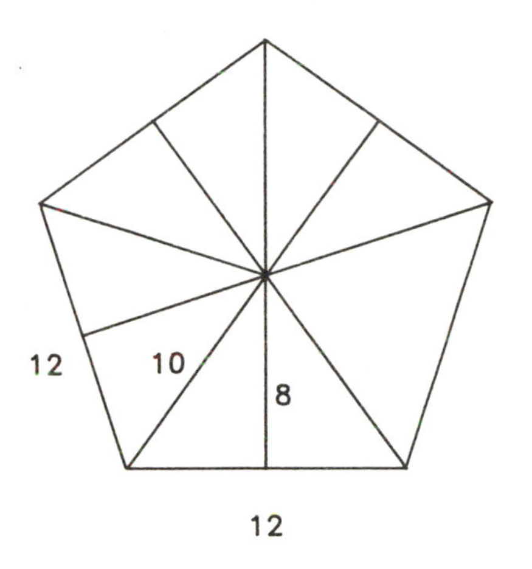

Source: Ramus [1599, 130]

FIGURE IX-24. Construction of the Pentagon

I. Stevin (1548-1620)

In the third book of his *Problemata geometrica* [1585], Stevin considers the construction of both regular and semi-regular polyhedra. As far as the regular polyhedra are concerned he simply states [pp. 241, 253] that one obtains the edge of the dodecahedron by dividing the edge of the cube in EMR and refers the reader to XIII,18 of the *Elements*.

When considering the semi-regular polyhedra, we find an avoidance of the direct use of the term DEMR and related properties. Thus when describing [p. 239, definition 20] the construction of the so-called truncated dodecahedron (i.e., twelve decagonal and twenty triangular faces $= 3.10^2 = \{12_{10}, 20_3\}$; see Williams [1972, 88] and Beard [1973, 60, 64]), he speaks of "[dividing the edges of a dodecahedron] into three parts in such a way that all the middle parts are to the two other parts of said edges as the chord of an arc of two-fifth of the circumference of a circle [i.e., the diagonal of the pentagon] to the chord of an arc of one-fifth of the same circumference [i.e., the side of the pentagon . . . .]"

For a comparison of Stevin's and Francesca's methods of constructing various semi-regular polyhedra, see Clagett [1978, 398].

In his definition [p. 229, def. 9; p. 243, section 9] of the so-called "small stellated dodecahedron" (i.e., a dodecahedron with pyramids built upon its 12 faces $= \{4/2,5\}$; see Williams [1972, 69] and Beard [1973, 74, 79]), he speaks of the altitude going from one vertex of the pentagon to the opposite side, but does not enter into details of the computation.

In the second book of the *Problemata geometrica*, we find the following problem [p. 215, third example]: "Let the line A be given, a perpendicular line of any unknown equilateral and equiangular pentagon, from an angle to the mid-point of the opposite side thereof. . . . Let it be required, if the kind of such a line A is given, to construct a pentagon equal and similar to the unknown pentagon in which A occurs."

The problem is solved by the method of false position ("regula falsi," see the introduction [p. 124] or Smith [1923, II, 437]), that is, essentially by using proportions. The concept of DEMR thus does not enter directly into the solution.

J. Pre-1600 Numerical Approximations to DEMR

i. Unknown Annotator to Paccioli's Euclid

In the margin of the Bibliothèque Nationale (Paris, côte Rés. V. 104) copy of Paccioli's 1509 edition of the *Elements* [Euclid–Paccioli], there are annotations relative to II,11. The book contains no indications of ownership or dates and the response to my inquiry to the Réserve stated that the Bibliothèque Nationale has no record of when this book entered the collection. However, it appears from the handwriting that the annotator was an Italian writing in the early part of the 16th century. The diagram is as follows:

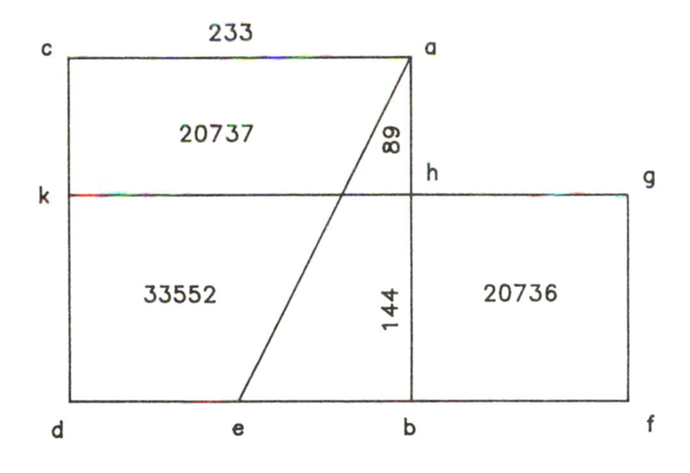

Source: Euclid–Paccioli

FIGURE IX-25. Annotated Diagram of II,11

A reproduction, transcription, translation, and discussion of this text appear in Curchin and Herz-Fischler [1985]. Although the text presents some difficulties, it is clear that the writer was well aware of the relationship between the Fibonacci sequence (see Subsection A,v and Section 32,B)—from which the terms 89,144,289 appear here—and the ratio determined by DEMR.

This text is the earliest unequivocal example known of the relationship. The appearance of 8 and 5 in the text of Hero (Section 25,A) has nothing to do with the Fibonacci sequence; recall also the discussion of Archimedes in Section 23,A. The question of early evidence for knowledge of the Fibonacci approximations to the ratio determined by DEMR was also discussed in Curchin and Herz-Fischler [1985].

I will discuss the relationship of Kepler to the above in Section 32,A,ii and the later history of the approximations in Section 32,B.

ii. Holtzmann (1562)

Despite its supposed aim of serving as a "practical" text, the 1562 edition of the *Elements* [Euclid–Holtzmann; see the description in Euclid–Heath, I, 107] does not give any actual numerical value in connection with VI,30.

The diagram (Fig. IX-26) is a variation on that of II,11. Instead of the hypotenuse $R(245)$ of the right triangle, with sides 14 and $1/2(14) = 7$, being swung down, the distance $R(245)$ is shifted over to the midpoint of the square of side 14. This gives the distance $R(245) - 7$ which is the length of the larger segment.

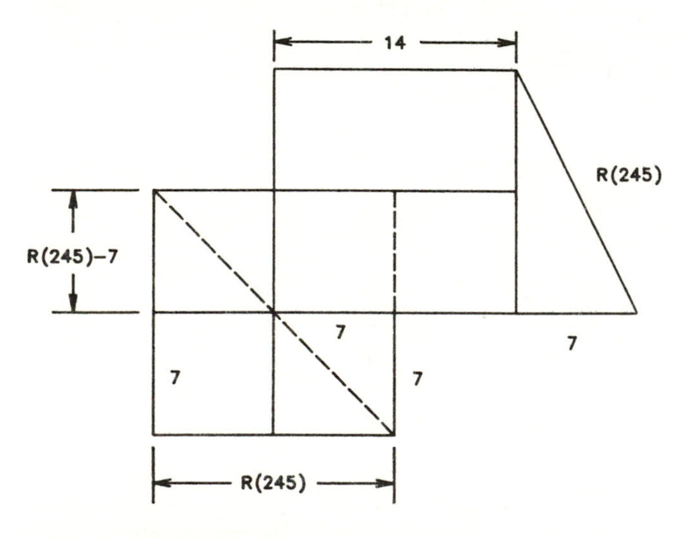

Source: Euclid–Holtzman

FIGURE IX-26. Diagram with VI,30

iii. Mästlin (1550-1631)

In October 1597 Kepler wrote a letter to his former professor, Michael Mästlin, in which he gives the statement and proof of a theorem that I shall discuss in more detail in Section 32,A,i. For our purposes here, it suffices to say that, relative to Figure IX-27, if line AE is divided in EMR at F, then one shows that $ED = AF$ and $S(AE):S(AD):S(ED) = AE:AF:EF$.

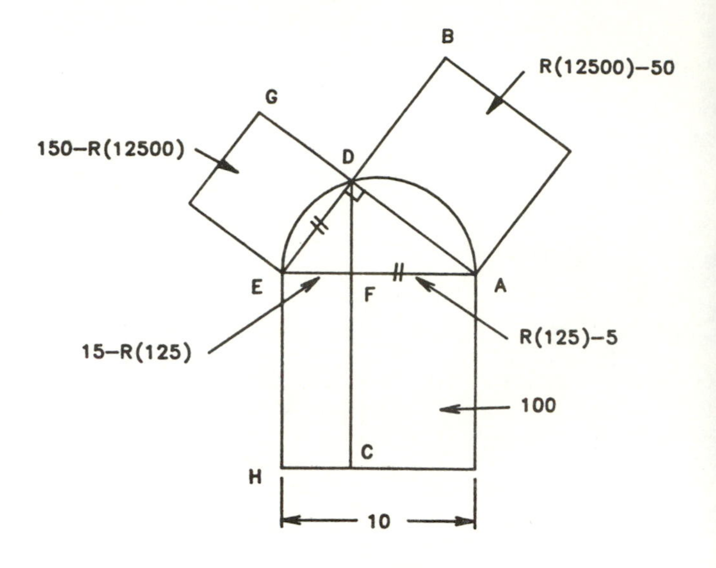

Source: Diagram and annotations in Kepler [1597, 144]

FIGURE IX-27

On the original letter [Kepler, 1597, 144], Mästlin wrote the following numerical verification of the result: "If EA is 10, ED will be R(125) – 5 and EF 15–R(125). And since [S(AE)] = 100 and [S(ED)] = 150 – R(12500), [S(AD)] will be R(12500) – 50. Its [i.e., $S(AD)$] side is AD. Therefore the following things are proportional: 1) ED [=] R(125) – 5, AD [=] R(R(12500) – 50). 3) AE [=] 10. Thus just as EA = 10000000 so ED is about 6180340 and AD is about 7861514."

Recall that we saw the division of 10 in EMR and the same values for the segments in connection with Paccioli's *Divina proportione* in Subsection D.

The interest in the annotations lies in the fact that we have here, to my knowledge, the earliest attested example of somebody actually finding a decimal number for the numerical value associated with DEMR. Recall also the various sexagesimal calculations, starting with Ptolemy (Section 26,A), for a_{10} and a_5 that we have seen. Mästlin's value for ED agrees with the first seven places of the exact value $(R(5)-1)/2 \cdot 10^7$.

K. Approximate Constructions of the Pentagon

Since their origin does not seem to be directly related to DEMR as such, I shall simply give biographical details of various approximate constructions of the pentagon.

i. Geometrica Deutsch *(Anonymous, 1484)*

The text and construction are given by Wieleitner [1927, II, 27]; see also Coolidge [1949, 67] and Hallerberg [1959, 234].

ii. *Da Vinci (1452-1519)*

The approximate construction is given in da Vinci–de Toni [73, plate of 17V; the picture is reproduced in Beaujouan, 1975, 458, fig. 11; see also Coolidge, 1949, 55].

iii. *Dürer (1525)*

In his *Underweysung der Messung mit dem Zirkel und Richtscheit* (*Treatise on Mensuration with Compass and Ruler*) [1525, Book II, no. 15,16], we find the same construction as in *Geometrica Deutsch*, as well as another one; see also Wieleitner [1927, II, 44] and Coolidge [1949, 67].

iv. *Benedetti (1585)*

In his *Book of Various Mathematical and Physical Speculations* (text given by Wieleitner [II, 44]), Benedetti computes the accuracy of the construction in *Geometrica Deutsch*.

Section 32. The 17th and 18th Centuries

A. Kepler (1571-1630)

As with so much of his work, Kepler's connection with DEMR is multifaceted and we see serious mathematics intertwined with, at least from our point of view, mysticism (see Appendix II). For summaries of Kepler, see Hofmann [1972] and Gingerich [*DSB*].

i. *Magirus—The Right Triangle with Proportional Sides (1597)*

In a letter Kepler wrote to his former professor Michael Mästlin in October 1597, we read the following [Kepler, 1597]:

"Now furthermore so that the Royal professor may achieve something even worthy of his title he proposes to you most renowned D. [Doctor] a geometrical problem; to build a right angled triangle all of whose three sides are mutually and continuously proportional so that just as the lesser side is to the greater around the right triangle so is the latter to the one subtended by the right right angle [i.e., the hypotenuse]. I do not know the distinctive use of this discovery except that it more greatly enhances geometry. For if you think about it, you will see that there can only be one form of it not many [i.e., only one solution to the problem?]. This discovery is due to Magirus wherefore I courteously greet him and show myself thankful to him because of this, that as by his very pleasant theorem he has pleased me among others with a new enthusiasm for geometry. The proof is easy on the basis of his discovery; because I have changed it into another form such that I think I will easily persuade even Magirus himself to think that it is entirely mine."

I have not been able to identify who Magirus was. What are we to credit him with? The discovery of the statement and a proof of some or all of what follows? The fact that there can only be one solution without determining what the solution is?

Kepler now gives the statement of the result: "If on a line which is divided in e.m.r. one constructs a right angled triangle such that [the vertex of the] right angle is on the perpendicular constructed at the section point then the smaller leg [literally, the smaller side around the right angle] will equal the larger segment of the divided line."

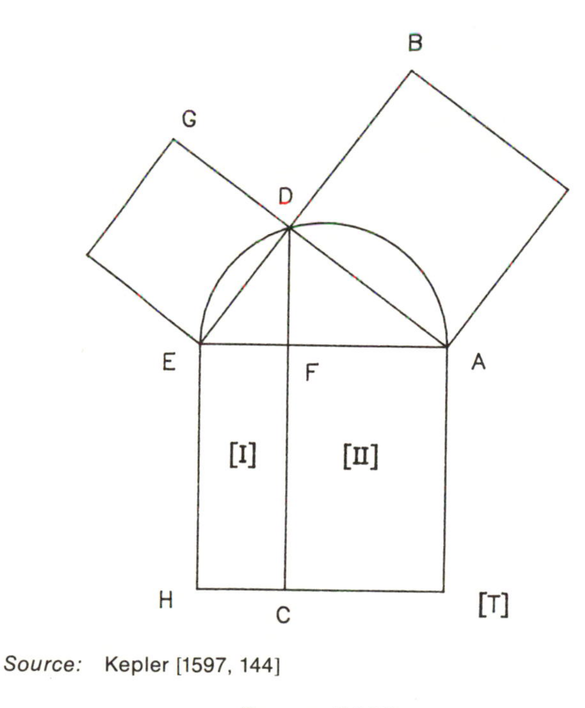

Source: Kepler [1597, 144]

FIGURE IX-28

Proof: Let the line *AE* be divided in EMR as in VI,30 [with *FA* the larger segment]. Draw the perpendicular at F and construct a semi-circle on AE so that in triangle EDA and ADE will be a right triangle. We must show that $DE = EA$.

By VI,8, $\Delta EFD \sim \Delta DFA \sim \Delta ADE$ and $AE{:}ED = ED{:}EF$. But since AE is divided in EMR at F, we also have $AE{:}AF = AF{:}EF$. These two proportions together imply (V,9) $AF = ED$.

"COROLLARY I. The sides of the triangle are proportional."

Proof: By VI,8 $AF{:}FD = FD{:}FE$ and since $ED = AF$, $ED{:}FD = FD{:}FE$, thus the sides of ΔEFD are proportional. Consequently, also by VI,8, the sides of ΔEDA are proportional, that is, $ED{:}DA = DA{:}EA$.

Note: Since both the statement and proof of the next corollary present certain difficulties I will leave them as they appear, elucidating matters where I can.

"COROLLARY II. The proportion of the sides is given. Moreover I consider the proportion to be determined if we express how it relates to another known proportion in particular a section according to e.m.r."

Proof: We construct squares on the three sides of triangle AED and draw altitude FC to form $R(EFCH) = \text{I}$ and $R(CFAT) = \text{II}$. Since these two rectangles and $S(AE)$ have the same altitude, we have $\text{I}{:}\text{II}{:}S(AE) = EF{:}FA{:}AE$. Furthermore, $\text{I} = S(ED)$ and $\text{II} = S(DA)$ (Kepler invokes the Pythagorean Theorem here, but this is not what is needed; rather by VI,8 $ED{:}EF = AE{:}ED$ or $S(ED) = R(EF,AE) = \text{I}$). If we substitute for I and II in the previous string of proportions we obtain:

$$(1) \qquad S(ED){:}S(DA){:}S(AE) = EF{:}FA{:}AE.$$

Note: It seems to me that, since $EF{:}FA = FA{:}AE$ (by the definition of EMR), Kepler could have stopped right here, except for the immediate corollary which follows the Q.E.D. at the end of this proof. What he has shown can be illustrated numerically as follows:

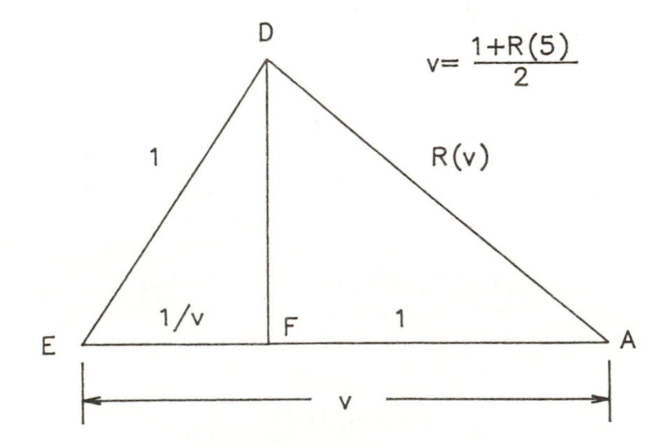

FIGURE IX-29. Kepler's (Magirus' ?) Triangle

If we let v be the numerical value, $(1 + R(5))/2$, associated with DEMR and let this be the length of AE, then $AF = 1$. By the result of the theorem itself, we have $ED = AF = 1$. If now we let a be the other leg DA then (1) gives $1{:}a^2{:}v^2 = EF{:}1{:}v$ so that $a = R(v)$ and $EF = 1/v$.

Note that when we look at the square of the sides we obtain $v^2, v, 1$, and thus by the Pythagorean Theorem $1 + v = v^2$, something that of course can be easily checked numerically. The result in this numerical form has been rediscovered many times; for some of its pseudo-mathematical history see Fischler [1979]. Recall also Mästlin's numerical remarks on Kepler's letter (Section 31,J,ii).

However, Kepler now continues: "But by [VI,19] amongst [S(ED), S(DA), S(AE)] which are similar figures there is the 'duplicate ratio' to the sides ED, DA, AE."

Note: The term "duplicate ratio" is introduced in V,def.9. In its simplest form, it occurs when three straight lines are proportional, as is the case with EF, FA, AE, since F divides AE in EMR; see the proof of VI,19 [Euclid–Heath, II, 233, line 21]. From the corollary to VI,19, we learn that, since EF, FA, AE are proportional, we have $S(AE){:}S(FA) = EF{:}AE$. The terminology here too is that $S(EF)$ and $S(FA)$ are in duplicate ratio [see line 26ff.]. I do not understand how Kepler obtains his statement involving the sides of the big triangle ED, DA, AE. Kepler now writes: "In fact it is the proportion of a section in e.m.r. Therefore amongst the sides ED, DA, AE there is 'dimidia proportio' of a section [divided] in e.m.r. Therefore the proportion is expressed by a well determined ["certo nomie"] proportion. Q.E.D. And therefore AE, AD, DE, DF, FE are in continuous proportion."

Note: $AE{:}AD = AD{:}DE$ is the result that Kepler announced at the beginning of the letter and would have followed immediately from the main statement of the theorem and similar triangles.

ii. Fibonacci Approximations to DEMR (1608)

We saw in Section 31,J,i that an unknown annotator to Paccioli's edition of the *Elements* used the division of a line of length 233 into segments of length 144 and 89 to obtain an approximation of the ratio determined by DEMR. We now find a more complete discussion of these approximations in Kepler's letter [1608b], dated May 12, 1608, to Joachim Tanckius, a professor in Leipzig. This is a rather amazing letter combining mathematics, mysticism, and music. I shall present other parts of this letter in Appendix II,g, but for now what interests us are the following segments [lines 116-42]:

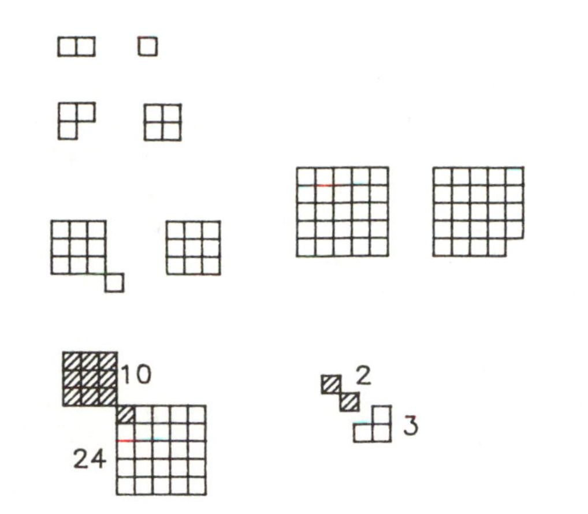

Source: Kepler [1608b]

FIGURE IX-30

"Now the divine proportion [i.e., DEMR; this terminology comes from Paccioli's book, Appendix I,C,2] cannot, however, be perfectly expressed in numbers; it can nevertheless be expressed in such a way that through an infinite process we come closer and closer to it, and in delineating the square we are never more than a unity away. Let us begin with the smallest numbers. Let the smallest divisible number be 2; it has the parts 1 and 1; from these, one would form the smaller, the other the larger part so that one obtains the terms 1, 1, 2. If the divine proportion were thereby perfectly expressed, the rectangle [i.e., product] of the extremes would have to be equal to the square of the middle. It is, however, too big by one. I therefore continue, and add the larger part 1 to the whole 2; so that it makes 3, and so the terms are 1, 2, 3. Here the rectangle [i.e., the product] of the outside terms is 3, the square of the middle terms 4. So, I again go further and add the larger part 2 to the whole 3 so that the new sum is 5; the terms are 2, 3, 5. The rectangle [i.e., product] of the outer terms 2, 5 is 10, the square of the middle terms 9. Likewise add 3 and 5 to make 8. Five times 5 is 25, three times eight is 24. There I continue in this way so the shortage of the one is always equal to the surplus of the other. I believe I can't make this any clearer or more evident than when I say: you see there the diagram of the masculine, here of the feminine term. And so either the surplus in the one case, or the shortage from the rectangle [i.e., product] in the next, makes it so that we can never express the divine proportion perfectly. And yet we approach it ever more closely, the more often we take up the combination. And so when one adds 5 and 8 one obtains 13. 13 and 5 multiplied work out to 65. Eight times 8, however, is 64. (See, above I mentioned that 13 rotations of Venus come close to 8 years.) The shortage of one rectangle [i.e., product] compensates, then, for the surplus of the other, so that in both cases we almost have two divine proportions."

As the first line of the text indicates Kepler is well aware of two fundamental properties of the Fibonacci sequence (see Subsection B); not only do the ratios of the terms of the Fibonacci sequence approximate the numerical value $(1 + R(5))/2$ determined by DEMR, but also that the square of any term f_n differs by at most one from the product of the two adjacent terms in the sequence. Kepler starts off with $f_3 = 2 = 1 + 1 = f_2 + f_1$, but finds that $2 \cdot 1 = 1^2 + 1$, that is, the product of the extremes is too big by 1. So now he tries $f_4 = f_3 + f_2 = 2 + 1 = 3$ and finds that $3 \cdot 1 = (2)^2 - 1$, which results in the product of the extremes being too small by 1. He continues in this way obtaining $5 = 3 + 2$, $8 = 5 + 3$, and $13 = 8 + 5$. Note that Kepler has not really shown that no integers work but rather that we will not be able to find numbers that work from this sequence.

A shorter version, which however leaves out the explanation of how Kepler arrived at the sequence, appears in his *Six-Cornered Snowflake* [Kepler–Hardi, 21; Kepler–Halleux, 65] which dates from 1611:

"Of the two regular solids, the dodecahedron and the icosahedron, the former is made up precisely of pentagons, the latter of triangles but triangles that meet five at a point. Both of these solids, and indeed the structure of the pentagon itself, cannot be formed without this proportion that the geometers of today call divine. It is so arranged that the two lesser terms of a progressive series together constitute the third, and the two last, when added, make the immediately subsequent term and so on to infinity, as the same proportion continues unbroken. It is impossible to provide a perfect example in round numbers. However, the further we advance from the number one, the more perfect the example becomes. Let the smallest numbers be 1 and 1, which you must imagine as unequal. Add them, and the sum will be 2; add to this the greater of the 1's, result 3; add 2 to this, and get 5; add 3, get 8; 5 to 8, 13; 8 to 13, 21. As 5 is to 8, so 8 is to 13, approximately, and as 8 to 13, so 13 is to 21, approximately."

See also the quotation from Book V of *Harmonice mundi* (1619) given in Appendix II,9.

B. *The Fibonacci Sequence*

The Fibonacci numbers are the elements of the sequence 1,1,2,3,5,8,13,21,34,55,89,144. . . . If we designate the nth term by f_n then the rule of formation is:

(2) $$f_{n+2} = f_{n+1} + f_n.$$

This sequence has many interesting mathematical properties and there are two that are of particular interest to us; for modern proofs see Hardy and Wright

[1938] and Niven and Zuckerman [1960]. The first states the relationship between the Fibonacci numbers and the numerical value determined by DEMR:

(3) f_{n+1}/f_n is an increasingly good approximation to $(1+R(5))/2$.

An indication of how good the approximation is was given in Section 23,A.

The second result is:

(4) $$(f_{n+1})^2-(f_{n+2})(f_n)=\pm 1.$$

In Section 31,A,v I discussed Fibonacci's "rabbit problem" and how it gave rise to the Fibonacci sequence. Further I noted that Fibonacci's discussion was completely verbal and that there is no indication that Fibonacci was aware of either (3) or (4).

We also saw in Sections 31,J,i and 32,A,ii that both an anonymous 16th-century annotator to Paccioli's *Euclid* and Kepler in 1608 were aware of relationships (3) and (4). Since Fibonacci's works were not printed until the second half of the 19th century it seems unlikely that either Kepler or the two authors whom I shall now discuss knew of the Fibonacci text.

After Kepler the relationship (3) next appears in a work by Girard [1634, 20]. It appears that Girard discovered it for himself for, he says, "I will add two or three items of special interest hitherto unknown" ("particularitez non encor par cy devant practiquées . . ."). He then goes on to say that the ratio determined by EMR can be obtained by looking at successive ratios of members of the 0,1,1,2,3,5,8,13,21, etc. (note the term 0 added in front). Then Girard says that the isosceles triangle of sides 13,[21],21 will accurately approximate the 36°–72°–72° triangle of IV,10 and also gives the terms 59,475,986 and 96,234,155. Maupin, the commentator on the text [p. 207], points out that these figures are incorrect and indicates how the error may have crept in. I note that Girard does not indicate how he obtained the series; see the comments by Maupin.

Girard's work was taken up by Simpson [1753] whose approach is very close to that of Kepler. Simpson says that the fact that the ratios are all approximately the same will follow if one can show (4) and this he proceeds to do, essentially by the principle of mathematical induction. Simpson also states that the number towards which the ratios tend is $(1+R(5))/2$ (i.e., that (3) holds), but he does not give a proof of this fact.

For more historical references to the Fibonacci sequence and in particular to relationship to DEMR, see Dickson [1981, chap. XVII; note the difference with the above description of the work of Girard and Simpson].

C. *Fixed Compass and Compass Only Constructions*

In Sections 28,C and 31,A,6 I mentioned the work of Abu'l-Wafa' and a group of 16th-century Italian mathematicians, involving constructions with a compass with a fixed opening and a straight edge. We find a renewal of interest in this problem in the 17th century in Mohr's *Euclidis Curiosi*, and then in a further work of Mohr. In the 18th-century study of Mascheroni the constructions are performed with a variable compass but without a straight edge; see Hlavaty [1957] and Kostovski [1959] for a discussion of the mathematics involved.

i. Mohr (1640-1697)

The *Euclidis Curiosi*, published in Dutch in 1673 and in English in 1677, has been recently reprinted in Mohr–Meyer. This work and Mohr's life are discussed in Hallerberg [1959, 1960].

In problem XXV [Mohr–Meyer, p. 17 of the reproduction of the English version] one is required to inscribe a regular pentagon in a circle whose radius is the opening of the fixed compass. The text is defective (the location of point F is not described), but from what is said it is clear that Ptolemy's construction (Section 26,A) is involved. In proposition XXVI one is required to circumscribe a pentagon and this construction simply uses the previous one; see IV,12.

Mohr's *Euclides Danicus* was only "rediscovered" in 1928 [Mohr–Hjelmslev and Pal] and was found to have anticipated Mascheroni's work on constructions using only a variable compass; the two authors are compared in Seidenberg [*DSB*].

The constructions of interest to us [p. 19], namely, problem 39—division of a line in EMR using the area formulation of II,11; problem 40—the construction of the 72°–72°–36° triangle; problem 41—inscription of the regular pentagon, are all based on the construction of the *Elements*. No proofs are given and the reader is presumably supposed to be familiar with Euclid; the emphasis of course is on the actual technique of the construction using the compass rather than on the method. Note how the approach to the construction of the pentagon differs in the *Euclidis Curiosi* and *Euclides Danicus*.

ii. Mascheroni (1750-1800)

On Mascheroni, see Seidenberg [*DSB*]. The proofs in *Geometry of the Compass* are "algebraic" in nature, that is, the various lengths are actually calculated. I have used both the German and French editions [Mascheroni–Gruson and Mascheroni–Carette].

In Book II, section 40, the pentagon is constructed by a method inspired by Ptolemy's construction (Section 26,A) and then in section 41 the decagon is constructed using XIII,10 rather than bisection.

Of potential great interest is section 97 of Book V where one is to divide a given line in EMR. We are told that this construction is "once more one of those that one solves more simply by means of a compass alone than with both a compass and straight edge." Unfortunately it appears that there is an error. I present the construction as given in both the German and French (Fig. IX-31).

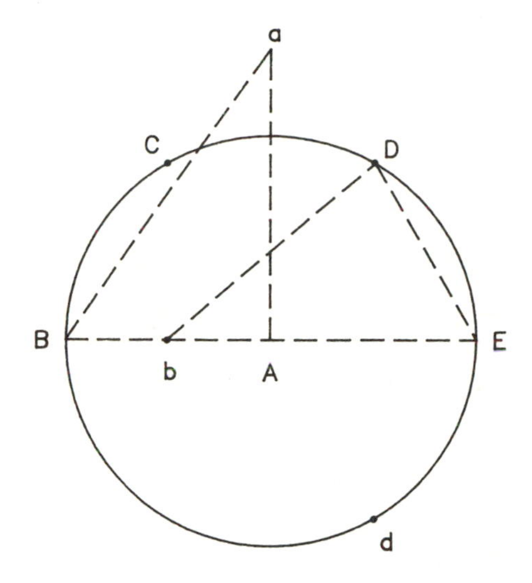

Source: Mascheroni–Gruson [fig. 41]

FIGURE IX-31 (dotted lines added)

Given line AB one draws the circle of radius AB and then finds C,D,E,d such that $BC = CD = DE = Ed = AB$, that is, each of the corresponding arcs is one-sixth of the circle. Next one finds point a equidistant from B and E—and therefore on the perpendicular bisector of the diameter BE—such that $Ba = Ea = BD$. Finally one finds point b such that $Eb = db = Aa$. Point b, which is supposed to lie on the diameter BE, is claimed to divide AB in EMR.

Note that if $Eb = bd$ then b will lie on the perpendicular bisector of Ed which passes through the centre A. Furthermore since $\measuredangle DBE = 30°$, we have $aB = BD = R(3) \cdot AB$ and consequently $aA = R(2) \cdot AB$. Thus even the assumption that b is the point on the diameter BE, such that either bE or bd is equal to aA, will not lead to the proper division point. I have not been able to think of a way of emending the text.

In Book IX, section 180 one is to construct a pentagon whose side is given and two solutions are given.

Recall that in Subsection C we saw Kepler's construction of a right triangle whose sides are in geometric proportion. This appears in Mascheroni's work in section 180 of Book XI. Mascheroni starts off by using the same construction that I discussed above for dividing a line in EMR. Then by swinging an arc to the circumference, one has constructed a right triangle, one of whose sides is the larger segment when the hypotenuse is divided in EMR. This is the triangle of Figure IX-29, with the diameter here corresponding to v and the leg corresponding to 1.

Since Kepler's letter was not published until the second half of the 18th century [Kepler–Frisch], this letter could not be the direct source of Mascheroni.

The constructions of Book XIII of the *Elements* are used in Book XI, section 181 to find the radius of a sphere which corresponds to a given edge of a polyhedron.

Also in Book II, sections 186 and 187 we find two more problems involving the pentagon. In the first one must locate the points of intersection of the diagonals. Numerically this corresponds to XIII,8, but the construction given by Mascheroni does not use properties of DEMR.

In the next problem, one is required to inscribe six regular pentagons in a circle of a given radius. One of the pentagons is in the centre and the five others share an edge with the interior one and also have a vertex that touches the circumference.

By Way of a Conclusion

We have now arrived at the end of the 18th century. A description of all the results obtained from then on would constitute, not a history, but a textbook. Already almost 150 years ago in Chasles [1837, 513], we find the suggestion that somebody should write a book about the properties of DEMR. Concerning that possibility, I refer the reader, for the second time (see Section 10,A) in this book, to Csörgő and Révész [1981, 19]. As I sit here in July 1982 finishing this book, I am waiting for "A Generalisation of the Golden Number" [Fowler, 1982b] to be printed/and or arrive, and one can be sure that other results related to DEMR, presently unknown to me, are in the process of being printed or at least written up. Paccioli (see the quotation on the title page) was right when he said, "Excellent Duke, it does not seem to be appropriate to me to continue further on its infinite effects because the paper would not be sufficient for the ink to express them all. . . ."

APPENDIX I

"A PROPORTION BY ANY OTHER NAME": TERMINOLOGY FOR DIVISION IN EXTREME AND MEAN RATIO THROUGHOUT THE AGES

In Parts A and B of this appendix I present examples of the terminology that has been used in connection with DEMR. This is followed by a discussion of some of the names that have been associated with DEMR.

When it comes to the terminology involved in the definition of DEMR (VI,def.3), there are two traditions. The first one that we find in Greek manuscripts, translations from the Greek, and certain Latin manuscripts (see Section A,2 on Fibonacci) which adopt a Greek terminology, use the expression "division in extreme and mean ratio" (Greek: akros kai mesos logos).

The question arises as to the origin of this expression. If we look at VI,16 we see that when we have lines A,B,C,D such that $A:B = C:D$—the lines are said to be proportional—then A and D are called the extremes and B and C are called the means. In VI,17 there are only three lines involved and $A:B = B:C$. Again the three lines are said to be proportional; A and C are called the extremes and B is called the mean. Thus in the expression "extreme and mean ratio," the words "extreme" and "mean" are simply technical terms for the lines involved in the definition. Note that if three lines A,B,C are proportional in the sense of VI,17 and if in addition $B + C = A$, then the line A will be divided in EMR. As to the word "ratio" (vaguely stated to be "a kind of relation with respect to size" in V,def.3), I suspect that the proportion involved in the definition of DEMR was considered to have defined a ratio so that the term "extreme and mean ratio" was considered more appropriate than, say, "extreme and mean proportion." See, however, Section 25,A, on the "Ratio Lemma." For various examples and explanations of the Greek words involved in the definition of DEMR, see Mugler [1958, 274, 287, 413, 425]. There are discussions of the meaning of DEMR and the correct way of translating it in Terequem [1838, 1853] and Vincent [1844], but I found the arguments difficult to follow and I am not quite sure what the problem is. See also the citation from Ramus' *Geometry* in Appendix II,6.

The second tradition, found in Arabic texts and Latin and Hebrew translations from the Arabic, employs the expression "proportion having a middle and two ends" (Arabic: nisbah ḏāt wasaṭ wa-ṭarafayin; usually "nisbah" is translated as "ratio," but since this would not seem to make any mathematical sense I have translated it as "proportion"). Presumably this terminology came directly from the definition of DEMR and the terminology of VI,17.

In Subsections A and B I give examples of texts, and their translations, which illustrate the two traditions. The statement of XIV,** (in Section 24) in various texts is given in Herz-Fischler [1985].

A. "Extreme and Mean Ratio"

1. *Critical Edition* [Euclid–Heiberg]

VI,def.3: "Akron kai meson logon eutheia tetmēsthai legetai hotan ēi hōs hē holē pros to meizon tmēma, houtōs to meizon pros to elatton" (for the text in Greek letters, see the title page).

"A straight line is said to be divided in extreme and mean ratio when as the whole is to the greater segment so is the greater to the less."

VI,30: "Tēn dotheisan eutheian peperasmevēn akron kai meson logon temein."

"To cut a finite straight line in extreme and mean ratio."

2. *Bibliothèque Nationale (Paris), ms. Lat. 10257* (12th century) [Goldat, 1957]

VI,29 [= VI,30]: "Datam lineam dividere secundum proportionem medium et extrema continentem."

"To divide a given line according to a proportion containing a mean and extremes."

3. *Bibliothèque Nationale (Paris), ms. Lat. 7373* (12th century)

VI,def.3: "Secundum extremam et mediam proportionem recta secari dicitur quando fuerit ut tota ad maiorem sectionem ita maior ad minorem."

"A straight line is said to be cut according to extreme and mean proportion when it happens [literally, when it shall have been] that the whole [line] to the larger segment is as the larger to the smaller."

VI,30: "Datam rectam terminatam secundum extremam et mediam proportionem secare."

"To cut a given, finite straight [line] according to extreme and mean proportion."

On this manuscript, see Murdoch [1966].

4. *Fibonacci* (1220) [Fibonacci–Boncompagni, II, 196]

"Modus dividendi lineam media et extrema proportione."

"The method of dividing a line in extreme and mean proportion."

Fibonacci consistently uses the Greek expression even in problems taken from Arabic texts; see Herz-Fischler [1985, section 5] where it is also pointed out that Fibonacci used Book XIV of the manuscript of 3 or its prototype.

5. *Zamberti* (1516) [Euclid–Zamberti; the same wording is used in Fine, 1556; Euclid–Ramus]

VI,def.3: "Per extremam et mediam rationem recta linea dividi dicitur quando fuerit sicut tota ad maius segmentum sic maius ad minus."

"A straight line is said to be divided through mean and extreme ratio when the whole is to the larger segment as the larger to the smaller."

VI,30: "Datam rectam linear terminatam per extremam ac mediam rationem dispescere."

"To divide a given straight finite line into extreme and mean ratio."

Zamberti used a "Theonian" manuscript; see Murdoch [*DSB* 1, 437].

6. *Gryaneus* (1533) [Euclid–Gryaneus]

VI,def.3 is exactly as in the critical edition.

7. *Candalla* (1566) [Euclid–Candalla]

VI,def.3: "Extrema et media ratione recta linea dividi dicitur, quando fuerit sicut tota ad maius segmentum, sic maius ad minus."

"A line is said to be divided in mean and extreme when the whole is to the greater segment as the greater to the less."

VI,30: "Datam rectam lineam terminatam, extrema at media ratione secare."

"To cut a given straight finite line in extreme and mean ratio."

8. *Billingsley* (1570) [Euclid–Billingsley]

VI,def.3: "A right line is said to be divided by an extreme and mean proportion, when the whole is to the greater part, as the greater part is to the less."

VI,30: "To divide a right line given by an extreme and mean proportion."

9. *Commandino* (1572) [Euclid–Commandino]

VI,def.3: "Extrema ac media ratione secari recta linea dicitur, quando sit ut tota ad maiorem protionem, ita maior portio ad minorem."

"A straight line is said to be cut in extreme and mean ratio when the whole is to the greater part as the greater part is to the smaller."

10. *Clavius* (1579) [Euclid–Clavius]

VI,def.3: "Secundum extremam, et mediam rationem recta linea secta esse dicitur, cum ut tota ad maius segmentum, ita maius ad minus se habuerit."

"A line is said to have been cut according to extreme and mean ratio when the whole is to the larger segment as the larger to the smaller."

11. *Barrow* (1722) [Euclid–Barrow]

VI,def.3: "A right line AB is said to be cut according to extreme and mean proportion, when as the whole AB is to the greater segment AC, so is the greater segment AC to the less CB."

VI,30: "To cut a finite right line given AB according to extreme and mean ratio."

12. *L'Encyclopédie* (1756) [Diderot, D'Alembert, 1756, tome 16, 338; article "extrême"]

"Quand une ligne est divisée, de manière que la ligne enti[è]re est à l'une de ses parties, comme cette même partie est à l'autre, on dit en Géométrie que cette ligne est divisée en moyenne et extrême raison. Voici comme on trouve cette division. . . ."

B. *"Middle and Two Ends"*

1. *Abu Kamil* (c. 850-930) [Abu Kamil ms., problem 17, 76b; see Section 28,B]

"Mima d̲akara Euclid wa-d̲alika ānahu qāla 'idā qusima h̲aṭa 'alay nisbah d̲āt wasaṭ wa-ṭarafayin. . . ."

"It is clear from what Euclid said that if we divide a line according to the proportion which has a middle and two ends. . . ."

2. *Al-Biruni* (973-1050) [al-Biruni–Wright, 17, section 54]

"Ma nisbah d̲āt wasaṭ wa-tarafayin yakun fi h̲aṭṭ mustaqim yuqasim taqsimin nisbat aṣghar mima ili a'aẓamaha kanisbat had̲ā al-a'atar ili majmūahā i'ani likhaṭṭ kuluhu mā takā fi al-nisbah hia an takun al-thani wa-althalith fi junba wāḥida. . . ."

"What is the proportion with a middle and two ends? [It is the one] which divides a straight line in two [in such a way that] the ratio of the smaller is to the larger as the ratio of the greater is to the whole, that is the whole line. . . ."

3. [*Euclid–"Rome Taḥrīr," 131*]

VI,def.5: "Al-h̲aṭṭ al-maqsūm 'ala nisbah d̲āt wasaṭ wa-ṭarafayin hūa al-h̲aṭṭ al-maqsum bi-muktalafin takūn nisbat al-h̲aṭṭ kula ili aṭwal qismiya kanisbat aṭwal qismiya ili aṣghar humā."

"A line is divided in the proportion with a middle and two ends if the line is divided into two parts such that the ratio of the whole line to the greatest segment is the same as the ratio of the greater segment to the smaller."

4. *Mordechai Finzi* (c. 1460) [Abu Kamil–Sacerdote, 177]

". . . yḥs b'l 'mṣ' wšty qṣwwt" [Hebrew].

". . . proportion with a middle and two ends."

5. *Gerard of Cremona* (12th century) [Euclid–Gerard of Cremona]

VI,def.3: "Linea recta dicitur dividi secundum proportionem habentem medium et duo extrema, quando fuerit proportio totius linee ad maiorem eius sectionem sicut proportio maioris divisionis ipsius ad minorem eius sectionem."

"A straight line is said to be divided according to a proportion having a mean and two extremes, when the proportion of the whole line to its larger section is as the proportion of its larger division to its smaller section."

VI,30: "Lineam datam secundum proportionem habentem medium et duo extrema dividere."

"To divide a given line according to a proportion having a mean and two extremes."

6. *"Adelard I"* (12th century) [Adelard I–Busard]

VI,29 [= VI,30]: "Nunc demonstrandum est quomodo lineam assignatam supra proportionem continentem medium atque duas extremitates dividamus."

"It must now be demonstrated how we may divide the line assigned above according to a proportion containing a mean and two extremes."

For a discussion of the various manuscripts attributed to Adelard of Bath, see Clagett [1953a,b] and Adelard I–Busard].

7. *"Adelard II"* (12th century) [Goldat, 1957]

VI,29 [= VI,30]: "Quamlibet lineam propositam proporcionem habentem medium, duoque extrema secare."

"To cut any proposed line [into] a proportion having a mean and two extremes."

8. *"Adelard III"* (12th century) [Euclid–Adelard III]

There is no VI,def.3 even though definition 1 appears [299v = p. 594].

VI,30: "Quamlibet lineam propositam secundum proportionem habentem medium duoque extrema secare."

"To cut any given line according to the proportion having a mean and two extremes."

9. *Hermann of Carinthia* (fl. 1140) [Euclid–Hermann of Carinthia, 36]

The definition of DEMR does not appear at the beginning of Book VI.

VI,39: "Datum lineam medii duorumque extremorum proporcione secamus."

"We cut a given line according to the proportion of one mean and two extremes."

10. *Campanus of Novara* (13th century) [Euclid–Campanus of Novara]

VI,def.3: "Linea dicitur dividi secundum proportionem habentem medium et duo extrema quando eadem est proportio totius ad maiorem sui sectionem que est maioris ad minorem."

"A line is said to be divided according to a proportion having a mean and two extremes, when the proportion of the whole to its greater section is the same as that of the greater to the lesser."

VI,29[= VI,30]: This is the same as the statement of Adelard III.

11. *Billingsley* (1570) [Euclid–Billingsley]

We saw in Section A,8 that Billingsley used the "extreme and mean proportion" terminology. However, in a comment he says, "commonly it is called a line divided by proportion having a mean and two extremes."

C. Names for DEMR

In addition to the two descriptions "division in extreme and mean ratio" and "proportion having a middle and two ends," there have been other designations that have been used. This topic has already been extensively investigated by Tropfke [1903, 101] and Archibald [1918]. I have, however, rechecked their references and added some new references and material.

1. *"The section" (?)*

In Section 22,A I discussed the views of several authors who believed that when Proclus used the expression "the section" in reference to Eudoxus (Q.22) he was referring to DEMR. In Section 22,C I expressed the opinion, also held by others, that Proclus did not have DEMR in mind.

2. *"Divina proportione"*

This was the title of Paccioli's book [Paccioli, 1509]. I have already discussed this work in Section 31,D, and in Appendix II I will give quotations from this work which illustrate why Paccioli chose this name. We saw in Section 32,A,ii that Kepler used this expression in his 1608 letter. He probably obtained this expression from the works of Petrus Ramus or Clavius (Appendix II, 6,8). Ramus' *Scholarum mathematicarum* is mentioned in the letter to Mästlin of Section 32,A,i and the index to the *Collected Works* of Kepler has several references to Clavius as well as to Ramus.

3. *"Proportionally divided"*

We find this in Clavius' 1574 edition of the *Elements* (Section A,10) in connection with his comments on VI,30, where the term "divisa proportionaliter" is used; see Appendix II,8.

Again in the *Scholarum mathematicarum* of Ramus [Ramus, 1627, 306], we find: "Si latus sexanguli secetur proportionaliter. . . ." The original version dates from 1569 but the 1627 edition that I used is posthumous and this terminology may be due to Lazarus Schoner the editor. Kepler (Appendix II,9, on fn. 17 to *Mysterium cosmographicum*) employed "proportional division" ("sectione proportionali").

4. *"Continuous proportion"*

This expression, and its different variations in several languages, probably came from V,def.10, where Euclid speaks of four magnitudes being continuously proportional, that is, $A:B = C:D$, in connection with VI,16,17 mentioned at the beginning of the appendix. Recall that it is VI,17 that is used in XIII,1-5 to go from the proportion definition of DEMR to the area definition.

We find this expression in Kepler's 1597 letter to Mästlin, discussed in Section 32,A,i, where we read: "triangle all of whose three sides are mutually and continuously proportional. . ." ("triangulum rectangulum constituere, cujus omnia tria latera sint invicem continue proportionalis . . .").

5. *"Medial section"*

This is used by Leslie [1820, 62] in connection with II,11.

6. *"Golden Number (Section, etc.)"*

This is the "popular" name for DEMR. Tannery [1882, 399] used "section d'or" and Cantor [1894, I, 151] used "goldene Schnitt."

The origins of the name "golden number" as a designation for DEMR had been the object of much inquiry and searching even before the previously mentioned studies by Tropfke [1903] and Archibald [1918]. In 1881, only forty-six years after Ohm [1835] (which will be discussed presently), Sonnenburg [1881, 3; repeated in Fink, 1900] stated, without however mentioning Ohm or any other mathematician by name, that the expression "golden number" was recent in origin and was not found in the more widely circulated mathematical compendiums of the 19th century. Even earlier Zeising [1854, 163]—this was in fact one of two works which started the golden number genre—stated explicitly that he did not know where the expression had come from. My readings of other early "golden numberists" indicate that they did not know the source either.

A measure of the interest in the matter is given by the number of times that authors have broached the subject. We find the question "Who first used the expression 'golden number'?" asked by Beman [1895] in the *Intermédiare des mathématicians*. The same inquiry was reprinted in 1904, then made again by Russo [1912] and answered by Quint in both [Quint, 1905] and [Quint, 1913] and by Brocard [1913]! However, none of the authors was able to supply any information beyond references to Paccioli and the German "golden numberists" of the second half of the 19th century. Then after almost a half a century the query was brought up again in Sarton [1951] and now the matter has been codified in perpetuity by its appearance in the *Oxford English Dictionary* [1972].

I too have entered the fray, with the help of various correspondents and libraries, and the following is the result of my investigations.

The earliest reference mentioned in any of the sources is the second edition (1835) of Martin Ohm's *Die reine Elementar-Mathematik* and indeed on page 194 the footnote to proposition 5 (which turns out to be none other than XIV,**—see Section 24,A—one of the other great mysteries of the golden number world) reads: "Diese Zertheilung eine beliebigen Linie r in 2 solche Theile, nennt man wohl auch den goldenen Schnitt; auch sagt man in diesem Falle zuweilen: die Linie r werde in stetige Proportion getheilt" ("One is also in the habit of calling this division of an arbitrary line in two such parts the golden cut [or "section"]; one sometimes also says in this case: the line r is divided in continuous proportion"). In the statement of the proposition itself we only find a description of DEMR; i.e., the segment (which turns out to be the side of the decagon) is the mean proportional between the radius and the smaller segment, and no name is given to the division of the line.

I will return to this text after answering the obvious question of what expression Ohm used in the first (1826) edition. If we check this out [Ohm, 1826a, 188], we find that the statement of no. 157 is just the definition of DEMR and that the term "stetige Proportion" is used. This latter term, as we saw in Subsection 4 above, goes back at least as far as Kepler and was used in a late 18th-century German edition of the *Elements* by J. Lorenz [Euclid–Lorenz*]; furthermore as we shall see below it was a very common term in 18th- and 19th-century German textbooks. Then in no. 159 which corresponds to page 194, no. 5 [i.e., XIV,**] of the second edition, although the proof is not the same, the term "stetige Proportion" is again used. I note that in the first edition there is no footnote which discusses the terminology.

The situation is thus that in the first edition Ohm uses, directly in the text, a term ("continuous proportion") which had already appeared in print in Germany, whereas in the second edition no name is used in the text and the term "continuous proportion" is relegated to the "one sometimes also says in this case" position. Further the statement "one is also in the habit of calling this division . . . golden cut . . ." indicates that Ohm was not making up the name on the spot and that it had gained at least some, and perhaps a great deal of currency, by 1835.

Just before this book went into the final stages, I came across a review of, and with the aid of Professor Albert Lewis was able to obtain a copy of, an article [Schubring, 1983; Lewis, 1985] dealing with the development of Ohm's career in the period 1807 to 1835. One does indeed find the standard initials G.S. for "goldene Schnitt," but unfortunately for this investigation they refer in this case to Ohm's brother and the author of the article. I then wrote to Dr. Schubring who kindly not only answered my questions concerning the possibility of a mention of "goldene Schnitt" elsewhere in the collection of Ohm's letters, but also did additional research and made some very pertinent remarks that I will discuss in my next book on the golden number. Dr. Schubring informed me that there is no mention of the "goldene Schnitt" in any of the letters so that no direct information is available from the existing Ohm archival material. Furthermore no relevant information is forthcoming from other books by Ohm.

In lieu of being able to provide a definite answer, what I will now do, as a possible aid to future research, is to provide a combined list of those German works that were looked at by Dr. Schubring or myself or for which I have found an indication in the literature.

Lorenz [1781?; Euclid–Lorenz*]: "stetige Teilung."

Kastner [1796; see p. 417, 435, 445 on Paccioli]: "divina proportione."

Klügel [1803-1831; vol. 4, 363*]: "sectio divina."

Ohm [1819]: "stetige Proportion."

Moebius [1824]: no name is used.

Ohm [1826a = *Die reine Elementar-Mathematik*, first edition]: "stetige Proportion."

Ohm [1826b = *Die analytische und höhre Geometrie*]: "stetige Proportion."

Ohm [1835 = *Die reine Elementar-Mathematik*, second edition]: "goldene Schnitt."

Ohm [1841? = the date of the third edition; I could not find the date of the first edition): "stetige Proportion."

Kunze [1842*; second edition 1851, par. 170, p. 231]: The main text uses the term "stetiger Proportion," but in the remark we read "Euclides . . . äusserem und mittlerem Verhältniss geschnitten. Sonst wird eine solche theilung auch Sectio divina oder Sectio aurea genannt.'' Archibald [1918, 235] gives a reference to an 1839 book entitled *Lehrbuch der Planimetrie* and states that there is a mention of a ''rechteck der schönsten Form'' on page 124. I was unable to find a bibliographic reference to such a work and Dr. Schubring assures me that there is no work by Kunze written in that year. From the given title it seems certain that [Kunze, 1842] is meant. I could not find the quotation in the second edition. However, page 124 falls in chapter 6 which deals with regular polygons including the square, but not the pentagon (which can only be constructed in chapter 9 where the above-mentioned section 170 is to be found). Thus the ''rechteck der schönsten Form'' is, I suspect, the square and has nothing to do with the ''golden rectangle'' as one might suspect from the inclusion of this reference in Archibald's article.

Helmes [1844]: ''sectio aurea.'' Helmes also wrote an expanded version of this for a ''Schulprogramme'' in 1844, but this does not contain any historical remarks; see Schubring [1986].

Wiegand [1847, 1848, 1849]: ''goldene Schnitt.'' Both the terms ''goldene Schnitt'' and ''äusseren und mittleren Verhältnissen'' (i.e., the technical term EMR) are used in [1847, 14]. And on pages 14, 90, 142 we find the rediscovered statement and proof of the Magirus-Kepler theorem discussed in Section 32,A,i. I further note that this theorem was also, apparently, rediscovered by the other co-founder of ''golden numberism,'' F. Röber, in 1855 (he does not use the term ''golden number'') as well as by other ''golden numberists''; see Fischler [1979b].

Zeising [1854]: ''goldene Schnitt.''

Schöttler [1857]: ''goldene Schnitt.''

In addition Dr. Schubring checked a large number of 18th- and 19th-century German textbooks at the Institut für Didaktik der Mathematik at the Universität Bielefeld and found no trace of the term ''goldene Schnitt''; the terms that he did find were all variations on the terms ''stetige Proportion'' and ''mittlere Proportion.''

In my original manuscript, I stated that based on what we find and do not find in Ohm and other German works it seemed that the term ''golden section'' developed or at least became acceptable between 1826 and 1835 and that Ohm did not make it up, at least not just for the 1835 edition. My own personal feeling, based on the lack of any written sign and the omission of the term in Ohm's 1826 edition, was and still tends to be that the term came into existence in the first quarter of the 19th century. Dr. Schubring, however, believes that the term comes from a well-established tradition, possibly from the Latin term ''sectio aurea.'' He points out that according to a count that he had made there were at least 665 (not including revised editions!) geometry textbooks published in Europe between 1775 and 1829, so that we are far from a complete analysis. Furthermore he suggests that it is quite possible that the term ''goldene Schnitt'' existed only orally, perhaps among artisans or engineers, outside the general school system. In particular he points out that Ohm would have had the opportunity to become acquainted with such an oral tradition after 1826 while teaching at the Vereinigte Artillerie- und Ingenieurschule or at the Bau-Akademie.

In addition to the problem of when the term ''golden number'' came into being, there is also the question of why the adjective ''golden'' came to be attached to the concept of DEMR. With respect to this the following examples of the use of the term ''golden'' should be noted.

a. The so-called ''rule of three'' for solving proportion problems of the form $a:b = x:c$ (see Sanford [1927, 17]; Smith [1923, II, 483]) was often called the ''golden rule.'' The earliest use of this term seems to be by Johann Böschensteyn in 1514; see Weitletner [I, 3].

b. Cardano (see Section 31,E) uses the expression ''On the Golden Rule'' as the title of chapter xxx of his *The Great Art* (1545) [Cardano–Witmer, 182] which deals with the approximate solutions of polynomial equations.

c. The ''golden number'' refers to the year in the nineteen-year cycle used in computing the date of Easter; see Neugebauer [1975, 1063] and Smith [1923, II, 651]. This number appears on most calendars printed in France under the name ''nombre d'or.''

d. Maistrov [1967, 73] says that Jakob Bernoulli (1654-1705) spoke of the main theorem (the so-called weak law of large numbers of probability theory) of his *Ars conjectandi* (The Art of Conjecturing, published in 1713) as the ''golden theorem.''

e. The expression ''Eine Buch mit goldem Schnitt'' (see Saunders Wörterbuch [1860, vol. 1, 608]) was used in early 19th-century Germany. Here, however, the expression means ''gilt edged'' in reference to a book.

I, of course, do not know if any of the above examples influenced the choice of ''goldene Schnitt'' in connection with DEMR. Examples *a* and *e* were surely widely known and *c* was most likely widely known. Sonnenburg [1881, 5] suggests in connection with *a* that the expression ''golden'' became attached to the geometrical concept to replace a usage in connection with an arithmetic technique that was disappearing from textbooks, but no precise works are cited to support this hypothesis. It may just be that since the German language used the adjective ''golden'' for so many things at that time (see Saunders Wörterbuch [1860] or the various 19th-century editions of the Brockhaus and Meyers dictionaries) that the term was created ''ab initio.''

Whatever the origin of the term ''goldene Schnitt,'' it seems to have become commonly accepted in the German mathematical literature shortly after Ohm [1835], for Helmes [1841] could entitle his article "A Simpler Solution of the Golden Section [sectio aurea] based on a New Analysis, Together with a Critical Elucidation of the Usual Solution of this Problem and a View of its Pedagogical Value," and Wiegand [1849] could use the expression "the generalized golden cut" in the title of his booklet. Furthermore by 1854 Zeising who was not a mathematician had, as mentioned earlier, become aware of the term.

In the discussions on the meaning and correct translations of the term DEMR by Terequem [1838, 1853] and Vincent [1844], the expression "golden number" does not appear and the same is true of Chasles [1837], so that the usage of this term was probably restricted to Germany. The earliest employment of the term that I have seen in an English-language mathematics book (there are earlier occurrences in the "golden number" literature) as in Ackermann [1895], which we are told is based directly on a German article. The adjective "golden" is also used in Beaman and Smith [1899, 196] and therefore may have been used in the 1895 edition, but I was unable to locate the latter.

7. *Medieval Expressions*

In addition to the above, I mention that Beaujouan [1973, 453] says that in a manuscript of the *Mathematica* by Philippus Elephantis (14th century) the expression "Victoria" is used in connection with the diagram accompanying II,11, the term "figura Demonis sive intelligentis" ("the figure of the devil" or "he who understands") with IV,10, and "Faratra" (= Pharetra, quiver) with XIII,18.

APPENDIX II

"MIRABLIS . . . EST POTENTIA . . .": THE GROWTH OF AN IDEA

1. *Campanus* (13th century)

In Campanus' 13th-century edition of the *Elements* [Euclid–Campanus], we read the following commentary at the end of his XIV,10 (= XIV,8′; see Section 24):

"Mirabilis itaque est potentia linee secundum proportionem habentem medium duoque extreme divise. Cui cum plurima [sic] pholosophantium [sic] admiratione digna conveniant hoc principium vel praecipuum ex superiorum principiorum invariabili procedit natura ut tam diversa solida tum magnitudine tum basium numero tum etiam figura irrationali quadam simphonia rationabiliter conciliet."

"Wonderful therefore is the power of a line divided according to a ratio having a mean and two extremes: since most things worthy of the philosophers' admiration accord with it, this foundation or pre-eminence proceeds from the invariable nature of higher foundations, that a certain harmony can rationally unite solids that are so diverse, first in magnitude, then in the number of bases, then too in their irrational shape."

I have checked Book XIV in various Latin and Arabic texts and did not find any trace of a related statement. It would thus appear that this commentary originated with Campanus. The nature of Campanus' remarks are discussed in Murdoch [1968].

In this appendix we shall follow the spread of the idea that DEMR was worthy of the philosopher's (i.e., mathematician's) admiration. It is important to emphasize that all these statements refer to the strictly mathematical properties of DEMR. They do not refer to any supposed superiority of DEMR in aesthetics, etc. Only in the letter of Kepler to Tanckius do we find a suggestion that DEMR is somehow connected to natural phenomena. Kepler says that he is playing "with symbols," but the question of whether he "believed" these statements is a difficult one. But even in Kepler's case the statements glorifying DEMR are always related to the mathematical properties. It may be, however, that Campanus' statement was a factor in the spread or even the start of "golden numberism" whose origins can be traced to 1854. For Campanus' statement appears on page 512 of Chasles' book [Chasles, 1837] and at least some of the early "golden numberists" were acquainted with Chasles' history of geometry. The Tanckius letter was not published until 1858-1871 [Kepler–Frisch]. These matters will be discussed in another book.

2. *Paccioli* (1509)

I have already discussed Paccioli's *Divina proportione* in Section 31,D. Let us read in Paccioli's own words why he chose the name that he did.

On the Appropriate Title of the Present Treatise

Chapter V

Excellent Duke, it seems to me that the proper title for this treatise must be the Divine Proportion. This is because there are very many similar attributes which I find in our proportion—all befitting God himself—which is the subject of our very useful discourse. Of the many I will take four as being sufficient to our purpose. The first is that it is one only and not more. And it is not possible to grant to it different species, nor differences. This unity, according to all the theological and the philosophical schools, is the supreme epithet of God himself. The second attribute is that of the Holy Trinity. That is, just as in the divine there are three persons in the same [Italian: Medisma] substance: Father, Son and Holy Ghost, likewise a [Italian: Medisma; Winterberg translates it as "one and the same"] proportion of this kind always involves three terms, and never more nor less is needed to find it, as we shall see later on. The third attribute is that just like God cannot be properly defined, nor can be understood through words, likewise this proportion of ours cannot ever be designated through intelligible numbers, nor can it be expressed through any rational quantity, but always remains occult and secret, and is called irrational by the mathematicians. The fourth attribute is that just as God can never change and is all in all, and all everywhere, likewise our present proportion is always in every quantity, be it big or small, is one and always invariable and there is no way that it can be changed, nor can it be learned otherwise by our intellect, as our sequel will show. A fifth [literally, "the fifth"] attribute can be added to the above not undeservedly. This is: just as God has conferred being to the heavenly virtue by another name, called the fifth essence, and through it on to the other four simple bodies, that is the four elements: earth, water, air and fire, and through these [He conferred] being to every other thing in nature, likewise this holy proportion of ours gives the formal being, (according to the ancient Plato in his Timaeus [see Section 16,D,i]) to heaven itself conferring to it the figure of the body called Dodecahedron, or the body with twelve pentagons. Which, as will be shown below, is impossible to form without our proportion. And similarly assigns [it is not clear if the subject is intended to be Timaeus, proportion or God] to each of the other elements its own specific form, bodies not at all similar to each other: that is, to fire the pyramidal figure called Tetrahedron; to earth the cubic figure called Hexahedron; to air the figure called Octahedron; and to water the one called Icosahedron. These forms and figures are all called regular bodies by the sages, as they will be treated individually below. Without our proportion it is not possible to compare them with each other, nor can they be inscribed in a sphere; all of this will appear below. Let these features be sufficient to the introduction of the present compendium, although many more could be mentioned.

On Its Worthy Naming

Chapter VI

Excellent Duke, this proportion of ours is of such quality, and of an excellence worthy as much as one could ever say because of its infinite power; so that without its knowledge very many things worthy of admiration could not have come to light either in philosophy [the Italian makes clear that philosophy is the main concern] or in any other science. This gift has been granted to it by the invariable nature of the superior principles, as our famous mathematician and great philosopher Campanus said about the 10th of the 14 [book of Euclid]. Especially when one sees that it is that which brings together, with a certain irrational symphony, such diverse bodies in their sizes, multitudes of surfaces and shapes and forms. As will be understood in our sequel by bringing forth the stupendous effects which (in a line which is proportioned according to it) are not to be called natural, but truly divine. Of which the first according to sequence is this. [Now follow the chapters discussed in Section 31,D.]

The mentioning of Campanus' remark on XIV,10 leaves no doubt as to the source of Paccioli's feelings towards DEMR. What is curious is that when we look at Paccioli's "corrected" version of Campanus [Euclid–Paccioli] published in the same year 1509 there is no "castigator" [= corrector] statement with XIV,10. Why did Paccioli not refer to his *Divina proportione*? He certainly was not shy about mentioning his work, for at the beginning of Book V is the text of a talk that he had given on the theory of proportions. When we look at the VI,29 (= VI,30), we only find the following under "castigator": "a division of this kind is very much prized by all the philosophers."

As to the Platonic element of the above text, Guzzo [1955, 155; see also Garin, 1965, 186] considers Paccioli to be more Platonistic than Plato in the sense of the degree of belief that Plato's views actually mirrored the system according to which God had created the universe.

Rather than give more quotations of the above nature, I shall simply give the names that Paccioli associates with the "effects" 2 to 8 and 10 to 13 (i.e., the results from Books XIII and XIV discussed in Section 31,D; these correspond to chapters 11 to 16 and 18 to 22 in the text): essential, singular, ineffable, admirable, unnamable, inestimable, surpassing all others, most excellent, almost incomprehensible, most worthy.

3. *Tartaglia* (1543)

This edition of the *Elements* [Euclid–Tartaglia] is to a large extent an Italian version of Campanus' edition. In XIV,10 we find essentially Campanus' statement given in (1).

4. *Fine* (1556)

Fine's *De rebus mathematicis* [Fine, 1536] deals with various geometrical propositions and constructions. Immediately following the table of contents, Fine has the following:

"De Divine proportione, quae in linea recta per mediam et extrema rationem divisa continetur." ("Concerning divine proportion which is contained in a straight line divided through mean and extreme ratio.")

"authoris distichon
Si quid divinum condebat pulchra Mathesis
Quod Geometra colat: haec tibi sola dabit"

("A couplet by the author: If beautiful education founded something divine which geometry deals with; she [i.e., education] alone will give it to you.")

Underneath this couplet there is the following picture, surrounded by a floral design:

A B C
Divina pro portio

Source: Fine [1536]

FIGURE App.-1

5. *Candalla* (1566; see Section 31,B) [Euclid–Candalla]

This author is rather restrained; in connection with VI,30 we read:

"This section worthy of being admired ["admiranda"] could be demonstrated at greater length. . . ."

There is no comment on XIV.

6. *Ramus* (1569, 1599; see Section 31,H)

The *Scholarum mathematicarum* dates from 1569. In the discussion of II,11 [p. 191] Ramus speaks of its use in IV,10 and by Ptolemy and then says: "but especially in all the mysteries of the regular bodies which [depend] principally on that proportional section. Finally a certain divine proportion was noticed here by certain Christians so that from it one trinity and a threefold unity were conceived which is complete in its entirety and in any part; a single principle complete in large [part], complete in small [part] which is most beautiful and most blessed."

From the content and the following citations, it is clear that the above is ultimately based on Paccioli's text.

In connection with VI,def.3 [p. 222] he speaks of: "the divine (as I would call it) proportion. . . ."

In the *Geometry* [Ramus, 1599, 94] proposition 1 of Book XIV is just the definition of DEMR. The commentary reads:

> This definition is wholly of a geometrical sort, yet by a certain quirk of expression the mean and extreme ratio is spoken of instead of the mean and extreme terms of the proportion. Namely, this line is so cut that [the line] itself with the two segments forms three terms of a proportion, and it itself is the first term, the greater segment is the middle, the smaller is the third. It may be stated, briefly and accurately, that it is proportionally divided. Moreover, a marvelous power is attributed to the proportion of this division by Campanus (10 p., 14) in his descriptions of regular solids. Since [he says] so many things worthy of the admiration of philosophers come together in it, this principle, or pre-eminence, arises from the unchanging nature of superior principles so that it unites rationally, in a sort of symphony [?], so many solids differing first in size, then in the number of faces, then even in their irrational figures. This [is what] Campanus [says], a speaker in other respects very sparing of praise. This same proportion of the division is called divine by Luca Pacioli in the book written on this subject, and indeed its unique uses are revealed in the *Elements* of Euclid, on the construction of the pentagon, icosahedron, and dodecahedron. In addition even the sublime mysteries of heavenly matters are deduced from this source by Ptolemy.

Hookyas [1958, 317, 394] considers the excursions by Ramus into "mysticism" rather extraordinary in particular in contrast to Kepler.

7. *Billingsley* (1570)

In the introduction to Book XIII [Euclid–Billingsley] we find the following statement: "In this thirteenth booke are set forth certayne most wonderfull and excellent passions of a lyne divided in extreme and meane proportion: a matter undoubtedly of great and infinite use in Geometry, as ye shall both in thys booke, and in the other bookes following most evidently perceave." There are also "Platonic" statements about the regular polyhedra in this introduction.

Then in the introduction to Book XIV we find: "Which things also undoubtedly for the woorthiness and hardnes thereof (for thinges of most price are most hardest) were first searched and found out of Philosophers, not of the inferior or meane sort, but of the depest and most grounded Philosophers, and best exercised in Geometry."

8. *Clavius* (1579) [Euclid–Clavius]

Following VI,def.3 we find the following commentary: "Moreover there are almost countless dignities and uses of a line divided in this way as will be established from the books on stereonometry especially book 13 so that it is not unworthily called by some [authors] divine proportion in which a line is divided in that way."

Then his comments on VI,30 (which he proves in three ways), we read: "Moreover this admirable ['admiranda,' instead of Campanus' 'mirabilis'] sectioning of a line in e.m.r. has important uses and properties as will be made clear in the books on stereometry, so that a line so divided is said by a great many mathematicians, not without reason, to have some sort of divine proportion because of its admirable strength and nature: but by some it may simply be called proportionally divided [divisa proportionaliter]."

9. *Kepler* (texts 1608-1621)

As I indicated in Appendix I,C,2, Kepler most probably obtained the expression "divine proportion" from the works of Ramus and Clavius, and the same is probably true about Kepler's knowledge of the statements of Paccioli.

We saw in Section 32,A,ii in connection with the 1608 letter to Tanckius [Kepler, 1608b] how Kepler combines mathematics and mysticism. Let us start off by considering other parts of the same letter (lines 85-116, 142-171):

> In geometry, a ratio is defined by two terms; a proportion consists in likeness of two terms. Thus, a proportion is comprised of four terms. If the two middle terms are equal, a continuous proportion is obtained; in reality this proportion comprises three terms, although four according to definition. Among the continuous proportions there exists one particu-

larly excellent kind: the divine proportion, when out of the three quantities the two smaller ones added total the larger quantity, or when a whole is divided into two parts such that between the parts and the whole there exists a continuous proportion. A peculiarity of this proportion lies in the fact that a similar proportion can be constructed out of the larger part and the whole; what was formerly the larger part now becomes the smaller, what was formerly the whole now becomes the larger part, and the sum of these two now has the ratio of the whole. This goes on indefinitely; the divine proportion always remaining. I believe that this geometrical proportion served as idea to the Creator when He introduced the creation of likeness out of likeness, which also continues indefinitely. I see the number five in almost all blossoms which lead the way for a fruit, that is, for creation, and which exist, not for their own sake, but for that of the fruit to follow. Almost all tree-blossoms can be included here; I must perhaps exclude lemons and oranges; although I have not seen their blossoms and am judging from the fruit or berry only which are not divided into five, but rather into seven, eleven, or nine cores. But in geometry, the number five, that is the pentagon, is constructed by means of the divine proportion which I wish (to assume to be) the prototype for the creation. Furthermore, there exists between the movement of the sun (or, as I believe, the earth) and that of Venus, which stands at the head of generative capability [''facilitati seminali''] the ratio of 8 to 13 which, as we shall hear, comes very close to the divine proportion. Lastly, according to Copernicus, the earth-sphere is midway between the spheres Mars and Venus. One obtains the proportion between them from the dodecahedron and the icosahedron, which in geometry are both derivatives of the divine proportion; it is on our earth, however, that the act of procreation takes place.

Now see how the image of man and woman stems from the divine proportion. In my opinion, the propagation of plants and the progenitive act of animals are in the same ratio as the geometrical proportion, or proportion represented by line segments, and the arithmetic or numerically expressed proportion.

Lines 116-142 of the letter are given in Section 32,A,ii. The letter continues:

If one were to ask a country satyr [''Satyro rusticano''] where the difference between the propagation of plants and the procreation of man lies, he would answer that the difference lies in the fact that the plant carries its seeds within itself and therefore a single plant suffices for propagation, whereas where humans are concerned, a male and a female member must unite; that which in one is a minus and is receptive, is in the other a plus and projects, so that a third likeness comes about only through the uniting of humans, through the joining of plus and minus.

Enough of this digression which was sparked by an interest in Rheinhard's [author of a book on music, 1604; see Kepler, 1608, 429, letters 492, 493] speculations. I also play with symbols; I have started a small work: ''Geometrical Cabbala''; it deals with the ideas of the things of nature in geometry. Only I play in such a way as to never forget that I am playing. For nothing is proven with symbols; in the philosophy of nature no secrets are unveiled by geometrical symbols, only things already known are put together, in case it is not definitely proven that we are not only concerned with symbols, but with a portrayal of the kind of and the reason for the relation between these two areas. . . .

I do not doubt that Pythagoreans thought of something similar as I have, in representing shortage and surplus on squares, when they named numbers masculine and the even numbers feminine. Thus Platon interpreted the first uneven number, that is, unity, as ''idea or form'' which is considered masculine. But the first even number, that is, two, he made matter which has the image of [?, correspond to ?] the feminine. For in an odd number, compared with an even, there appears also a deviation. Therefore Pythagorean males have their penises but Pythagorean females are missing their vaginas.

An earlier occurrence of the expression ''divine proportion'' is found in another letter written in 1608 to Georg Brengger [Kepler, 1608a, 140]. In the middle of some calculations (see XIII,9) we read: ''Thus since the chord of 1/10 [of a circle, i.e., a_{10}] was in the same proportion; its noble proportion, called divine, to which the highest honour is paid, was added to the radius.''

In the *Epitome Astronomiae Copernicae* (*Epitome of Copernican Astronomy*) 6.143 [Kepler, 1618, 50] we find the following:

[Question:] How will you prove that the sphere is sufficient for itself and others [i.e., the polyhedra].

[Answer:] Indeed, a comparison of the side of a figure [i.e., a polyhedra inscribed in a sphere], both to the other measurements of a body and, of most bodies, among themselves, is made only from the fixed sphere in which all bodies are contained, and from that sphere's diameter, divided into fixed ratios. The sphere is therefore the cause and model of definition, or scientific description, or proportions, for the other figures.

But even those admirable qualities of such proportions [i.e., of the polyhedra] cannot be determined and understood except with the help of the circle, and your men of Geometry call this proportion ''divine.''

In Book I, section XXVI of *Harmonice mundi* (*Harmonies of the World*) (1619) [Kepler–Caspar 2, 27] we read:

When the given line is the whole and its two parts are sought, mathematicians speak of the ''division according to the extremes and the mean''. This designation is intended to express the following: when one undertakes, on the one hand, a common division of the whole according to any ratio, but on the other hand, wants to add a segment to the whole which forms the same ratio with the whole as the smaller part does with the larger, then in the former case there are four terms, two outer and two inner, and in the latter case only three terms: the whole and the smaller part as the two extremes and the larger part as the one inner term.

For the same reason one also speaks of continuous division. Today one uses the word divine in reference to division as well as to ratios, due to its wonderful nature and its various special qualities; the most important of these is the fact that one repeatedly obtains a line divided in the same way when one adds the larger part to the whole, whereby the larger part now becomes the smaller and what was formerly the whole becomes the larger part, according to Euclid XIII,5.

Again in Book V of *Harmonice mundi* [Kepler–Wallis, 1013] we find the following statement:

The second digress of kinship, which is genetic, is to be conceived as follows: First, some harmonic ratios of numbers are akin to one wedding or family, namely, perfect ratios to the single family of the cube; conversely, there is the ratio which is never fully expressed in numbers and cannot be demonstrated by numbers in any other way, except by a long series of numbers gradually approaching it; the ratio is called divine, when it is perfect, and it rules in various ways throughout the dodecahedral wedding. Accordingly, the following consonances begin to shadow forth that ratio: 1:2 and 2:3 and 2:3 and 5:8. For it exists most imperfectly in 1:2, more perfectly in 5:8, and still more perfectly if we add 5 and 8 to make 13 and take 8 as the numerator, if this ratio has not stopped being harmonic.

The following is found in chapter 12 of the 1621 edition of Kepler's *Mysterium cosmographicum* (*Cosmic Mystery*) [Kepler–Caspar 2, 73; this is not in the original 1596 edition; see Kepler, 1596, 41, line 26; Kepler–Duncan, 133].

Therefore perfect harmonies are to be associated with the cube, the pyramid and the octahedron; imperfect harmonies with the dodecahedron and the icosahedron. Also entering into this is a notion which indeed provides a hint of the deeply concealed basis of (reason behind) this matter—we shall discuss this in the next chapter: geometry possesses two treasures: the ratio of the hypotenuse in the right triangle to the legs, as well as division in extreme and mean ratio.[17] Out of this ratio comes the construction of the dodecahedron and the icosahedron. It is for this reason that the pyramid can be inscribed into the cube, and the octahedron into both of these solids, so easily and simply. How the particular harmonies are to be related to the particular solids is, however, not so readily apparent.

Footnote 17 reads [Kepler–Caspar 2, 81; Kepler–Duncan, 143]:

Two propositions of unlimited use and consequently highly valuable; there exists, however a big difference between them. The former, which states that in a right-triangle the sum of the legs squared is equal to the hypotenuse squared, I would like to compare to a gold-nugget; the other, relating to proportional division, I would like to call a gemstone. This one is indeed nice in itself, but it is of no value without the other. When the former, after we have first made the initial step with it, leaves us in the lurch, this one leads us further into science: that is, it leads us to the calculation and construction of the side of the decagon and related quantities.

CORRECTIONS AND ADDITIONS

Minor typographical errors are not included in the following list. The location of the errors are indicated in the form: [p. 5, c. 2, l. 5 (l.-5)]; i.e. page 5, column 2, line 5 from the top (line 5 from the bottom). For their comments and corrections I wish to thank Messrs. Artmann, Bien, Burckhardt, Curchin, Fowler, Kutler, Vajda and Van der Schoot. The reviews cited in the preface contain suggested changes and purported errors in my text and these should be consulted by the reader. I have not done a survey of the literature so the additional references listed here should not be considered as complete.

As Professors Pottage and Høyrup correctly pointed out, not all the diagrams are correctly proportioned. I have not however commented on the diagrams unless an error would affect the proof.

title page: Ἀκρον καὶ

[p. xi, c. 2, l.-3] The original family name of my wife Eliane was Herz. She used Fischler (as I did) after our marriage, but wrote Eliane Herz Fischler (without a hyphen) on her thesis. Then in 1982, after eighteen years of marriage, we all changed our name to Herz-Fischler.

[p. 2, c. 2, l.-10; fig.]: ". . . proof of II,6, whose . ."; in the figure replace IV by V.

[p. 11, c. 2, l. 7; 12]"*A,C,D* . . point *B*; $\sphericalangle BDC = \sphericalangle DAC$

[p. 18] For the "converse" of XIII,5 see "Theorem XIV**", fn. 57. Also Compare XIII,5 with II,11′ on p. 31.

[p. 24, c. 2, l. 2] The quotation ends after 'lengths'. The precise relationships are not, as indicated by the brackets, given in Euclid's statement of XIII,18.

[p. 34, discussion, diagram] XIII,8 is used in XIII,17[a], corollary; see p. 23.

[p. 36] See W. Knorr, *The Ancient Tradition of Geometric Problems*, revised edition, New York: Dover, 1996; A. Wasserstein, "Some Early Greek Attempts to Square the Circle", *Phronesis* 4(1959), pp. 92–100.

[p. 43, G] See J. Bidwell, "A Babylonian Geometrical Algebra", *College Mathematics J.,* 17(1986), pp. 22–31.

[p. 44 H, J] See Appolonius, *On Cutting Off a Ratio*, E. Macierowski trans., Cumberland, Maryland: Golden Hind Press, 1988; Euclid, *Recipients, Commonly Called the Data, With a Commentary by Marinus*, R. Schmidt trans., Cumberland, Maryland: Golden Hind Press, 1988; C. Taisbak, "Elements of Euclid's *Data*" in I. Mueller ed., *Peri Tōn Mathēmatōn [On Mathematics]/Essays on Ancient Mathematics and its Later Development*, Edmonton: Academic Printing, 1992, pp. 135-71, in particular p. 168.

[p. 46, c. 1] See the review of [Herz-Fischler, 1984] by C. Fletcher, *Mathematical Reviews*, review 85e:01006.

[p. 50, c. 2] See J. Høyrup, "*Dýnamis*, the Babylonians, and *Theaetetus* 147c7–148d7", *Historia Mathematica*, 17(1990), 201–22; B. Artmann, "A Proof of Theodorus' Theorem by Drawing Diagrams", *J. of Geometry*, 49(1994), pp. 3–5.

[p. 51, c. 2, l.-18] 21/13.

[p. 55, c. 2, l.20] ÁR-NIGIN is example 5.

[p. 60, C; c. 2, l.26] See B. Artmann, "Mathematical Motifs on Greek Coins", *Mathematical Intelligencer*, 12(1990), pp. 43–50. The discussion of (19) ends after "real objects."

[p. 61, D, E] I shall not be publishing this material. Mssrs. Artmann, Bien, Curchin sent me a great deal of material, but the following only includes examples, or possible examples, that may predate -400. The literature on the dodecahedra includes material that I was unaware of when I did research for Section D. A complete catalogue of these dodecahedra and similar objects, and a complete survey of the literature would be most useful.

The most important text (provided by Len Curchin) is H. Murray, *A History of Board-Games Other than Chess*, Oxford: Clarendon, 1952, p. 18. The author describes game boards which were incised on the roofing slabs of the temple (c.-1400) at Kurna (Thebes) and one of these game boards is a pentagram. Further Murray suggests that a Greek game, the only mention of which is an obscure line from Sophocles, is the same as the Egyptian game. If Sophocles, who lived in the vth century, is indeed describing a pentagram game then—unless of course someone wishes to speculate about "secret games"—the pentagram was a widely known figure in vth century Greece.

Other texts of interest are: S. Cook *Religion of Ancient Palestine in the Light of Archaeology*, London: Oxford, 1930, p. 214; H. May, *Material Remains of the Megiddo Cult*, Chicago: Univ. Chicago, p. 6; W. Petrie, *Tannis, Part II*, London: Trubner, 1888, plate xxxvi, item 18; M. Cameron, "'Stars of David' on a Mural Plaster Fragment from Knossos", *Kadmos*, 1979, pp. 40-6. E. Budge, *Amulets and Superstitions*, London: Oxford, 1930; reprint Dover 1978, p. 209.

In response to an inquiry, Dr. A. Sherrat of the Department of Antiquities, Ashmolean Museum referred me to the book J. Evans, *Ancient Stone Implements of Great Britain*, London: Longmans-Green, 1897, pp. 420-24. The objects come from Scotland and Northern Ireland and, from the ornaments on some of them, they may date from c. -3000. From the photographs sent by Dr. Sherratt and from figures 351 and 352 in the Evans book, I would say that purported associations of the markings with "geometry" are completely unwarranted. Paul Bien has informed me of a numbered metal icosahedron from India. Since the numbers are in Arabic, the icosahedron is evidentally not ancient; the date and tradition behind this object is unknown. That the pentagram/pentagon is a universal sign is again suggested by the following works: A. Leroi-Gourhan, *The Dawn of European Art/ An Introduction to Paleolithic Cave Painting*, Cambridge: Cambridge U., 1982, p. 56; M. Robertson, *Rock Drawings of the Micmac Indians*, Halifax: Nova Scotia Museum, 1973, fig. 3, 5, 89; K. Wellman, *A Survey of North American Indian Rock Art*, Graz: Akademische Druck, 1970, fig. 905.

[p. 63, c. 2, l.-17] Aristotle, -384 to -322.

[p. 64, i] See L. Zhmud, "Pythagoras as a Mathematician", *Historia Mathematica* 16(1989), pp. 249–68.

[p. 66] See B. Artmann, "Hippasos und das Dodekaeder", *Mitteilungen aus dem mathematischen Seminar Giessen*, 163(1984), pp. 103–21.

[p. 81, c. 2, l.5] In Q.3 . . . .

[p. 82, c. 2, l.18] See B. Artmann, L. Schäfer, "On Plato's 'Fairest Triangles' (*Timaeus* 54A)", *Historia Mathematica* 20(1993), pp. 255–64.

[p. 83, c. 1, l.12; Q.12] V 1003; "says that earth arose. . .

[p. 84, ii] See Y. Balashov, "Should Plato's Line be Divided in the Mean and Extreme Ratio?", *Ancient Philosophy*, 14(1994), pp. 283–95.

[p. 86, c. 1, l.-9] H. Thesleff in *Studies in Platonic Chronology*, Helsinki: Societas Scientiarum Fennica, 1982 and in "Platonic Chronolgy", *Phronesis*, 34(1988), p. 18, fn. 67 argues that Theatetus died in -390.

[p. 88, c. 2, l.11] See also Section 11,B.

[p.100, c. 2, (1)] $OD^2 / AD^2 = \{(CO + OA)^2 + CA^2\} / CA^2$. See W. Knorr, "Archimedes and the *Elements*: Proposal for a Revised Chronological ordering of the Archimedean Corpus", *Archive History of Exact Sciences*, 19(1978/79), pp. 211–90.

[p. 102, Section 24] Professor Hogendijk's study of the al-Maghribī version of the *Elements* that is discussed in section 3G of "XIV**" has now appeared: "An Arabic Text on the Comparison of the five Regular Polyhedra: 'Book XV' of the *Revision of the Elements* by Muhyī al-Dīn al-Maghribī", *Zeitschrift für Geschichte der Arabisch-Islamischen Wissenschaften* 8(1993), pp. 133–233. Unfortunately the Hebrew text discussed in section 3F does not appear to have been published. This latter text may shed more light on the question of whether the Arabic source is really based on a Greek text, or whether—as Professor Hogendijk now suggests is a possibility—it was of Arabic-Islamic origin.

[p. 105, c. 2, theorem 8] For a modern discussion, see the very interesting book by J. Pottage, *Geometrical Investigations*, Reading, Mass.: Addison-Wesley, pp. 379, 451.

[p. 107, c. 2, l. 3] The two Aristaeus are not the same.

[p. 111, c. 2] The letter e is missing from the end of arc 3.

[p. 113, A] G. Toomer, *Almagest*, New York: Springer, 1984; see also the enlightening study of the construction of the tables: G. Van Brummelen, "Lunar and Planetary Interpolation Tables in Ptolemy's *Almagest*", *J. History of Astronomy*, 25(1994), 297–311.

[p. 114, B] O. Neugebauer, (*A History of Ancient Mathematical Astronomy*, New York: Springer, 1975, pp. 733) says that significance of 36° may be related to a division of the quadrant of the circle into 15 parts.

[p. 129, c. 2, problem 10] See J. Berggren, *Episodes in the Mathematics of Medieval Islam*, New York: Springer, 1986.

[p. 136, iii] See R. Franci, L. Toti Rigaletti, "Towards a History of algebra from Leonardo of Pisa to Luca Pacioli." *Janus* 72 (1983), pp. 17–82.

[p. 149, c. 2, problem 21] The references to [Chuqet–L'Huillier] and [Chuquet–Flegg] were omitted. A comparison of Fig. IX-17 with those of related problems in [Chuqet–Flegg, 266, 268] suggests that there was a common source.

[p. 157, J] Thanks to the good offices of David Fowler, several people have commented on the marginal note. Because of this Len Curchin and I hope to publish a further note in the near future.

[p. 158, c. 1, l. 1; c. 2, l. 4] . . . (see Section 32,B) . . . 89,144,233. The text omits mention of the Mästlin's letter of October 30, 1597, [Mästlin, 1597], in which just the triangle is drawn and the numbers of Fig. IX-27 are given in the text.

[p. 159, c. 1, l.-3] Samuel Magirus (d. 1631) was one of Kepler's music teachers; see M. Dickreiter, *Der Musiktheoretiker Johannes Kepler*, Bern: Franck, 1973, pp. 126, 145, 164. For a continuation of the story of this result, see R. Herz-Fischler, "A 'Very Pleasant' Theorem", *College Mathematics Journal*, 24(1993), pp. 318–324.

[p. 160, c. 1, l. 1] ". . . so that $\sphericalangle EDA$ will be a right angle. We must show that $AF = ED$.

[p. 160, c. 2] See the discussion in Dickreiter, pp. 167–70; Tanckius was a doctor.

[p. 162, c. 1] Relationship (3) was known to the anonymous commentator, generally presumed to be William Brouncker, of the 1653 English translation of Descartes *Compendium musicae*. See *Renaturs Des-Cartes Excellent Compendium of Musick: With Necessary and Animadversions thereupon by a Person of Honour*, London: Humphrey Mosley, 1653, p. 87 ff. I shall discuss this example, and others from the world of music, in *Golden Numberism*.

Relationship (4) was known to Cassini in 1680. This is pointed out in R. Graham, D. Knuth, O. Patashnik, *Concrete Mathematics*, Reading, Mass., 1989, p. 280. Those historians of mathematics who are surprised at the appearance of such detailed historical statements in a book entitled *Concrete Mathematics* should consult the second author's "Ancient Babylonian Algorithms", *Communications of the Association of Computing Machinery*, 15(1972), pp. 671–77; 19(1976), p. 108. I am pleased to announce that I was recipient of the US$2.56 prize, which, as the introduction of the above book stated, was to be awarded to the first person who pointed out a historical mistake. These authors, [p. 285], also point out that Euler in 1765 was quite familiar with the Fibonacci sequence and knew how to express them in terms of positive and negative powers of (1 + R(5)) / 2 (the so-called Binet Formula). S. Vajda, *Fibonacci and Lucas Numbers, and the Golden Section*, Chichester, England: Horwood, 1989, p. 52, points out that the Binet Formula was already known to DeMoivre in 1718. P. Scholfield, *The Theory of Proportion in Architecture*, Cambridge: Cambridge Univ., 1958, p. 101, fn. 6 attributes the formula to Daniel Bernoulli in 1732.

P. Singh, "The So-called Fibonacci Numbers in Ancient and Medieval India", *Historia Mathematica* 12 (1985), pp. 229–244, points out that the Fibonacci sequence appears in the treatises on poetry of Virahanka (between 600 and 800), Gopala (prior to 1135) and Hemicandra (c. 1150). As was the case with Fibonacci (Section 31,B,v), there is no indication of a knowledge of the relationship with DEMR.

[p. 164, Appendix I] See "Theorem XIV**" for the texts in Greek, Arabic and Hebrew characters.

[p. 165, items 3,4,10 and p. 173, 8] The manuscript has now been published: H. Busard, *The Latin Translation of Euclid's 'Elements' Made Directly from the Greek*, Stuttgart: Steiner, 1987. In appendix II to the introduction, Busard suggests that Fibonacci himself was the author of the combined books XIX–XV. Clavius (1574)

[p. 167, items 1,6] Section 20,A; Section 20,C. The matter of the name "golden section" will be discussed further in my forthcoming book *Golden Numberism*.

[p. 169, item a; item e] In Ulrich Wagner's 1483 Rechenbuch the title of chapter 10 is "Von der guilden Regel"; see the facsimile edition by J. Burckhardt, Munich: Graphos Verlag, 1966. Wieleitner [1927, I, 3]. "Ein Buch mit goldenem Schnitt".

[p. 170, c. 1, l. -2; c. 2, 3] I now know that the earliest English usage of "golden number" dates to 1864. I shall discuss this and other examples in *Golden Numberism*. (there . . literature) in Ackermann. . . . "Beman".

[p. 171, Title] Mirabilis.

[p. 173, c. 1, l.-1] Hooykaas; R. Weitman, *Vistas in Astronomy* 18(1975), pp. 723, suggests that Kepler used the Candalla edition (item 5). See also [Kepler-Duncan, fn. 2].

[p. 175, c. 1, l.-3] It is footnote 17 which is not in the 1596 edition. See also A. Segonds, *Jean Kepler / Le secret du monde*, Paris: Les Belles Lettres, 1984, p. 80.

[p. 176, c. 1, l. -5] mfi = microfiche.

[p. 179] Curchin , 1980 "Sumer"!

[p. 181] [Fowler, 1980] . . . Section 20,A,vii.

[p. 182, c. 2] Goldat, G. 1957. *The Early Medieval Traditions of Euclid's 'Elements'*. Thesis, University of Wisconsin (see *Dissertation Abstracts* 17(1957), 1319). Appendix I, 2,7; Girard . . . Section 32,B.

[p. 183] For [Herz-Fischler, 1985], see the new preface.

[p. 184, c. 2, l.-6] Klein, J.; E. Brann.

[p. 185] [Langermann, Hogendijk, 1984] This is discussed in "Theorem XIV**"; see on p. 102 above.

[p. 189] [Struick, 1969] . . 1200–1800.

[p. 191] [Zeising, 1854] *Neue Lehre* . . . Albert Van der Schoot of the Department of Philosophy of the University of Amsterdam will be publishing a book on the golden number in 1998: *De ontstelling van Pythagoras - over de geschiedenis van de goddelijke proportie*, Kampen: Kok Agora. I understand that this will contain an analysis of Zeising's work, principally from a philosophical viewpoint. My forthcoming book *Golden Numberism* will also contain an analysis of Zeising, but with an emphasis on the "sociological" aspects of his work; see *The Shape of the Great Pyramid* for a partial discussion.

BIBLIOGRAPHY

In addition to providing bibliographic information, this bibliography has two other functions.

(1) To indicate the library from which I was able to obtain a particular work. This will enable readers to use interlibrary loan services to locate material. In the case of Canadian and American libraries, I have given the library's identification letters in square brackets; a list of the identification codes can be found in the Library of Congress *National Union Catalogue*. As there is only one possible case of conflict, I have left off the Ca prefix for Canadian universities. Note also the following:

B.N. = Bibliothèque Nationale (Paris).
B.L. = British Library (London).

Other European locations are written out.

(2) To serve as an index to references to the works in this book. I have provided the section and subsections immediately after the entry.

Each entry is preceded by a short form which is the same, except for the addition of the initial and perhaps an alternate name or spelling, as the short form used in the text to cite the work. For a discussion of the short forms, see A Guide for Readers.

Each bibliographical entry has been made as complete as I thought necessary to enable the reader and/or librarian to accurately identify the work. On some occasions I have added comments and/or catalogue references.

The following special abbreviations are used:

DSB *Dictionary of Scientific Biography*. Edited by C. Gillespie. New York: Scribner, 1970-80. 16 vols.

OCD *The Oxford Classical Dictionary*. 2nd ed. Edited by N. Hammond and H. Scullard. Oxford: Clarendon Press, 1970.

NUC *National Union Catalogue* (pre-1956). Washington: Library of Congress.

mf microfilm

mc microcard

ms manuscript

An asterisk (*) after an entry indicates that I did not consult the work.

Abraham bar Ḥiyya–Curtze. "Der *Liber Embadorum* des Abraham bar Chijja Savasorda in der Übersetzung des Plato von Trivoli." Edited by M. Curtze. In *Urkunden zur Geschichte der Mathematik im Mittelalter und der Renaissance*, pp. 1-183. Edited by M. Curtze. Leipzig: Teubner; New York: Johnson Reprint, 1968. [Bibliotheca Mathematica Teubneriania, Band 45.] [OHM.] Section 31,A,i.

Abraham bar Ḥiyya–Vallicrosa. *Libre de geometria hibbur hameixiha uehatixboret*. Edited by I. Vallicrosa (based on the text by M. Guttman). Barcelona: Editorial Alpha, 1931. [MiU.] Section 31,A,i.

Abu Kamil–ms. *On the Pentagon and Decagon*. Kara Mustafa, library number 379 (Istanbul), ff. 67r.-78v. [mf UPB.] Section 28,B,i; Appendix I,B.

Abu Kamil–Levey. *The Algebra of Abu Kamil "Kitab fi al-jabar wa'l-mugabala" in a Commentary by Mordecai Finzi*. Translated by M. Levey. Madison: Univ. of Wisconsin Press, 1966. [OTY.] Sections 5G; 28,B,ii.

Abu Kamil–Sacerdote. "Il trattato del pentagono e del decagono di Abu Kamil Shogia' ben Aslam ben Muhammed." Translated by G. Sacerdote. In *Festschrift zum achtzigsten Geburtstage Moritz Steinschneiders*, pp. 169-94. Leipzig: Otto Harrassowitz, 1896. [OTY.] Sections 28,B,i; 31,B,iii.

Abu Kamil–Suter. "Die Abhandlung des Abu Kamil Shogia' b. Aslam 'über das Fünfeck und Zehneck'." Translated by H. Suter. *Bibliotheca mathematica* 3 (1909-10), pp. 15-42. [IaU.] Sections 28,B,i; 31,B,iii.

Abu Kamil–Weinberg. *Die Algebra des Abu Kamil Soga ben Aslam*. Translated by J. Weinberg. München: Druck der Salesianischen Offizin, 1935. Inaugural Dissertation, Ludwig-Maximilian Universität, Philosophische Fakultät II, München, September 7, 1934. [CRL.] Section 28,B,ii.

Abu Kamil–Yadegari and Levey. *Abu Kamil's "On the Pentagon and Decagon"*. Translated by M. Yadegari and M. Levey. Japanese Studies in the History of Science, Supplement 2. Tokyo: History of Science Society of Japan, 1971. [OTU.] Section 28,B,i.

Abu'l-Wafa'–Suter. "Das Buch der geometrischen Konstruktionen des Abu'l Wefâ'." Edited by H. Suter. *Abhandlungen zur Geschichte der Naturwissenschaften und der Medizen* 4 (1922), pp. 94-109. [RPB.] Section 28,C.

Ackermann, E. 1895. "The Golden Section." *American Mathematical Monthly* 2, pp. 260-64. Appendix I,6.

Adelard I–Busard. *The First Latin Translation of Euclid's 'Elements', Commonly Ascribed to Adelard of Bath*. Edited by H. Busard. Toronto: Pontifical Institute of Mediaeval Studies, 1983. [OOCC.] Appendix I,B.

Agostini, Amedeo. "Sopra un preteso plagio di Luca Pacioli e sopra un incunabulo italiano." *Arch. Stor. Sci.* 6 (1925), pp. 115-20. Section 31,D.

Alberti–Leoni. *Ten Books on Architecture by Leone Batista Alberti*. Edited by J. Leoni. London: Tiranti, 1965 (original 1726). [OOCC.] Section 31,A,iii.

Alberti–Orlandi. *Leon Battista Alberti L'Architectura (De Re Aedificatoria)*. Edited by G. Orlandi. Milano: Edizioni il Polifilio, 1966. [OOCC.] Section 31,A,iii.

Albinus–Burges. "The Introduction of Alcinous to the Doctrines of Plato." In *The Works of Plato*, vol. VI, pp. 241-314. Translated by G. Burges. London: Bell, Bohs Classical Library, 1908. [On Albinus vs Alcinous, see Dodds–*OCD*.] [OOCC.] Section 16,D,i.

al-Biruni–Schoy. *Die trigonometrischen Lehren des Persischen Astromen Abu'l-Raihan Muh. Ibn Ahmed al-Biruni*. Translated by C. Schoy. Hannover: Heinz la Faire, 1927. [IU.] Section 28,E.

al-Biruni–Suter. "Das Buch der Auffindung der Sehnen im Kreise von Abul-Raihan Muh. el-Biruni." Translated by H. Suter. *Bibliotheca mathematica* 11 (1910), pp. 11-78. [IaU.] Section 28,E.

al-Biruni–Wright. *The Book of Instruction in the Elements of the Art of Astrology*. Translated by R. Wright. London: Luzac, 1934. [QMM.] Appendix I,B.

al-Karaji–Woepcke. *Extrait du 'Farki' traité d'algèbre par Abou Bekr Mohammed Ben Alhacon Alkarhi*. Edited by F. Woepcke. Paris: Imprimerie Imperiale, 1853. [KU.] Section 28,intro.,i.

al-Khayyami–Kasir. *The Algebra of Omar Khayham*. Translated by D. Kasir. New York: Columbia Univ., Contribution to Education, no. 385, 1931. [OLU.] Section 28,intro.,i.

al-Khayyami–Woepcke. *L'Algèbre d'Omar Alkhayyami*. Translated by F. Woepcke. Paris: Duprat, 1851. [KU.] Section 28,intro.,i.

al-Khwarizmi–Rosen. *The Algebra of Mohammed Ben Musa*. Translated by F. Rosen. London: Oriental Translation Fund, 1831. [OTU.] Section 28,A,i.

al-Nayrizi–Curtze. *Anaritii in decem libros priores Elementorum Euclidis commentarii ex interpretatione Gheradi Cremonensis in codice Gracoviensi 569 servata*. Edited by M. Curtze. Leipzig. Teubner, 1899 (Supplement to vol. 8 of Euclid–Heiberg). [OKQ.] Section 25,B.

al-Sama'wal–Ahmad and Rashed. *Al-Bahir en algèbre d'al-Sama'wal*. Edited by S. Ahmad and R. Rashed. Damascus: Ministère de l'Enseignement Supérieur, R.A.S.-imp. de l'Université de Damas, 1972. [OTU.] Section 28,intro.,i.

Alexander, C. 1929. "Miscellaneous Accessions in the Classical Department." *Bulletin of the Metropolitan Museum of Art* 24, pp. 201-204. Section 10,D.

Alexander, C. 1937. "Accessions of Greek and Roman Antiquities." *Bulletin of the Metropolitan Museum of Art* 32, pp. 175-77. Section 10,D.

Allan, D.–*DSB*. "Plato." *DSB*, XI, pp. 22-31. Section 16,B.

Allman, G. 1889. *Greek Geometry from Thales to Euclid*. Dublin: Hodges, Figgis. [OLU.] Sections 12; 13; 14; 18,A,B; 20,B,ii; 24,C.

Amma, T. Sarasvati. 1979. *Geometry in Ancient and Medieval India*. Delhi: Motilal Banarsidass. Section 29.

Anbouba, A. 1978. "L'Algèbre arabe aux IXe et Xe siècles. Aperçu général." *Journal for the History of Arabic Science* 2, pp. 66-100. [OON.] Section 28,intro.,i.

Anson, L. 1910. *Numismata Graeca/Greek Coin Types Classified for Immediate Identification*. Section 1 (inanimate). London, 1910-16; Bologna: Forni, 1967 (2 vols.). [OLU.] Section 10,C.

Apostle, H. 1952. *Aristotle's Philosophy of Mathematics*. Chicago:Univ. of Chicago Press. [OOCC.] Section 16,A,C.

Archibald, R. 1918. "The Golden Section—Fibonacci Series." *American Mathematical Monthly* 25, pp. 232-37. Appendix I,C.

Archimedes–Heath. *The Works of Archimedes with 'The Method' of Archimedes*. Translated by T. Heath. Cambridge: Cambridge Univ. Press, 1912; New York: Dover, 1950. [OOCC.] Section 23,A.

Arias, P. 1962. *A History of 1000 Years of Greek Vase Painting*. New York: M. Abrams. [OOCC.] Section 10,A.

Aristophanes–Dubner. *Scholia in Aristophanem*. Edited by F. Dubner. Paris: Didot, 1877; Hildesheim: G. Olms, 1969. Section 11,A.

Aristotle–Hett. *On the Soul*. Translated by W. Hett. London: Heinemann, Loeb Classical Library, 1957. [OOCC.] Section 16,C.

Aristotle–Hicks. *De anima*. Translated by R. D. Hicks. London: Cambridge Univ. Press, 1907; Amsterdam: Hakkert, 1965. [OOCC.] Section 16,C.

Aristotle–Tredennick. *Aristotle—The Metaphysics*, Books I-X. Translated by H. Tredennick. London: Heinemann, Loeb Classical Library, 1968. [OOCC.] Sections 11; 22,C.

Arrighi, G. 1968. "La matematica a Firenze nel Rinascimento, Il codice Ottoboniano Latino 3307 della Biblioteca Apostolica Vaticana." *Physis* 10, pp. 70-82. [OON.] Section 31,A,iii.

Ast, F. 1835. *Lexicon Platonicum*. Göttingen: Dieterich; New York: Franklin, 1969 (3 vols.). [OOCC.] Section 16,C.

Avery, C. 1962. *The New Century Handbook of Classical Geography*. Edited by C. Avery. New York: Appleton Century Crofts. Chapter III,intro.

Avigad, N. 1966. "Aramaic Writing on a Bowl from Tel-Dan" [Hebrew]. *Yediot* 30, pp. 209-19. [OTU.] Section 9,E.

Avigad, N. 1974. "More Evidence on the Judean Post-Exilic Stamps." *Israel Exploration Journal* 24, pp. 52-58. Section 9,E.

Avigad, N. 1976. *Bullae and Seals from a Post-Exilic Judaean Archive* (*QEDEM* 4, 1976). Jerusalem: Hebrew Univ. Section 9,E.

Babelon, E. 1901. *Traité des monnaies grecques et romanes*. 2nd part, vol. III. Parigi, 1901-16 (9 vols.). Section 10,C.

Badrikian, A. 1968. "Les espaces prénuclèaires à l'époque d'Euclide." To appear.

Bandinelli, R., A. Giuliano. 1973. *Les Etrusques et l'Italie avant Rome*. Paris: Gallimard. [OOC.] Section 10,A.

Banu Musa–Curtze. "Verba filiorum Moysi, Fillii Sekir, id est Maumeti, Hameti et Hassen/Der Liber trium fratrum de Geometria/Nach der Lesart des Codex Basileensis F.II.33." *Nova Acta Academiae Caesarae Leopoldino-Carolinae Germanicae Naturae Curiosorum* 89, Nr 2 (1885), pp. 109-67. [OON.] Section 28,intro.,i.

Barr, S. 1961. *The Will of Zeus*. Philadelphia: Lippincott. [OOCC.]

Baumgartel, E. 1955. *The Cultures of Prehistoric Egypt II*. Oxford: Oxford Univ. Press. [OOCC.] Section 9,A.

Beard, R. 1973. *Patterns in Space*. Palo Alto: Creative Publications. Section 31,C,I.

Beaujouan, G. 1973. "Réflexions sur les rapports entre théorie et pratique au moyen-âge." In *The Cultural Context of Medieval Learning*, pp. 437-77. Edited by J. Murdoch and E. Sylla. Boston: Reidel, Boston Studies in the Philosophy of Science, vol. 26, 1975. [OOCC.] Appendix I,C.

Becker, Oskar. 1957. *Das mathematische Denken der Antike*. Göttingen: Vandenhoeck und Ruprecht. [OOUM.] Section 12,viii.

Beman, W. 1895. "Question 658." *L'Intermédiaire des Mathématiciens* 2, p. 317. Reprinted in *L'Intermédiaire des Mathématiciens* 11 (1904), p. 210. Appendix I,C.

Beman, W., D. Smith. 1899. *New Plane and Solid Geometry*. Boston: Ginn. (Original 1895?) [OCU.] Appendix I,C.

Benfey, O. 1974. "Dodecahedral Geometry in a T'ang Era Incense Burner Preserved in the Shoshin." In *Proceedings, No. 3*. XIVth International Congress of the History of Science. Tokyo and Kyoto, Japan, August 1974, pp. 273-77. Tokyo: Science Council of Japan. [InRe.] Section 30.

Berggren, J. 1984. "History of Greek Mathematics: A Survey of Recent Research." *Historia Mathematica* 11, pp. 394-410. Section 7.

Berggren, J. 1985. "History of Mathematics in the Islamic World: The Present State of the Art." *Middle East Studies Association Bulletin* 19, pp. 9-33. Section 28,intro.,i.

Berggren, J. 1986. *Episodes in the Mathematics of Medieval Islam*. New York: Springer. Section 28,intro.,i.

Bhaskara–Colebrooke. *Algebra with Arithmetic and Mensuration from the Sanscrit of Brahmagupta and Bhaskara*. London: John Murray, 1817. [OTU.] Section 29.

Boethius–Folkerts. *"Boethius" Geometrie II*. Edited by M. Folkerts. Wiesbaden: Franz Steiner, 1970. [OOCC.] Section 31,A,i.

Bombelli–Bortolotti. *Rafael Bombelli da Bologna/L'algebra*. Edited by E. Bortolotti. Milano: Feltrinelli, 1966 (original Books I,II,III, published 1572; IV,V, 1929). [NcRS.] Section 31,F.

Bonitz, H. 1870. *Index Aristotelicus*. 2nd ed. Berlin; Graz: Akademische Druck- u. Verlagsanstalt, 1955. [OOCC.] Section 16,C.

Bortolotti, E. 1930. "Le fonti arabe di Leonardo di Pisano." *Memorie della R. Accademia delle scienze dell' Istituto di Bologna classe di scienze fisiche*, Ser. 8,7 (1929-39), pp. 1-30, 39-49. [OON.] Section 31,B.

Bose, D. 1971. *A Concise History of Science in India*. New Delhi: Indian National Science Academy. [OOCC.] Section 29.

Boyance, P. 1939. "Sur la vie pythagoricienne de Iamblichus." *Revue des études grecques* 52, pp. 36-50. Section 11,B.

Boyne, W. 1968. *A Manual of Roman Coins*. Reprint. Chicago: Ammon Press. [OOCC.] Section 10,C.

Brentlinger, J. A. 1963. "The Divided Line and Plato's Theory of Intermediates." *Phronesis* 7, no. 2, pp. 146-66. Section 16,D.

Bretschneider, C. 1870. *Geometrie und die Geometer vor Euklides*. Leipzig. Teubner; Wiesbaden: Sandig, 1968. [OOUM.] Sections 14; 20,A,B; 24,C.

Brisson, L. 1977. "Platon 1959-1975." *Lustrum* 20, pp. 5-304. Section 16,intro.

British Museum, 1873. *British Museum, Dept. of Coins and Medals. A Catalogue of the Greek Coins in the British Museum/Italy*. London: British Museum. [NcDU.] Section 10,C.

British Museum. 1892. *A Catalogue of the Greek Coins in the British Museum/Mysia*. London: British Museum; Bologna: Arnaldo Fourni, 1964. Section 10,C.

Brocard, H. 1913. "[Response to Question 4149]." *L'Intermédiaire des Mathématiciens* 20, pp. 119-20. Appendix I,C.

Broughton, T.–*OCD*. "Heraclea (3) Pontica." *OCD*, p. 498. Section 18,A.

Brown, M. 1967a. Introduction to *Plato's "Meno"*. Edited by M. Brown. Indianapolis: Bobbs-Merrill, 1971.

Brown, M. 1967b. "Plato Disapproves of the Slave-boy's Answer." In *Plato's "Meno"*. Edited by M. Brown. Indianapolis: Bobbs-Merrill, 1971. [OHM.] Section 22,C.

Brown, M. 1969. "*Theaetetus*: Knowledge as Continued Learning." *Journal of the History of Philosophy* 7, pp. 359-79.

Bruins, E. 1950a. "Quelques textes mathématiques de la mission de Suse." *Koninklije Niederlandsche Akademie Van Wetenschappen. Proceedings of the Section of Sciences, Amsterdam* 53, pp. 1025-33. [OON.] Section 9,D.

Bruins, E. 1950b. "Aperçu sur les mathématiques babyloniennes." *Revue d'histoire des sciences et de leurs applications* 3 (1956), pp. 301-14. [OON.] Section 9,D.

Bruins, E., M. Rutten. 1961. *Textes mathématiques de Suse*. Paris: Geuthner. [QQL.] Section 9,D.

Bruins, E. 1976. "The Division of the Circle and Ancient Arts and Sciences." *Janus* 63, pp. 61-84. Section 23,A.

Bruins, E. 1979. "On Interpretation in the History of Mathematics." *Janus* 66, pp. 83-129.

Brumbaugh, R. 1954. *Plato's Mathematical Imagination, The Mathematical Passages in the Dialogues and Their Interpretation*. Bloomington: Indiana Univ. Press; Millwood, N.Y.: Krauss, 1977. [OOCC.] Sections 5,D; 6; 16A,C.

Brun, V. 1968. "A quelle époque a-t-on observé pour la première fois les rapports irrationneles. In *Actes XIIIe congrès international de l'histoire de la scice*, Paris, 1968, vol. 4, pp. 27-30.

Brunes, T. 1967. *The Secrets of Ancient Geometry and Its Use*. 2 vols. Copenhagen: Rhodes. [OON.] Section 16,D.

Bulliet, R. 1975. *The Camel and the Wheel*. Cambridge, Mass.: Harvard Univ. Press. [OOCC.] Preface.

Bulmer-Thomas, I.–*DSB* 1. "Hypsicles." *DSB*, VI, pp. 616-17. Section 24,C.

Bulmer-Thomas, I.–*DSB* 2. "Pappus of Alexandria." *DSB*, X, pp. 293-304. Section 27.

Bulmer-Thomas, I.–*DSB* 3. "Theaetetus." *DSB*, XIII, pp. 301-307. Sections 7; 18,B; 21.

Bulmer-Thomas, I.–*DSB* 4. "Theodorus." *DSB*, XIII, pp. 314-19; XV, p. 503. Section 15.

Bulmer-Thomas, I.–*DSB* 5. "Euclid: Life and Works." *DSB*, IV, pp. 414-37. Section 21.

Buonamici, G. 1932. "Rivista di epigrafia Etrusca, 1931-1932." *Studi Etruschi* 6, pp. 459-96. Section 10,A.

Burkert, W. 1972. *Lore and Science in Ancient Pythagoreanism*. Cambridge, Mass.: Harvard Univ. Press. [OOCC.] Sections 7; 9; 11,intro.,B; 12; 16,B,C; 20,A.

Burnyeat, M. 1978. "The Philosophical Sense of Theaetetus' Mathematics." *Isis* 69, pp. 489-513. Section 7.

Burnyeat, M. 1979. "Reply by M. F. Burnyeat." *Isis* 70, pp. 569-70. Section 7.

Busard, H. 1968. "L'Algèbre au moyen âge: Le 'Liber mensurationum' d'abû Bekr." *Journal des Savants* (avril-juin), pp. 65-124. Section 31,A,i.

Busard, H. 1974. "Uber einige Euklid-Scholien, die den *Elementen* von Euklid, übersetzt von Gerard von Cremona, angehängt worden sind." *Centaurus* 18, pp. 97-128. Section 31,A.

Cajori, F. 1924. "Notes on Luca Pacioli's 'Summa'." *Archivio di storia della scienze* 5, pp. 125-30. [NfSM.] Section 31,D.

Cajori, F. 1928. *A History of Mathematical Notations*, Vol. 1: *Notations in Elementary Vathematics*. LaSalle, Ill.: Open Court. [OOCC.] Guide for Readers,D.

Cantor, M. 1892. *Vorlesungen über Geschichte der Mathematik*. 3rd ed. Stuttgart: Teubner, 1907; 1965. [OTU.] Section 16,D.

Cardano–Witmer. *The Great Art or the Rules of Algebra*. Edited by R. Witmer. Cambridge, Mass.: M.I.T. Press, 1968 (original 1545). [OOCC.] Section 31,D,E; Appendix I,C.

Carruccio, E. 1964. *Mathematics and Logic in History and in Contemporary Thought*. Translated by I. Quigly. Chicago: Aldine. [OOCC.]

Carson, R. 1970. *Coins, Ancient, Mediaeval and Modern*. 2nd ed. London: Hutchinson. [OOCC.] Section 10,C.

Carson, R. 1979. *Principal Coin Types of the Romans*, Vol. 2: *The Principate 31 BC-AD 296*. London: British Museum. [National Numismatic Museum, Ottawa.] Section 10,C.

Chase, G. 1902. "The Shield Devices of the Greeks." *Harvard Studies in Classical Philology* 13, pp. 61-127. Section 10,B.

Chasles, M. 1837. *Aperçu historique sur l'origine et le développement des méthodes en géométrie*. Paris: Gauthiers-Villars, 1889. [OOUM.] Conclusion.

Cherniss, H. 1945. *The Riddle of the Early Academy*. Berkeley: Univ. of California Press; New York: Russell and Russell, 1962. [OOCC.] Section 16,B.

Cherniss, H. 1951. "Plato as Mathematician." *Review of Metaphysics* 4, pp. 395-425. [OOCC.] Section 16,intro.,B,C.

Cherniss, H. 1959. "Plato (1950-1957)." *Lustrum* 4 (1959), pp. 5-308; 5 (1960), pp. 321-648. Section 16,intro.

Chicago Oriental Institute. 1950. *The Assyrian Dictionary of the Oriental Institute of the Univ. of Chicago*. Chicago: Oriental Institute, 1956. [OOCC.] Section 9,C.

Chuqet–Flegg. *Nicolas Chuqet, Renaissance Mathematician*. Edited by G. Flegg, C. Hay, and B. Moss. Dordrecht: Reidel, 1985. Section 31,A,i.

Chuqet–L'Huillier*. *Nicolas Chuqet, La géométrie. Première géométrie algébrique en langue française (1484)*. Paris, 1979. Section 31,A,i.

Clagett, M.–*DSB*. "Archimedes." *DSB*, I, pp. 213-31. Section 23.

Clagett, M. 1953a. "The Medieval Latin Translations from the Arabic of the *Elements* of Euclid with Special Emphasis on the Versions of Adelard of Bath." *Isis* 44, pp. 16-41. Appendix I,B.

Clagett, M. 1953b. "Medieval Mathematics and Physics: A Check List of Microfilm Reproductions." *Isis* 44, pp. 378-81. Appendix I,B.

Clagett, M. 1978. *Archimedes in the Middle Ages*, vol. 3. Philadelphia: American Philosophical Society. [OOCC.] Section 31,C,D,I.

Clark, K. 1969. *Piero della Francesca*. London: Phaidon. [OOCC.] Section 31,C.

Cook, R. 1961. *The Greeks till Alexander*. London: Thames and Hudson. [OOCA.] Section 10,A.

Cook, R. 1972. *Greek Painted Pottery*. 2nd ed. London: Methuen. Section 10,A.

Coolidge, J. 1949. *The Mathematics of Great Amateurs*. Oxford: Clarendon Press. [OOCC.] Section 31,K.

Cornford, F. 1932. "Mathematics and Dialectic in the *Republic* VI-VII." In *Studies in Plato's Metaphysics*, pp. 61-95. Edited by R. Allan. London: Routledge & Kegan Paul, 1965. [OOCC.] Section 16,A.

Cornford, F. 1948. *Plato's Cosmology*. London: Routledge. [OOCC.] Section 16,D.

Crawford, M. 1974. *Roman Republican Coinage*. Cambridge: Cambridge Univ. Press. Section 10,C.

Csörgő, M., P. Révész. 1981. *Strong Approximations in Probability and Statistics*. Budapest: Akademiai Kiado. Section 10,A; Conclusion.

Curchin, L. 1977. "Eannatum and the Kings of Adab." *Revue d'assyriologie* 71, pp. 93-96. Preface.

Curchin, L. 1979a. "A Greek Epitaph of 'Syrian' Type." *Zeitschrift für Papyrologie und Epigraphik* 36, pp. 135-36, pl. V. Preface.

Curchin, L. 1979b. "Minoans at Chalcis?" *Quaderni di storia* 9, pp. 271-78. Preface.

Curchin, L. 1980. "Old Age in Summer: Life Expectancy and Social Status of the Elderly." *Florilegium: Carleton University Papers on Classical Antiquity and the Middle Ages* 2, pp. 61-70. Preface.

Curchin, L., R. Fischler. 1981. "Hero of Alexandria's Numerical Treatment of Division in Extreme and Mean Ratio and Its Implications." *Phoenix* 35, pp. 129-33. Section 25,A.

Curchin, L., R. Herz-Fischler. 1985. "De quand date le premier rapprochement entre la suite de Fibonacci et la division en extrème et moyenne raison?" *Centaurus* 28, pp. 129-38. Section 31,J,i.

da Vinci–de Toni and Corbeau. *Le manuscrit A de l'institut de France*. Transcription by N. de Toni; translated by A. Corbeau. Grenoble: Roissard, 1972. [B.N.] Section 31,K.

Davis, M. 1977. *Piero della Francesca's Mathematical Treatises, The "Trattato d'abaco" and "Libellus de quinque corporibus regularibus."* Ravenna: Longo. [OOCC.] Section 31,C.

de Clavasio–Busard. "The *Practica geometriae* of Dominicus de Clavasio." Edited by H. Busard. *Archive for History of Exact Sciences* 2 (1962), pp. 520-75. Section 31,A,i.

Deimel, A. 1922*. *Schultexte aus Fara*. Leipzig: Deutsche Orient-Gesellschaft, 1923. Vol. 2 of *Die Inschriften von Fara*, 1922-24. Section 9,C.

Deonna, W. 1954. "Les dodécaèdres ajourés et bouletés; à propos du dodécaèdre d'Avenches." *Bulletin, Association Pro Aventico (Avenches, Switzerland)* 16, pp. 19-89. Section 10,D.

de Stefani, S. 1885. "Intorno un dodecaedro quasi regolare di pietra a facce pentagonali scolpite con cifre scoperto nelle antichissime capanne di pietra del Monte Loffa." *Atti del reale istituto Veneto delle scienze, lettere ed arti*, 6th ser., 4 (Nov. 1885-Oct. 1886), pp. 1437-60, plate. [CRL.] Section 10,D.

Dickson, L. 1918. *History of the Theory of Numbers*, vol. 1. Reprint. New York: Chelsea, 1952. [OOCC.] Section 32,B.

Diels, H. 1934. *Die Fragmente der Vorsokratiker*, Bd. 1. 10th ed. Edited by W. Kranz. Berlin: Weidmann, 1961. [OOCC.] Section 4.

Dies, A. 1959. Introduction to *Platon oeuvres complètes*, t. VI (*La république*). Edited by E. Chambry. Paris: Les Belles Lettres. [OOCC.] Section 16,D; 22,C.

Diogenes Laertius–Hicks. *Lives of Eminent Philosophers*, vol. 1. Translated by R. D. Hicks. London: Heinemann, Loeb Classical Library, 1966. [OOCC.] Section 17.

Diophantus–Heath. *Diophantus of Alexandria: A Study in the History of Greek Algebra*. Translated by T. Heath. Cambridge: Cambridge Univ. Press, 1910; New York: Dover, 1964. [OOCC.] Section 5,F.

Diringer, D. 1934. *Le iscrizioni antico-ebraiche palestinesi*. Florence: Felice le Monnier, 1934. Section 9,E.

Dittenberger, W. 1898. *Sylloge inscriptionum Graecarum*. Leipzig: Hirzel, 1898-1901. [OOCC.] Section 11,A.

Dodds, E.–*OCD*. "Albinus." *OCD*, p. 34. Section 16,D.

Drachmann, A.–*DSB*. "Hero of Alexandria." *DSB*, VI, pp. 310-14. Section 25.

Dürer, A. 1525. *Unterweisung der Messung mit dem Zirkel und Richtscheit*. Zurich: Stocke-Schmid, 1966. [OOCC.] Section 31,K.

Duval, P. 1981. "Comment décrire les dodécaèdres Gallo-Romans, en vue d'une étude comparée." *Gallia* 39, pp. 195-200.

Eckhel, J. 1792. *Doctrina numorum veterum*. Vindobonae: Josephi Vincenti Degen. [American Numismatic Society, New York.] Section 10,C.

Einarson, B. 1936. "Mathematical Terms in Aristotle's Logic." *American Journal of Philology* 57, pp. 33-54, 151-72. Section 16A,C.

Euclid. For the early editions of Euclid and for the manuscripts, only a short English form of the title is given; consult the reference indicated for the complete title. The library indicated is the source of the microfilm. In parentheses is the date of the first edition or approximate date of composition in the case of manuscripts. The early editions are described in Euclid–Heath [I,97] and Murdoch [*DSB*].

Euclid–Adelard I. See Adelard I–Busard.

Euclid–Adelard III (12th century). *Elements*. [B. L. Burney 275; see Clagett, 1963, pp. 18 fn. 5, 22 fn. 8, 25 fn. 6.] Guide for Readers,C; Appendix I,B.

Euclid–al Tusi (13th century). *Elements ("Tahrir")*. Constantinople, 1801 (1809?). [B.N., vol. 48; 690-V.6063; also B.N., fnds Arab mss 2466.]

Euclid–Archibald. *Euclid's 'Book on Division of Figures' with a Restoration Based on Woepcke's Text and on the 'Practica Geometriae' of Leonardo Pisano*. Edited by R. Archibald. Cambridge: Cambridge Univ. Press, 1915. [OWTU.] Section 5,H.

Euclid–Barrow (1660). *Elements*. London, 1722. [B.L., vol. 69, 277-52.a.11.] Appendix I,A.

Euclid–Billingsley (1570). *Elements*. London: John Daye, 1570 (mc New York: Readex, n.d.; Landmarks of Science Series). [OTY.] Sections 31,G; Appendix I,A,B.

Euclid–Campanus of Novara (13th century). *Elements*. Venice: Erhardus Radtdolt, 1482. [B.L., vol. 69, 270-G.7837.] (Note: This was used for Books II,IV,XIV; for the other books use was made of Euclid–Paccioli, which is an emended version of the Campanus text; see Murdoch–*DSB* [448].) Section 31,A,ii; Appendix II.

Euclid–Campanus, Zamberti (1516). *Elements*. Paris: H. Stephani, 1516. [B.L., vol. 69, 272-47.e.2.] Appendix I,A.

Euclid–Candalla (1566). *Elements*. Paris: Johannem Royerium, 1566. [B.L., vol. 69, 273-530.m.10(1); see also B.N. vol. 23, 234-Res. V. 108.] Section 31,G; Appendix I,A; Appendix II.

Euclid–Clavius (1574). *Elements*. Rome: B. Grassium, 1589. 2 vols. [B.N., vol. 48, 686-V.18130-18131.] Appendix I,A; Appendix II.

Euclid–Commandino (1572). *Elements*. Pisa: C. Francischinum. [B.L., vol. 69, 273-8534.g.9.] Appendix I,A.

Euclid–Fine (1536). *Elements*. Paris. [B.L., vol. 69, 270-716.i.3,(2).] Appendix II.

Euclid–Frajese. *Gli Elementi di Euclide*. Edited by A. Frajese and L. Maccioni, Torino: Unione tipografico-editrice Torinese, 1970. [WU.] Guide for Readers,D; Section 1.

Euclid–Gerard of Cremona. *Elements*. [B.N., ms fnds Latin 7216.] (See Clagett [1963, 27].) Appendix I,B.

Euclid–Gryaneus (1533). *Elements*. Basel: I. Hervagium. [B.L., vol. 69, 269-47.f.11.] Appendix I,A.

Euclid–Heath. *The Thirteen Books of Euclid's Elements*. 2nd ed. Edited by T. Heath. Cambridge: Cambridge Univ. Press, 1926; New York: Dover, 1956. [OOCC.]

Euclid–Hermann of Carinthia. *The Translation of the 'Elements' of Euclid from the Arabic by Hermann of Carinthia (?) I-VI*. Edited by H. Busard. Leiden: Mathematisches Centrum, 1968 (Books VII-IX, 1972; X-XII, 1977). [OLU.] Section 31,A,ii; Appendix I,B.

Euclid–Heiberg. *Euclidis opera omnia*. Edited by J. Heiberg and H. Menge. Leipzig: Teubner, 1883-1916 (9 vols.). New edition of Books I-XIII edited by E. Stamatis. Leipzig: Teubner, 1970 (4 vols.). [OkQ; OOCC.]

Euclid–Holtzmann (1562). *Elements*. Basel: Jacob Kundig. [B.L., vol. 69, 297-529.m.19.] Section 31,J,ii.

Euclid–Ito. *The Medieval Latin Translation of the Data of Euclid*. Edited by S. Ito. Tokyo: Univ. of Tokyo Press, 1980. [OTU.] Section 5,J.

Euclid–Kayas. *Euclide/Les Eléments*. Translated by G. Kayas. Paris: Editions CNRS, 1978. [QMU.]

Euclid–Lorenz*. See NUC, vol. 341, 364. Appendix I,6.

Euclid–Paccioli (1509). *Elements*. Venice: Paganium de Paganinis. [B.N., vol. 48, 684-Res. V. 104; mf OOCC.] Section 31,J,i; Appendix II.

Euclid–Peyrard. *Les oeuvres d'Euclide*. Translated by F. Peyrard. Paris: Patris, 1816 (3 vols.). [OOUM.] Section 5,J; 24,A.

Euclid–Ramus (1545). *Elements*. Paris: Thomae Richardi, 1558. [B.L., vol. 69, 530.h.5.] Appendix I,A.

Euclid–"Rome Tahrir." [Elements. Rome: Medicea, 1594. [B.L., vol. 69, 295-G.7840.] (On this work, see Murdoch–*DSB* [453, no. 5].) Appendix I,B.

Euclid–Tartaglia (1543). *Elements*. Vinegia: V. Roffinella. [B.L., vol. 69, 298-8534.ee.32, shelf mark 1605/116.] Appendix II.

Euclid–Thaer. *Euklid/Die Elemente*.Translated by C. Thaer. Darmstadt: Wissenschaftliche Buchgesellschaft, 1969 (original Leipzig, 1933-37; Ostwald's Klassiker der exakten Wissenschaften nr. 235,236,240,241,243). [NFSM.]

Euclid–Woepcke. "Notice sur des traductions Arabes de deux ouvrages perdus d'Euclide." Edited by F. Woepcke. *Journal Asiatique*, 4th ser., (sept.-oct.), pp. 232-47. Section 5,H.

Euclid–Zamberti (1505). *Works*. Venice: J. Tacuini. [B.L., vol. 69, 267-531.m.11.] (This is identified as Zamberti's edition in Murdoch–*DSB* [448] and Euclid–Heath [I,98].) [mf OOCC.] Appendix I,A.

Eudoxos–Lasserre. *Die Fragmente des Eudoxos von Knidos*. Edited by F. Lasserre. Berlin: Gruyter, 1966. [OOCC.]

Evans, G. 1976. "The 'Sub-Euclidean' Geometry of the Earlier Middle Ages, up to the Mid-Twelfth Century." *Archive for History of Exact Sciences* 16 (1976-77), pp. 105-18. Section 31,A,i.

Fibonacci–Boncompagni. *Scritti di Leonardo Pisano matematico del secolo decimoterzo*. Edited by B. Boncompagni. Rome: Tipografia delle scienze matematiche e fisiche, 1857, 1862 (2 vols.). [mc New York: Readex, 1970; Landmarks of Science Series.] [Text IU; Cards OTY.] Section 31,B; Appendix I,A.

Field, G.–*OCD*. "Academy." *OCD*, p. 1. Section 16,intro.

Field, G. 1930. *Plato and His Contemporaries: A Study in Fourth-Century Life and Thought*. London: Methuen. [OOCC.] Section 16,intro.

Fine, 1556. *Orontij Finaei... De rebus mathematicis hactenus desideratis Libri IIII. Paris: Michaelis Vascosani, 1556*. [B.L., vol. 73, 209, C.54 + 8; this work should not be confused with *De re et praxi geometricae* (1556); see B.N., V.51, 1067.] Appendix II.

Fink. 1900. *A Brief History of Mathematics*. Chicago: Open Court. [OOCC.]

Fischler, R. 1974. "Quelques théorèmes du calcul des probabilités dont la valeur limite dépend d'une variable aléatoire." *Annales institut Henri Poincaré* 9, pp. 345-49. Preface.

Fischler, R. 1976. "Convergence faible avec indices aléatoires." *Annales institut Henri Poincaré* 12, pp. 391-99. Preface.

Fischler, R. 1979a. "A Remark on Euclid II,11." *Historia Mathematica* 6, pp. 418-22. Section 2,C.

Fischler, R. 1979b. "What Did Herodotus Really Say? or How to Build (a Theory of) the Great Pyramid." *Environment and Planning B* 6, pp. 89-93. Preface.

Fischler, R. 1979c. "The Early Relationship of Le Corbusier to the Golden Number." *Environment and Planning B* 6, pp. 95-103. Preface.

Fischler, R. 1981a. "How to Find the 'Golden Number' Without Really Trying." *Fibonacci Quarterly* 19, pp. 406-10. Preface.

Fischler, R. 1981b. "On Applications of the Golden Ratio in the Visual Arts." *Leonardo* 14, pp. 31-32, 262-64, 349-51. Preface.

Folkerts, M. 1970. *"Boethius" Geometrie II; ein mathematisches Lehrbuch des Mittelalters*. Wiesbaden: Steiner, 1977. [OOCC.] Section 31,A,1,A.

Forbes–*OCD*. "Suda." *OCD*, p. 1019. Section 18.

Fowler, D. 1979. "Ratio in Early Greek Mathematics." *Bulletin of American Mathematical Society* (new series), 1, pp. 807-46. Section 13,ii.

Fowler, D. 1980. "Book II of Euclid's *Elements* and a Pre-Eudoxean Theory of Ratio." *Archive for History of Exact Sciences* 22, pp. 5-36. Section 13,ii.

Fowler, D. 1981. "Anthyphairetic Ratio and Eudoxean Proportion." *Archive for History of Exact Sciences* 24, pp. 69-72. Section 13,ii.

Fowler, D. 1982a. "Book II of Euclid's *Elements* and a Pre-Eudoxean Theory of Ratio, Part 2: Sides and Diameters." *Archive for History of Exact Sciences* 26, pp. 193-209. Section 13,ii.

Fowler, D. 1982b. "A Generalisation of the Golden Section." *Fibonacci Quarterly* 20, pp. 146-58. Section 13,ii; Conclusion.

Fowler, D. 1982c. "Logos(ratio) and analogon(proportion) in Plato, Aristotle, and Euclid." Paper read at colloquium, "Logos et Théorie des Catastrophes (R. Thom)," Cerisy la Salle, France, September 17, 1982. Mathematics Research Centre, Univ. of Warwick. Section 13,ii.

Fowler, D. 1983a. Review of Taisbak [1982]. *Mathematical Intelligencer* 5, pp. 69-72.

Fowler, D., 1983b. Review of Mueller [1981]. *British Journal of Philosophy of Science* 34, pp. 57-70.

Fowler, D. 1983c. "Egyptian Land Measurement as the Origin of Greek Geometry?" *2-Manifold* 4, pp. 5-21. (Preprint: Mathematics Research Centre, Univ. of Warwick, 1983.)

Fowler, D. 1985. *The Mathematics of Plato's Academy: A New Reconstruction*. To be published by Oxford Univ. Press. (Preprint: Mathematics Research Centre, Univ. of Warwick.) Sections 5,H; 16,intro.

Frajese, A. 1963. *Platone e la matematica nel mondo antico*. Rome: Editrice Studium (Universale Studium, Testi e Documenti 4). [ICU.] Section 16,A.

Francesca–Arrighi. *Piero della Francesca, "Trattato d'abaco"*. Edited by G. Arrighi. Pisa: Domus Galilaenona, 1970. [ACU.] Section 31,C,D.

Francesca–Mancini. "L'opera 'De corporibus regularibus' di Piero Franceschi detto della Francesca, usurpata da Fra Luca Pacioli." Edited by G. Mancini. *Atti Accad. Lincei*

Mem. Cl. Sci. Morali Stor. Filol., 5th ser., 14 (1916), pp. 441-580 (+8 plates). [IU.] Section 31,C,D.

Franci, R., L. Rigatelli. 1981*. *La trattatisca matematica del Rinascimento senese*. Siena: Centro Studi della matematica mediovale. [UATL.] Section 31,A,i.

Frank, E. 1940. "The Fundamental Opposition of Plato and Aristotle." *American Journal of Philology* 61, pp. 34-53. Chapter IV,intro.

Fraser, P. 1972. *Ptolemaic Alexandria*. 2 vols. Oxford: Oxford Univ. Press. [OOCC.] Section 21.

Freeman, K. 1946. *The Pre-Socratic Philosophers*. 3rd ed. Oxford: Blackwell, 1953. [OOCC.] Section 4; Chapter IV,intro.; 18,B; 22,C.

Freeman, K. 1948. *Ancilla to the Pre-Socratic Philosophers*. Oxford: Blackwell. [OOCC.] Section 22,C.

Freudenthal, H. 1966. "Y avait-il une crise des fondements des mathématiques dans l'antiquité?" *Bull. S.M. Belg.* 18, pp. 43-55.

Freudenthal, H. 1976. "What Is Algebra and What Has It Been in History." *Archive for History of Exact Sciences* 16 (1976-77), pp. 189-200. Section 5.

Fritz, K. von. 1945. "The Discovery of Incommensurability by Hippasus of Metapontum." *Annals of Mathematics* 46, pp. 242-64. Sections 11,B; 12,viii.

Fritz, K. von. 1971. *Grundprobleme der Geschichte der antiken Wissenschaft*. Berlin: Gruyter. Section 14,vi.

Fritz, K. von–*DSB*. "Pythagoras of Samos." *DSB*, XI, pp. 219-25. Chapter IV,intro.

Fuÿe, A. de la. 1934. *Le Pentagramme Pythagoricien sa diffusion, son emploi dans le syllabaire cuneiforme*. Paris Geuthner. [OLU.] Sections 9,C,i; 12,i.

Gabrichi, E. 1913. "Cuma." *Monumenti Antichi* 22, pp. 5-871, plus plates. Reprint. Mainz: Von Zabern, 1979. [OOCC.] Section 10,A.

Gaiser, K. 1963. *Platons ungeschriebene Lehre*. Stuttgart: Ernst Klett. [WU.] Sections 16,B,C; 18,B;200.

Gandz, S. 1932. "The *Mishnat Ha Middot*, The First Hebrew Geometry of about 150 C.E. and the Geometry of Mohammad Ibn Muza Al-Khowarizmi, The First Arabic Geometry (c. 820) Representing the Arabic Version of the Mishnat Ha-Middot." *Quellen und Studien zur Geschichte der Mathematik, Astronomie und Physik*, Abteilung A (Quellen), 2, pp. 1-96, 2 plates. [OTU.] Section 28,A,i.

Gandz, S. 1936a. "Review of J. Weinberg's *Die Algebra des Abu Kamil Soga Ben Aslam*." *Isis* 23, pp. 145-47. [OOCC.] Section 28,B,ii.

Gandz, S. 1936b. "The Sources of al-Khowarzmi's Algebra." *Osiris* 1, pp. 263-77. Section 28,A.

Gandz, S. 1938. "The Origin and Development of Quadratic Equations in Babylonian, Greek and Early Arabic Algebra." *Osiris* 3, pp. 405-557. [OOUM.] Sections 5,B,F,G,K; 28,A.

Gardiner, A. 1957. *Egyptian Grammar*. 3rd ed. London: Oxford Univ. Press, 1973. [OOCC.] Chapter III,intro.

Gardiner, A. 1961. *Egypt of the Pharaohs*. London: Oxford Univ. Press. Section 9,A.

Garin, E. 1965. *Italian Humanism, Philosophy and Civic Life in the Renaissance*. Translated by P. Munz. New York: Harper & Row. [OOCC.] Appendix II.

Gazzaia–Nanni. 1982. *Praticha di geometria e tutte misure de terre*. Edited by C. Nanni. Siena: Centro studi della matematica medioevale. Section 31,A,ii,h.

Gherardi–Arrighi. "*Due tratti di Paolo Gherardi matematico fiorentino ....*" Edited by G. Arrighi. *R. Accademia delle scienze di Torino, classe di scienze* 101 (1967), pp. 61-82. [OTU.] Section 31,A,iii.

Gherardo-Arrighi. *Leonardo Fibonacci. La pratica di geometria, volgarizzata da Cristofano di Gherardo di Dino cittadino pisano. Dal codice 2186 della Biblioteca Riccardiana di Firenze*. Edited by G. Arrighi. Pisa: Domus Galilaeana, 1966. [OTU.] Section 31,A,iii.

Gherhard, O. 1830. "Vasi panatenaici." *Deutsches archaologisches Institut. Annali dell' instituto di corrispondenza archeologica* 2, pp. 204-24. [IaU.] Section 10,B.

Ghyka, M. 1927. *L'Esthétique des proportions dans la nature et dans les arts*. Paris: Gallimard. Sections 12,vi; 20,A,iv.

Ghyka, M. 1931. *Le nombre d'or*. Paris: Gallimard. Section 12,vi.

Giacardi, L. 1977. "On Theodorus of Cyrene's Problem." *Arch. Int. d'Hist. des Sciences* 27, pp. 231-36. Section 7.

Gibson, A. 1955. "Plato's Mathematical Imagination." *Review of Metaphysics* 9, pp. 57-70. Section 16,A,C.

Gille, B.–*DSB*. "Alberti." *DSB*, I, pp. 96-98. Section 31,A,iii.

Gingerich, O.–*DSB*. "Kepler." *DSB*, VII, pp. 289-312. Section 32,A.

Giorgio Martini–Arrighi. *Francesco di Giorgio Martini, La pratica di geometria dal Codice Ashburnham 361 della Biblioteca Medicea Laurenziano di Firenze*. Edited by G. Arrighi. Firenze: G. Barbera, 1970. [WUM.] Section 31,A,iii.

Girard, A. 1634. "Albert Girard tire parti des fractions continues." In G. Maupin, *Opinions et curiosités touchant les mathématiques*, pp. 203-209. 2nd ser. Paris: Naud, 1902. (The text of Girard is taken from his *Les oeuvres mathématiques de Simon Steven*. Leyde, 1634.) [CU.] Section 32,A,ii.

Glick, T. 1979. *Islamic and Christian Spain in the Early Middle Ages*. Princeton: Princeton Univ. Press. [OOCC.] Section 31,A.

Goff, B. 1963. *Symbols of Prehistoric Mesopotamia*. New Haven: Yale Univ. Press. [OOCC.] Chapter III,intro.; Section 9,B.

Goodenough, E. 1958. *Jewish Symbols in the Greco-Roman Period*. Vol. 7: *Pagan Symbols in Judaism*. New York: Bollingen, Pantheon Books. [OOCC.] Chapter III,intro.

Gottschalk, H. 1980. *Heraclides of Pontus*. Oxford: Clarendon Press. [OOCC.] Section 18,A.

Gow, S. 1884. *A Short History of Greek Mathematics*. Cambridge: Cambridge Univ. Press; New York: Chelsea, 1968. [OOCC.]

Grant, E.–*DSB*. "Jordanus de Nemore." *DSB*, VII, pp. 171-79. Section 31,A,i.

Grant, M. 1971. *Ancient Historical Atlas*. London: Weidenfeld. [OOCC.] Chapter III,intro.

Grose, S. 1923. *Catalogue of the McClean Collection of Greek Coins*. 3 vols. Cambridge: Cambridge Univ. Press, 1923-24; Chicago: OBOL Int., 1979. Section 10,C.

Grueber, H. 1970. *Coins of the Roman Republic in the British Museum*. Reprint. London, 1970. Section 10,C.

Gupta, R. 1975. "The Lilavati Rule for Computing Sides of the Regular Polygons." *Mathematics Education* 9, pp. 25-29. Section 29.

Gupta, R. 1976. "Sine of Eighteen Degrees in India up to the Eighteenth Century." *Indian Journal of History of Science* 11, pp. 1-10. Section 29.

Gupta, R. 1978. "Indian Values of the Sinus Totus." *Indian Journal of History of Science* 13, pp. 125-43. Section 29.

Guzzo, A. 1955. *La scienza*. Torino: Edizione di Filosofia. [OTY.] Appendix II.

Hadamard, J. 1945. *The Psychology of Invention in the Mathematical Field*. Princeton: Princeton Univ. Press; New York: Dover, 1954. Section 2,A.

Haeberlin, E. 1910. *Aes Grave/Das Schwergeld Roms und Mittelitaliens*. Frankfurt: Baer. Section 10,C.

Hallerberg, A. 1959. "The Geometry of the Fixed Compass." *The Mathematics Teacher* (Apr.), pp. 230-44. Sections 28,C; 31,A.

Hallerberg, A. 1960. "Georg Mohr and *Euclidis Curiosi*." *The Mathematics Teacher* (Feb.), pp. 127-32. Section 32,C,i.

Hammond, N. 1981. *Atlas of the Greek and Roman World in Antiquity*. Park Ridge, N.J.: Noyes Press. Chapter III,intro.

Hardy, G., E. Wright, 1938. *An Introduction to the Theory of Numbers*. 4th ed. Oxford: Clarendon Press, 1960. Sections 8; 25,A,iv; 32,B.

Hare, R. 1965. "Plato and the Mathematicians." In *New Essays on Plato and Aristotle*, pp. 21-38. Edited by R. Bambrough. London: Routledge & Kegan Paul. [OOCC.] Sections 16,A; 22,C.

Haskins, C. 1924. *Studies in the History of Mediaeval Science*. Reprint. New York: Ungar, 1960. [OOCC.] Section 31,A,J.

Hasse, H., H. Scholz. 1928. "Die Grundlagen-Krisis der griechischen Mathematik." *Kant-Studien* 33, pp. 4-34.

Head, B. 1910. *Historia Numorum, A Manual of Greek Numismatics*. Oxford: Clarendon Press; London: Spink, 1962. [OOCC.] Section 10,C.

Heath, T. 1921. *A History of Greek Mathematics*. 2 vols. Oxford: Clarendon Press. [OOCC.] Sections 4; 5; 6; 7; 11; 12; 16; 17; 18; 20; 22; 25; 26.

Heath, T. 1931. *A Manual of Greek Mathematics*. Oxford: Clarendon Press; New York: Dover, 1963. [OOUM.]

Heath, T. 1949. *Mathematics in Aristotle*. Oxford: Clarendon Press. [OOCC.] Section 16,C,iv.

Heiberg, J. 1925. *Geschichte der Mathematik und Naturwissenschaften im Alterum*. Munich: Beck. [Sainte Geneviève, Paris.] Section 14,viii.

Heidel, W. 1940. "The Pythagoreans and Greek Mathematics." *American Journal of Philology* 61, pp. 1-33. [OOCC.] Chapter IV,intro.; 11,B.

Heinevetter, F. 1912. *Würfel- und Buchstaben Orakel in Griechenland und Kleinasien*. Inaugural dissertation, Universität Breslau, Feb. 10. [CRL.] Section 10,D.

Heller, S. 1958. "Die Entdeckung der stetigen Teilung durch die Pythagoreer." *Abhandlungen der deutschen Akademie der Wissenschaften zu Berlin, Klasse für Mathematik, Physik, und Technik* 6, pp. 5-28. [OTU.] Sections 11,C; 12,ix.

Helmes, J. 1844. "Eine einfachere, auf einer neuen Analyse beruhende Auflösung der sectio aurea, nebst einer kritischen Beleuchtung der gewöhnlichen Auflösung dieses Problems und der Betrachtung ihres pädagogischen Wertes." *Archiv der Mathematik und Physik*, 1st ser., 4, pp. 14-22. [OWTU.] Appendix I,C.

Hendy, P. 1968. *Piero della Francesca and the Early Renaissance*. London: Weidenfeld and Nicolson. [OTY.] Section 31,C.

Herculaneum–Mekler. *Academicorum philosophorum index Herculanensis*. Edited by S. Mekler. Berlin: Weidmann, 1902. Section 16,B.

Hero–Bruins. *Codex Constantinopolitanus*. 3 vols. Edited by E. Bruins. Leiden: Brill, 1964. [OLU.] Sections 23,A; 25.

Hero–Heiberg. *Heronis Definitiones cum variis collectionibus Heronis quae ferunter Geometrica*. Vol. IV of *Heronis Alexandrini opera quae supersunt omnia*. Edited by J. Heiberg. Leipzig: Teubner, 1911 (mc New York: Readex, 1970). [OTY.] Section 25,A.

Herz Fischler, E. *La Dramaturgie de Thomas Corneille*. Thèse, doctorat de l'université (Paris III), 1977. Preface.

Herz-Fischler. See also Fischler and Curchin.

Herz-Fischler, R. 1984. "What Are Propositions 84 and 85 of Euclid's *Data* All About?" *Historia Mathematica* 11, pp. 86-91. Section 5,J.

Herz-Fischler, R. 1985. "Theorem XIV,** of the First 'Supplement' to the *Elements*." To appear in *Archives Internationales d'Histoire des Sciences*. Sections 24; 28, 31,A; Appendix I.

Hintikka, J., U. Remes. 1974. *The Method of Analysis/Its Geometrical Origin and Its General Significance*. Boston Studies in the Philosophy of Science, 25. Dordrecht: Reidel. [OOCC.] Section 17.

Hipparchus–Dicks. *The Geographical Fragments of Hipparchus*. Edited by D. Dicks. London: Univ. of London, Athlone Press, 1960. [OOCC.] Section 26,B.

Hlavaty, J. 1957. "Mascheroni Constructions." *The Mathematics Teacher* (Nov.), pp. 482-87. Section 32,C.

Hofmann, J. 1926. "Ein Beitrag zur Einschiebungslehre." *Zeitschrift für mathematischen und naturwissenschaftlichen Unterricht* 57, pp. 433-42. Section 12,v.

Hofmann, J. 1971. "Johannes Kepler als Mathematiker." *Praxis der Mathematik* 13, pp. 287-93, 318-24. Section 32,A.

Hogendijk, J. 1983. *Ibn al-Haytham's 'Completion of the "Conics" of Apollonius'*. Thesis, Utrecht, 1983. Publication forthcoming. (Sources in the History of Mathematics and Physical Sciences; see Toomer [1984, 38].) Section 28,E.

Hogendijk, J. 1985a. "Discovery of an 11th-Century Geometrical Compilation: The 'Istikmal' of Yusuf al Mu'taman ibn Hud, King of Saragossa. *Historia Mathematica* 13, pp. 43-52. Section 28,intro.

Hogendijk, J. 1985b*. "Al-Kūhī's Construction of an Equilateral Pentagon in a Given Square." *Zeitschrift für Geschichte der arabisch-islamischen Wissenschaften* 1, pp. 230-75. Section 28,intro.

Hooykaas, R. 1958. *Humanisme, science et réforme Pierre de la Ramée (1515-1572)*. Leiden: Brill. [OONL.] Section 31,A; Appendix II.

Høyrup, J. 1985. "Dynamis, The Babylonians and 'Theaetetus' 147c7-148d7." Preprint: Roskilde Univ. Centre, Institute of Educational Research, Media Studies and Theory of Science, Roskilde, Denmark. Section 7.

Hultsch, F. 1900. "Die pythagoreischen Reihen der Seiten und Diagonalen von Quadraten und ihre Umbildung zu einer Doppelreihe ganzer Zahlen." *Bibliotheca mathematica*, 1st ser., 3, pp. 8-12.

Huxley, G.–*DSB* 1. "Eudoxus." *DSB*, IV, pp. 465-67. Section 18,A.

Huxley, G.–*DSB* 2. "Theon of Smyrna." *DSB*, XIII, pp. 325-26.

Huxley, G. 1963. "Eudoxian Topics." *Greek, Roman and Byzantine Studies* 4, pp. 83-96. Section 20,intro.

Iamblichus–Albrecht. *Iamblichos/Pythagoras/Legende. Lehre. Lebensgestaltung* [*De vita pythagorica liber*]. Translated by M. von Albrecht. Zurich: Artemis, 1963. [OOU.] Chapter IV,intro.; Section 11,B.

Iamblichus–Festa. *De communi mathematica scientia*. Edited by W. Festa. Leipzig: Teubner, 1891. [OKQ.] Section 11,B.

Iamblichus–Nauck. *Iamblichi De vita pythagorica liber*. Edited by A. Nauck. St. Petersburg, 1884; Amsterdam: Hakkert, 1965. [OOCC.] Chapter IV,intro; Section 11,B.

Ibn Khaldun–Woepcke. "Traduction d'un chapitre des Prolégomènes d'Ibn Khaldoun, relatif aux sciences mathématiques." Edited by F. Woepcke. *Atti dell Accademia Pontificia de Nuovi Lincei* 10 (1856-57), pp. 236-48. [NNM.] Section 28,intro.,i.

Ibn Turk–Sayili. *"Logical Necessities in Mixed Equations" by 'Abd al Hamid Ibn Turk and the Algebra of His Time*. Edited by A. Sayili. Ankara: Turk Tarih Kurumu Basimevi, 1962 (Turk Tarih Kurumu Yayinlarindan, VII, ser. I, no. 41). [OTU.] Sections 5,B,G; 28,A,ii.

Isocrates–Norlin. "Antidosis." In *Isocrates*, vol. 2, pp. 261-62. Translated by G. Norlin. London: Heinemann, Loeb Classical Library, 1929. [OOCC.] Section 16,A.

Israel Museum. 1972. *Ketovot Mesaprot (Inscriptions Reveal)*. Jerusalem: Israel Museum. Section 9,E.

Itard, J. 1961. *Les livres arithmétiques d'Euclide*. Paris: Hermann. [OOU.] Sections 13,iv; 31,A,ii.

Jacobson, N. 1951. *Lectures in Abstract Algebra*. Toronto: Van Nostrand. Section 5,A.

Jahn, O. (ed.). 1854. *Beschreibung der Vasensammlung König Ludwigs in der Pinakothek zu München*. Munich: Lindauer. [IU.] Section 10,B.

Jayawardene, S. 1976. "The 'Trattato d'abaco' of Piero della Francesca." In *Cultural Aspects of the Italian Renaissance in Honour of Paul Oscar Kristeller*, pp. 229-43. Edited by C. Clough. Manchester. [OOCC.] Section 31,C.

Jayawardene, S.–*DSB*. "Pacioli." *DSB*, X, pp. 269-72. Section 31,D.

Johanne de Muris. *De arte mensurandi*. (See Clagett [1978, part I, chap. 3] for mss.) Section 31,A,i.

Joannes Philoponus–Rabe. *De aeternitate mundi*. Edited by H. Rabe. Leipzig: Teubner, 1899; Hildesheim: G. Olms, 1963. [OOCC.]

Junge, G. 1948. "Flächenanlegung und Pentagramm." *Osiris* 8, pp. 316-45. [OOUM.] Sections 11,B; 12,viii.

Junge, G. 1958. "Von Hippasus bis Philolaus, das Irrationale und die geometrischen Grundbegriffe." *Classica et mediaevalia* 19, pp. 41-72. [OOCC.] Section 5,L,ii.

Kapp, A. 1934. "Arabische Übersetzer und Kommentatoren Euclids." *Isis* 22 (1934), pp. 150-72; 23 (1935), pp. 54-99; 24 (1935), pp. 37-79. Section 28,intro., i.

Karpinski, L. 1914. "The Algebra of Abu Kamil." *American Mathematical Monthly* 21, pp. 37-48. Section 28,B,ii.

Kästner, A. 1796. *Geschichte der Mathematik*. Vol. 1. Göttingen: Rosenbusch. [B.N.] Appendix I,C,5.

Kent, R. 1905. "The Date of Aristophanes' Birth." *Classical Review* 19, pp. 153-55. Section 18,A.

Kepler, J. 1596. *Mysterium cosmographicum*. In *Gesammelte Werke*, vol. 1. Edited by M. Caspar and F. Hammer. Munich: Beck, 1938. [OON.] Appendix II,G.

Kepler, J. 1597. Letter to Mästlin. In *Gesammelte Werke*, vol. 1, p. 421; vol. 13, pp. 140-44, 394. Edited by M. Caspar and F. Hammer. Munich: Beck, 1938; 1945. [OON.] Sections 31,G,J,iii; 32,A,i.

Kepler, J. 1608a. Letter to Brengger. In *Gesammelte Werke*, vol. 16, pp. 137-49. Edited by M. Caspar and F. Hammer. Munich: Beck, 1955. [OON.] Appendix II.

Kepler, J. 1608b. Letter to Joachim Tanckius. In *Gesammelte Werke*, vol. 16, pp. 154-65, 429. Edited by M. Caspar and F. Hammer. Munich: Beck, 1955. A German translation of part of the letter appears in M. Caspar and W. von Dyck, *Johannes Kepler in seinen Briefen*, pp. 309-12. 2 vols. Munich, 1930. [OON.] Section 32,A,ii; Appendix II.

Kepler, J. 1618. *Epitome astronomiae Copernicae*. In *Gesammelte Werke*, vol. 7. Edited by M. Caspar and F. Hammer. Munich: Beck, 1953. [OON.] Appendix II.

Kepler, J.–Caspar 1. *Das Weltgeheimnis (Mysterium cosmographicum)*. Translated by M. Caspar. Munich: R. Oldenbourg, 1936 (original 1596; second version 1621). [WaU.] Appendix II.

Kepler, J.–Caspar 2. *Weltharmonik (Harmonicae mundi)*. Translated by M. Caspar. Munich: Oldenbourg, 1939; 1973 (original 1619). [OMU.] Appendix II.

Kepler, J.–Duncan. *Johannes Kepler. Mysterium Cosmographicum, The Secret of the Universe*. Translated by A. Duncan. New York: Abaris, 1981. [OON.] Appendix II.

Kepler, J.–Frisch. *Joannis Kepleri, astronomi opera omnia*. Edited by C. Frisch. Frankfurt, Erlangen, 1857-71; Hildesheim, 1971. (mc New York: Readex, 1967). [M.P.OTU.] Section 32,C,ii; Appendix I,C.

Kepler, J.–Hardi. *The Six-Cornered Snowflake*. Translated by C. Hardi. Oxford: Clarendon Press, 1966 (original 1611). [OOC.] Section 32,A,ii.

Kepler, J.–Halleux. *L'Etrenne ou La neige sexangulaire*. Translated by R. Halleux. Paris: Vrin, 1975 (original 1611). [OOU.] Section 32,A,ii.

Kepler, J.–Wallis. *The Harmonies of the World*. Translated by C. Wallis. In *Encyclopedia Britannica* (1952), pp. 1009-85. [ONL.] Appendix II.

King, D.–*DSB*. "Ibn-Yunus." *DSB*, 14, pp. 574-80. Section 28,D.

Klein, T. 1968. *Greek Mathematical Thought and the Origin of Algebra*. Translated by E. Brown. Cambridge, Mass.: M.I.T. Press. [OOUM.]

Klügel, G. 1803*. *Mathematisches Wörterbuch*. 5 vols. Leipzig: E. B. Schwickert, 1803-31. *Supplement*, 1833-36. (See NUC, vol. 299, p. 561). Appendix I,6.

Knorr, W. 1975a. "Archimedes and the Measurement of the Circle: A New Interpretation." *Archive for History of Exact Sciences* 15 (1975-76), pp. 115-40. Section 23,A.

Knorr, W. 1975b. *The Evolution of the Euclidean Elements: A Study of the Theory of Incommensurable Magnitudes and Its Significance for Early Greek Geometry*. Dordrecht: Reidel. [OOCC.] Sections 5; 6; 7; 8; 11; 13; 14; 15; 16; 18; 20; 23.

Knorr, W. 1978. "Archimedes and the Pre-Euclidean Proportion Theory." *Archive internationales d'histoire des sciences* 28, pp. 183-244. Section 12,vii.

Knorr, W. 1979. "Methodology, Philology and Philosophy." *Isis* 70, pp. 565-68. Section 7.

Kochavi, M. 1978. "Tell Esdar." In *Encyclopedia of Archaeological Excavations in the Holy Land*, vol. 4, pp. 1169-71. Edited by M. Avi-Yonah and E. Stern. Jerusalem: Massada Press. Section 9,E.

Kostovskii, A. 1959. *Geometrical Constructions Using Compasses Only*. New York: Blaisdell, 1961. Section 32,C.

Kraay, C. 1964. "The Melos Hoard of 1907 Reexamined." *The Numismatic Chronicle*, 7th ser., 4, pp. 1-20. [OOCC.] Section 10,C.

Kunze, C. 1842. *Lehrbuch der Geometrie*. Bd. 1: *Planimetrie*. 2nd ed. Jena: Friedrich Frommann, 1851. [PBM.] Appendix I,6.

Kuhn, F. von. 1887. "La Necropalidi Suessula." *Mitteilungen des Deutschen Archaeologischen Instituts. Roemische Abteilung*, pp. 235-75. [OHM.] Section 10,A.

Labat, R. 1952. *Manuel d'épigraphie Akkadienne*. Paris: Imprimerie Nationale de France. [OOCC.] Chapter IV,intro.; Section 9,C.

Langdon, S. (ed.). 1923*. *Cuneiform Inscriptions in the Ashmolean Museum, Oxford*, vol. 8. London: Oxford Univ. Press. [MB.] Section 9,C.

Langermann, Y., J. Hogendijk. 1984. "A Hitherto Unknown Hellenistic Treatise on the Regular Polyhedra." *Historia Mathematica* 11, pp. 325-26. Section 24,A.

Lasserre, F. 1964. *The Birth of Mathematics in the Age of Plato*. London: Hutchinson. [OOCC.] Sections 14,v; 16,C; 20,B,v.

Lasserre, F. 1966. *Die Fragmente des Eudoxus von Knidos*. Berlin: Gruyter. [OOCC.] Section 20,B.

Lee, J. 1978. "Tuleilat El-Ghassul." In *Encyclopedia of Archaeological Excavations in the Holy Land*, vol. 14, pp. 1203-13. Edited by M. Avi-Yonah and E. Stern. Section 9,E.

Leslie, J. 1820. *Elements of Geometry and Plane Trigonometry*. 4th ed. Edinburgh: Tait. Appendix I,C.

Levey, M.–*DSB*. "Abu Kamil." *DSB*, I, pp. 30-32. Section 28,A.

Leviant, C. 1973. *The Yemenite Girl*. New York: Avon. Guide for Readers.

Lewis, A. 1985. Review of Schubring [1983]. *Historia Mathematica* 12, p. 306.

Libri, G. 1840. *Histoire des sciences mathématiques en Italie*. Paris: Renouard; New York: Johnson Reprint, n.d. [OOCC.]

Liddell, H., R. Scott. 1968. *A Greek-English Lexicon*. Oxford: Clarendon Press. Section 18,A.

Lindemann, F. 1896. "Zur Geschichte der Polyeder und der Zahlzeichen." *Bayerische Akademie der Wissenschaften, Munich. Mathematisch-naturwissenschaftliche Klasse. Sitzungsberichte* 26, pp. 625-768. [OON.] Chapter III,intro.; Section 10,D.

Lonie, I.–*OCD*. "Heraclides(1) Ponticus." *OCD*, p. 500. Section 18,A.

Lucian–Kilburn. "A Slip of the Tongue in Greeting." In *Lucian*, vol. VI, pp. 171-89. Translated by K. Kilburn. Cambridge, Mass.: Harvard Univ. Press, Loeb Classical Library, 1959. [OOCC.] Section 11,A.

Mahoney, M. 1968. "Another Look at Greek Geometrical Analysis." *Archives for History of Exact Sciences* 5 (1968-69), pp. 318-48. Section 17.

Mahoney, M.–*DSB* 1. "Hero of Alexandria." *DSB*, VI, pp. 314-15. Section 25.

Mahoney–*DSB* 2. "Ramus." *DSB*, II, pp. 286-90. Section 31,H.

Mahoney, M. 1978. "Mathematics." In *Science in the Middle Ages*, pp. 145-78. Edited by D. Lindberg. Chicago: Univ. of Chicago Press. [OOCC.] Section 31,A,i.

Maistrov, L. 1967. *Probability Theory, A Historical Sketch*. Translated by S. Kotz. New York: Academic Press, 1974. [OOCC.] Appendix I,C.

Mainardi–Curtze. "Die 'Practica geometriae' des Leonardo Mainardi aus Cremona." Edited by M. Curtze. In *Urkunden zur Geschichte der Mathematik in Mittelalter und der Renaissance*, pp. 337-434. Edited by M. Curtze. Leipzig: Teubner, 1902; New York: Johnson Reprint, 1968. [Bibliotheca Mathematica Teubneriania, Band 45.] [OMU.] Section 31,A,i.

Manasse, E.–*DSB*. "Speusippus." *DSB*, XII, pp. 575-76. Section 19.

Marinus–Michaux. *Le commentaire de Marinus aux "Data" d'Euclide*. Edited by M. Michaux. Louvain: Bibliothèque de l'université, 1947. [OOLa.] Section 5,J.

Mascheroni, L.–Carette. *Géométrie du compass*. Translated by A. Carette. Paris: Duprat, 1798 (original 1797). [B.N.] Section 32,C,ii.

Mascheroni, L.–Gruson. *Gebrauch des Zirkels*. Translated by J. Gruson. Berlin: Schlesingerschen Buch- u. Musikhandlung, 1825 (original Pavia, 1797]. [IU.]

Mästlin, M. 1597. Letter to Kepler. In *Gesammelte Werke*, vol. 13, pp. 151-54. Edited by M. Caspar and F. Hammer. Munich: Beck, 1945. [OON.] Section 31,J,iii.

Mau, J. "Eudoxus." *Der kleine Pauly*, vol. II, pp. 408-10. Stuttgart: Druckenmüller, 1967. Section 20,intro.

McCabe, R. 1976. "Theodorus' Irrationality Proofs." *Mathematics Magazine* 49, pp. 201-203. Section 7.

McDiarmid, J.–*DSB*. "Theophrastus." *DSB*, XIII, pp. 328-34.

Meiggs, R.–*OCD*. "Herculaneum." *OCD*, p. 501. Section 16,B.

Michel, P. 1950. *De Pythagore à Euclide/Contribution à l'histoire des mathématiques préeuclidiennes*. Paris: Les Belles Lettres. [OOUM.] Sections 5,K; 12,vi; 13,i; 14,iii; 16,C,D; 18,A; 20,A,iv.

Mikami, Y. 1913. *The Development of Mathematics in China and Japan*. Leipzig. [OOCC.] Section 30.

Milhaud, G. 1916. Review of *De Theaeteto atheniensi mathematico* by Eva Sachs. *Revue des études grecques* 29, pp. 125-27.

Miller, G. 1926. "Weak Points in Greek Mathematics." *Scientia* 39, pp. 317-22. [OTU.]

Moebius, A. 1824. "Schreiben des Herrn Professors Moebius an den Herausgeber." *Astronomische Nachrichten* 3 (1825), pp. 132-36. Appendix I,C,5.

Mohr, G.–Hjelmslev, Pal. *Euclides Danicus*. Translated by J. Pal. Copenhagen: A. Host, 1928 (original Amsterdam, 1672). [IU.] Section 32,C,i.

Mohr, G.–Meyer. *Compendium Euclidis Curiosi*. Acta Historica Scientiarum Naturalium et Medicinalium, vol. 34. Introduction by H. Meyer. Copenhagen: Georg Mohr Foundation, 1982 (original Dutch edition, 1673; English edition, 1677). [OON.] Section 32,C,i.

Molland, A. 1978a. Review of "Neuenschwander–1973." *Math. Reviews* 56, 11686. Introduction,B.

Molland, A. 1978b. "An Examination of Bradwardine's Geometry." *Archive for History of Exact Sciences* 19, pp. 113-75. Section 31,A,i.

Montucla, J. 1758. *Histoire des mathématiques*. 2 vols. Paris: Jombert. [Bibliothèque Sainte Geneviève, Paris.] Section 31,D.

Montucla, J. 1799. *Histoire des mathématiques*. 4 vols. Paris: Agasse, 1799-1802; Paris: Blanchard, 1968; New York: Readex, 1971. According to *DSB*, volumes 1 and 2 were written by Montucla; Lalande finished volumes 3 and 4. [OTY.] Section 31,D.

Moore, F. 1981. "An Aramaic Ostracon of the Third Century B.C.E. from Excavations in Jerusalem." *Eretz-Israel* 15, pp. 67-69. Section 9,E.

Morison, S. 1933. *Fra Luca de Pacioli of Borgo S. Sepolcro*. New York: Grollier Club; New York: Kraus, 1969. [OONL.] Section 31,D.

Morrow, G. 1970. "Plato and the Mathematicians, An Interpretation of Socrates' Dream in the *Theaetetus* (201e-206c)." *Philosophical Review* 79, pp. 309-33. Sections 16A,B.

Mueller, I. 1981. *Philosophy of Mathematics and Deductive Structure in Euclid's 'Elements'*. Cambridge, Mass.: M.I.T. Press. Guide for Readers; Sections 1; 5.

Mugler, C. 1948. *Platon et la recherche mathématique de son époque*. Strasbourg: Heitz. [OOU.] Sections 12,vi; 16A,C.

Mugler, C. 1958. *Dictionnaire historique de la terminologie géométrique des Grecs*. Paris: C. Klincksieck. [OOUM.] Sections 5,A,J; 11,B; 15,ii; 20,A,B.

Muller, L. 1855. *Numismatique d'Alexandre le grand*. 2 vols. Copenhagen: Bianco Lun; Basil: Munzen und Medaillon, 1857. Translation *The Coinage of Alexander the Great, An Atlas*. New York: Attic Books, 1970. [Canadian Numismatic Museum, Ottawa.] Section 10,C.

Muller, L. 1860. *Numismatique de l'ancienne Afrique*. Vol. 1: *Les monnaies de la Cyrénaique, 1860-74*. Reprint. Chicago: OBOL Inter., 1977. [American Numismatic Society, New York.] Section 10,C.

Murdoch, J.–*DSB* 1. "Euclid: Transmission of the *Elements*." *DSB*, IV, pp. 437-59. Section 28,A,ii.

Murdoch, J.–*DSB* 2. "Al-Nayrizi." *DSB*, X, pp. 441-46. Section 25,B.

Murdoch, J. 1966. "Euclides Graeco-Latinus/A Hitherto Unknown Medieval Latin Translation of the *Elements* Made Directly from the Greek." *Harvard Studies in Classical Philology* 71, pp. 249-302. Appendix I,A,3.

Murdoch, J. 1968. "The Medieval Euclid: Salient Aspects of the Translations of the *Elements* by Adelard of Bath and Companus of Novara." *Revue de Synthèse* 89, pp. 67-94. Section 31,A,ii; Appendix II,1.

Museo etrusco Georgiano. 1842. *Musei etrusci quod Gregorius XVI . . . Part II*. Rome: ex aedibus Vaticanis. (See *NUC* pre-1956, v. 630, p. 594.) [mf IaU.] Section 10,B.

Nakayoma, S. 1965. *A History of Japanese Astronomy: Chinese Background and Western Impact*. Cambridge, Mass.: M.I.T. Press. [OOCC.] Section 30.

Nakayoma, S., W. Siven. 1973. *Chinese Science, Explorations of an Ancient Tradition*. Cambridge, Mass.: M.I.T. Press. [OOCC.] Section 30.

Needham, J. 1951. *Science and Civilization in China*. Vol. 3: *Mathematics and Sciences of the Heavens and the Earth*. Cambridge: Cambridge Univ. Press, 1959. [OOCC.] Section 30.

Neufert, E. 1960. *Bauordnungslehre*. Wiesbaden-Berlin: Bauverlag GMBH. French translation *La coordination dimensionelle dans la construction*. Paris: Dunod, 1967. Section 16,C.

Neugebauer, O., A. Sachs. 1945. *Mathematical Cuneiform Texts*. New Haven: American Oriental Society. [OWU.] Section 9,D.

Neugebauer, O. 1957. *The Exact Sciences in Antiquity*. 2nd ed. Providence: Brown Univ. Press. [OOCC.] Guide for Readers,E; Sections 5,C,H; 9,D; 11; 16,A,B.

Neugebauer, O. 1975. *A History of Ancient Mathematical Astronomy*. New York: Springer-Verlag, 1975. Sections 25,A; 26,A.

Neuenschwander, E. 1972. "Die ersten vier Bücher der Elemente Euklids." *Archive for History of Exact Sciences* 9 (1972-73), pp. 325-80. Sections 1; 12,x.

Neuenschwander, E. 1973. "Beiträge zur Frühgeschichte der griechischen Geometrie." *Archive for History of Exact Sciences* 11, pp. 127-33.

Neuenschwander, E. 1974. "Die stereometrischen Bücher der Elemente Euklids." *Archive for History of Exact Sciences* 14 (1974-75), pp. 91-125. Sections 1; 18,B,vii.

Neuenschwander, E. 1982. "Betrachtungen zu den Quellen der Arabischen Geometrie." Forthcoming. Section 28,A.

Nilson, M., J. Croon–*OCD*. "Panathenaea." *OCD*, p. 774. Section 16,B.

Niven, I., H. Zuckerman. 1960. *An Introduction to the Theory of Numbers*. New York: Wiley. Sections 8; 25,A,iv; 32,B.

Oates, J. 1979. *Babylon*. London: Thames, 1979. [OOCC.] Section 9,D.

Ohm, M. 1819*. *Elementar-Geometrie und Trigonometrie für Deutschlands Schulen und Universitäten*. Berlin: Mauer. (See *NUC*, vol. 428, 435.) Appendix I,C,6.

Ohm, M. 1826a. *Die reine Elementar-Mathematik*. Vol. 2. Berlin: Riemann. [PBL.] Appendix I,C.

Ohm, M. 1826b. *Die analytische und höhere Geometrie in ihren Elementen mit vorzüglicher Berucksichtigung der Theorie der Kegelschnitte . . . .* Berlin. [B.N.] Appendix I,C,6.

Ohm, M. 1835. *Die reine Elementar-Mathematik*. 2nd ed. Berlin: Jonas Verlags-Buchhandlung. [PSt.] Appendix I,C,6.

Ohm, M. 1841 (?). *Lehrbuch für den gesammten mathemati-*

schen Elementar-Unterricht. 5th ed. Leipzig: Graul, 1856. [B.N.] Appendix I,C,6.

Ong, W. 1958. *Ramus and Talon Inventory*. Cambridge, Mass.: Harvard Univ. Press. [OCC.]

Orlinsky, H. (ed.). 1970. *Notes on the New Translation of "The Torah"*. Philadelphia: Jewish Publication Society of America. Introduction.

Owen, G.–*DSB*. "Aristotle." *DSB*, I, pp. 250-58.

Paccioli, L. *Divina proportione* .... Venice: Paganinus de Paganinis, 1509. [B.N., B.L.; mc Zug: Switzerland, n.d., OHM; mf Institute of Chartered Accountants Collection of Rare Books on Accounting, World Microfilm Publications, n.d., CRL.] Section 31,D.

Paccioli–Bruschi. In *Scritti Rinascimentali di architectura*. Edited by A. Bruschi et al. Milano: Polifilio, 1978 (parts of *Divina proportione* are given on pp. 23-84, and the *Architectura* on pp. 85-144). [OOCC.] Section 31,D.

Paccioli–Duchesne. *Divina proportione*. Translated by G. Duchesne and M. Giraud. Paris: Librairie du Compagnonnage, 1980. [B.N.] Section 31,D.

Paccioli–Winterberg. *Fra Luca Pacioli 'Divina proportione', die Lehre vom goldenen Schnitt* .... Wien: Carl Graeser, 1896. [IU.] Section 31,D.

Pappus–Thomson. *The Commentary of Pappus on Book X of Euclid's Elements*. Edited by W. Thomson and G. Junge. Cambridge, Mass.: Harvard Univ. Press, 1930; New York: Johnson Reprint, 1968. [QMU.] Sections 7; 11,B.

Pappus–Ver Eecke. *Pappus d'Alexandre "La Collection mathématique"* (2 vols), vol. 1. Translated by P. ver Eecke. Bruges: Desclee de Brouwer, 1933. [QMU.] Sections 24,C; 27.

Pederson, O. 1974. *A Survey of the "Almagest"*. Odense: Odense Univ. Press. [OON.] Section 26.

Perls, H. 1973. *Lexikon der platonischen Begriffe*. Bern: Francke. [OOCC.] Section 16,C.

Philip, J. 1966. *Pythagoras and Early Pythagoreanism*. Toronto: Univ. of Toronto Press. [OOCC.] Sections 7; 11,intro.; 12.

Pingree, D.–*DSB*. "History of Mathematical Astronomy in India." *DSB*, XV, pp. 533-633. Section 29.

Plant, R. 1979. *Greek Coin Types and Their Identification*. London: Seaby. [OOCC.] Section 10,C.

Plato–Adam. *"The Republic" of Plato*. Translated by J. Adam. Cambridge: Cambridge Univ. Press, 1963 (original 1902). [OOCC.] Sections 16,D; 20,A.

Plato–Bluck. *Plato's "Meno"*. Translated by R. Bluck. Cambridge: Cambridge Univ. Press, 1961. [OONL.] Section 5,D.

Plato–Bury 1. *Laws in Plato*, vol. 9. Translated by R. Bury. London: Heinemann, Loeb Classical Library, 1926; 1961. Section 7.

Plato–Bury 2. *Timaeus* in *Plato*, vol. 6. Translated by R. Bury. London: Heinemann, Loeb Classical Library, 1966. [OOCC.] Section 16,D.

Plato–Chambry. *La république* in *Platon/Oeuvres complètes*, vol. 6. Edited by E. Chambry. Paris: Les Belles Lettres, 1959. [OOCC.] Sections 16,D; 20,H.

Plato–Dies. *Platon/Oeuvres complètes*, vol. 8, *Théetète*. Translated by A. Dies. Paris: Les Belles Lettres, 1963. [OOCC.] Section 18,H.

Plato–Fowler 1. *The Statesman* in *Plato*, vol. 3. Translated by H. Fowler. London: Heinemann, Loeb Classical Library, 1952. Section 16,B.

Plato–Fowler 2. *Greater Hippias* in *Plato*, vol. 6, pp. 333-424. Translated by H. Fowler. London: Heinemann, Loeb Classical Library, 1953. [OOCC.] Section 16,D.

Plato–Fowler 3. *Theaetetus* in *Plato*, vol. 7. Translated by H. Fowler. London: Heinemann, Loeb Classical Library, 1953 [OOCC.] Section 7; 18, A.

Plato–Hackforth. *Plato's 'Phaedo'* Translated by R. Hackforth. Cambridge: Cambridge Univ. Press, 1955. [OOCC.] Section 16,D.

Plato–Jowett. *The Dialogues of Plato*. Edited by B. Jowett. Oxford: Oxford Univ. Press, 1931. [OOCC.] Section 16,D.

Plato–Lamb. *Meno* in *Plato*, vol. 4, pp. 9-371. Translated by W. Lamb. London: Heinemann, Loeb Classical Library, 1952. [OOCC.] Section 5,D.

Plato–Shorey. *The Republic* in *Plato*, vol. 2.Translated by P. Shorey. London: Heinemann, Loeb Classical Library, 1963. [OOCC.] Sections 5,C; 6; Chapter IV,intro.; 16,A,C,D; 20,A,D.

Plato–Tarrant. *Hippias Major*. Edited by D. Tarrant. Cambridge: Cambridge Univ. Press, 1928. [OOCC.] Section 16,D.

Plato–Warrington. *Timaeus*. Translated by J. Warrington. London: Dent, Everyman's Library, 1965. [OOCC.] Section 16,D.

Plutarch–Babbit. "The Obsolescence of Oracles." In *Moralia*, vol. 5. Translated by F. Babbit. London: Heinemann, Loeb Classical Library, 1962. [OOCC.] Section 16,D.

Plutarch–Cherniss. "Platonic Questions." In *Moralia*, vol. 13, part 1. Translated by H. Cherniss. London: Heinemann, Loeb Classical Library, 1976. [OOCC.] Section 16,D.

Plutarch–Einarson. "A Pleasant Life Impossible." In *Moralia*, vol. 14. Translated by B. Einarson and P. de Lacy. London: Heinemann, Loeb Classical Library, 1967. [OOCC.] Section 9,D.

Plutarch–Minar. "Table Talk." In *Moralia*, vol. 9. Translated by E. Minar et al. London: Heinemann, Loeb Classical Library, 1969. [OOCC.] Section 9,D.

Porada, E. 1965. "The Relative Chronology of Mesopotamia/Part I: Seals and Trade (6000-1600 BC)." In *Chronologies in Old World Archaeology*. Edited by R. Ehrich. Chicago: Univ. of Chicago Press. [OOCC.] Section 9,B.

Poulle, E.–*DSB*. "Johannes de Muris." *DSB*, VII, pp. 128-33. Section 31,A,i.

Procissi, A. 1981. "Bibliografia matematica della Grecia classica e di altre civiltà antiche." *Bollettino di Storia delle Scienze Matematiche* 1, pp. 3-149.

Proclus–Festugière. *Commentaire sur le Timée*, Book 3, vol. 3. Translated by A. Festugière. Paris: Vrin, 1967. [OOCC.] Section 16,D.

Proclus–Friedlein. *Procli Diadochi in primum Euclidis Elementorum librum commentarii*. Edited by G. Friedlein. Leipzig: Teubner, 1873. [OKQ.] Sections 11,B; 16,B; 20,A,B,C.

Proclus–Kroll. *Procli Diadochi in Platonis Republicam commentarii*. Edited by G. Kroll. Leipzig: Teubner, 1901. [OOCC.] Section 6.

Proclus–Morrow. *Proclus/A Commentary on the First Book of Euclid's Elements*. Translated by G. Morrow. Princeton: Princeton Univ. Press, 1970. [OOCC.] Sections 4; 5,D,H; 6; Chapter IV,intro.; 11,B; 14; 16,B; 20,A,B,C.

Proclus–Ver Eecke. *Proclus de Lycie/Les commentaires sur le premier livre des Eléments d'Euclide*. Translated by P. Ver Eecke. Bruges: De Brouwer, 1948. [OOUM.] Sections 16,B; 20.

Pryce, N., B. Sparkes. *OCD*. "Pottery." *OCD*, pp. 870-71. Section 10,A.

Ptolemy–Manitius. *Ptolemaus, Handbuch der Astronomie*, vol. 2. 2nd ed. Translated by K. Manitius. Leipzig: Teubner, 1963. [OOCC.] Section 26,A.

Quint, N. 1905. "Response to Question 658." *L'Intermédiaire des Mathématiciens* 12, p. 53. Appendix I,C.

Quint, N. 1913. "Response to Question 4149." *L'Intermédiare des Mathématiciens* 20, p. 120. Appendix I,C.

Ramus (Ramée), P. 1569. *Petrus Rami Scholarum mathematicarum libri unus et triginta, dudum quidem a Lazaro Schonero recogniti et aucti....* Frankfurt: Andrea Wecheli, 1627. [This is item no. 706 in Ong, 1958, p. 439; mf OOCC.] Section 31,H.

Ramus (Ramée), P. 1599. [*Geometriae libri septem et viginti*]. In *Petri Rami Arithmeticae libri duo, geometriae septem et viginti, a Lazaro Schonero recogniti et aucti*. Frankfurt: Andrea Wecheli, 1627. [This is item no. 685 in Ong, 1958, p. 425; see also B.N. catalogue, vol. 140, p. 308, cote V6208; mf OOCC.] This work is difficult to trace because the two books on arithmetic and the twenty-seven books on geometry are bound together but have separate paginations; because *Scholarium mathematicarium* (Ramus, 1569) is sometimes bound with this work; and because there is another work (Ong, 427, no. 686) with a similar title. Section 31,H; Appendix I,C; Appendix II.

Rey, A. 1935. *Les mathématiques en Grèce au milieu du Ve siècle*. Paris: Hermann. [OOUM.]

Rey, A. 1939. *La maturité de la pensée scientifique en Grèce*. Paris: Albin Michel. [OOU.]

Robinson, R. 1936. "On Greek Analysis." *Mind* 45, pp. 464-73.

Robinson, R.–*OCD*. "Plato." *OCD*, pp. 839-42. Section 16,intro.,C.

Rose, P. 1975. *The Italian Renaissance of Mathematics, Studies on Humanists and Mathematicians from Petrarch to Galileo*. Geneva: Droz. [OOCC.] Section 31,A,D,E.

Rose, H.–*OCD*. "Hygeia." *OCD*, p. 533. Section 9,C.

Ross, W.–OCD. "Aetius." *OCD*, p. 20. Section 16,D.

Russo, G. 1912. "Question 4149." *L'Intermédiaire des Mathématiciens* 19, p. 271. Appendix I,C.

Sachs, E. 1914. *De Theaeteto atheniensi mathematico*. Berlin: Otto Schade. [IU.] Section 18,A.

Sachs, E. 1917. *Die fünf platonischen Körper*. Berlin: Weidmann. [OKQ.] Sections 2; 11,B; 14,i; 16,C; 18,B; 20,A,B.

Said, H. (ed.). 1979. *Al-Biruni Commemorative Volume (Proceedings of the International Congress (1973)*. Karchi: Hamdard Academy. Section 28,E.

Salmon, E.–*OCD* 1. "Cumae." *OCD*, p. 301. Section 10,A.

Salmon, E.–*OCD* 2. "Suessula." *OCD*, p. 1020. Section 10,A.

Salmon, E.–*OCD* 3. "Campania." *OCD*, p. 199. Section 10,A.

Sanford, V. 1927. *The History and Significance of Certain Standard Problems in Algebra*. New York: Columbia Univ. [OTY.] Appendix I,C,6.

Sarfatti, G. 1968. *Mathematical Terminology in Hebrew Scientific Literature of the Middle Ages* [Hebrew]. Jerusalem: Magnes Press. [WU.] Section 28,A,ii.

Sarton, G. 1936. *The Study of the History of Mathematics*. New York: Dover, 1954. [OOCC.]

Sarton, G. 1951. "Query no. 130: When did the term golden section or its equivalent in other languages originate?" *Isis* 42, p. 47. [OOCC.] Appendix I,C.

Saunders Wörterbuch. 1860. *Wörterbuch der Deutschen Sprache*. Leipzig: Wigand. [B.N.] Appendix,I,C.

Schliemann, H. 1878. *Mykenae*. Leipzig; Darmstadt: Wissenschaftliche Buchgesellschaft, 1964. [OOCC.] Chapter III,intro.

Schmidt, W. 1900. "Sind die heronischen Vielecksformeln trigonometrisch?" *Bibliotheca mathematica*, 3rd ser., 1, pp. 319-20. Section 6.

Schmitz, L. 1849. "Salus." In *Dictionary of Greek and Roman Biography and Mythology*, vol. 3, p. 702. Edited by W. Smith. London: Taylor, Walton and Waverly. [OOCC.] Section 9,C.

Schöttler. 1857. "Ueber eine mit dem goldenen Schnitt in Zusammenhang stehende Kreisgruppe." In *Schulprogramme Gütersloh*, pp. 3-10. [Göttingen.] Appendix I,C,6.

Schoy, C. 1923. "Beiträge zur arabischen Trigonometrie." *Isis* 5, pp. 364-99. [OOCC.] Section 28,D.

Schubring, G. 1983. "Das mathematische Leben in Berlin—Zu einer enstehenden Profession an Hand von Briefen des aus Erlangen stammenden Martin Ohm an seinen Bruder Georg Simon." *Erlanger Bausteine zur Fränkischen Heimatforschung* 30, pp. 221-49. Appendix I,C,6.

Schubring, G. 1986*. *Bibliographie der Schulprogramme*. In *Mathematik und Naturwissenschaften (1800-1875)*. Bad Salzdetfurth: Franzbecker-didaktischer dienst. Appendix I,C,6.

Schweitzer, B. 1955. "Zum Krater des Aristonothos." *Mitteilungen des Deutschen Archaeologischen Instituts Roemische Abteilung* 62, pp. 78-106 + Tafel 34-41. [OLU.] Section 10,A.

Scullard, H. 1967. *The Etruscan Cities and Rome*. Ithaca, N.Y.: Cornell Univ. Press. [OOCC.] Section 10,A.

Seidenberg, A. 1960. "The Ritual Origin of Geometry." *Archive for History of Exact Sciences* 1 (1960-62), pp. 488-527. Section 5,H.

Seidenberg, A. 1977. "Origin of Mathematics." *Archive for History of Exact Sciences* 18 (1977-78), pp. 301-42. Section 5,intro.,H.

Seidenberg–*DSB*. "Mascheroni." *DSB*, IX, pp. 156-58. Section 32,C,ii.

Sezgin, F. 1974. *Geschichte des arabischen Schriftums*. Band V: *Mathematik bis ca. 430 H*. Leiden: Brill. [OOCC.] Section 28,A,1.

Shorey, P. 1933. *What Plato Said*. Chicago: Univ. of Chicago Press. [OOCC.] Section 16,A.

Simpson, R. 1753. "An Explication of an Obscure Passage in Albert Girard's Commentary upon Simon Stevin's Work." In *Philosophical Transactions* 48, pp. 368-77. [OONL.] Section 32,B.

Slavutin, E. 1977. "On Euclid's *Data*" (Russian). *Istoriko-Mat. Issledovanija* 22, pp. 229-39. (See the review in *Zentralblatt für Mathematik* 401 (1979), no. 01002.) Section 5,J.

Smith, D. 1923. *History of Mathematics*. 2 vols. Boston: Ginn. [OOCC.] Section 31,A; Appendix I.

Sonnenburg, 1881. "Der goldene Schnitt. Beitrag zur Geschichte der Mathematik und ihrer Anwendung." In *Programm des Königlichen Gymnasiums zu Bonn, Schuljahr 1880-1881*, pp. 3-22. Bonn: Universitäts- Buchdruckerei von Carl Georgi. [Univ. de Strasbourg.]

Speziali, Pierre. 1953a. "Fra Luca Paciuolo et 'La Divina proportione'." *Les Musées de Genève* 10, no. 5, p. 1. Section 31,D.

Speziali, Pierre. 1953b. "La Divina proportione" de Paciuolo." *Stultifera Navis* 10, pp. 84-90, 10 fig. Section 31,D.

Speziali, Pierre. 1953c. "Leonard de Vinci et 'La Divina proportione' de Luca Paciuolo." *Bibl. Hum. Renaiss.* 15, pp. 295-305, 1 pl. Section 31,D.

Staigmüller, H. 1889. "Lucas Paciuolo. Eine biographische Skizze." *Zeitschrift für Mathematik und Physik* 34, pp. 81-102, 121-28. [OTU.] Section 31,D.

Stamatis, E. 1956. "A Contribution to the Investigation of the Geometrical Algebra of the Pythagoreans." *Platon* 8, pp. 144-57. [MiU.] Section 6.

Stapleton, H. 1958. "Ancient and Modern Aspects of Pythagoreanism." *Osiris* 13, pp. 12-53. [OOUM.] Section 9,D; 12,xi.

Steele, D. 1951. "A Mathematical Reappraisal of the Corpus Platonicum." *Scripta mathematica* 17, pp. 173-89. Section 16,A.

Steinschneider, M. 1893. *Die hebräischen Übersetzungen des Mittelalters und die Juden als Dolmetscher*. Berlin: Bibliographisches Bureau; Graz; Akademischer Druck, 1956. [OHM.] Section 28,intro.

Stenius, E. 1978. "Foundations of Mathematics: Ancient Greek and Modern." *Dialectica* 32, pp. 255-90. Section 22,C.

Stevin, S. 1585. *Problemata geometrica*. In *The Principal Works of Simon Stevin*, vol. IIa. Edited by D. Struik. Amsterdam: Swets and Zeitlinger, 1958 (original Antwerp, 1585). [OON.] Section 31,I.

Stillwell, R. 1976. *The Princeton Encyclopedia of Classical Sites*. Edited by R. Stillwell. Princeton: Princeton Univ. Press. Chapter III,intro.; 10,C.

Strong, D. 1968. *The Early Etruscans*. London: Evans. Section 10,A.

Struik, D. 1963. "On Ancient Chinese Mathematics." *Mathematics Teacher* 56, pp. 424-32. Section 30.

Struik, D. 1969. *A Source Book in Mathematics 1700-1800*. Cambridge, Mass.: Harvard Univ. Press. [OOCC.] Section 31,B,v.

Strycker, E. de. 1937. "Une énigme mathématique dans *L'Hippias Majeur*." *Annuaire de l'institut et d'histoire orientales et slaves* (Brussels), vol. V (*Mélanges Emile Boisacq*), pp. 317-26. [OTU.] Section 16,D.

Suidas–Adler. *Suidae Lexicon*. Edited by A. Adler. Leipzig: Teubner, 1929-35. (On the name of the work "Suidas," see P. Forbes–*OCD*.) Section 18,A.

Suter, H. 1900. *Die Mathematiker und Astronomen der Araber und ihre Werke*. Leipzig: Teubner; New York: Johnson Reprint, 1972. [OTU.] Section 28,intro.

Suter, H. 1902. "Uber die Geometrie der Söhne des Musa b. Schakir." *Bibliotheca Mathematica* 3, pp. 259-72. Section 28,A,1.

Swetz, F., T. Kao. 1977. *Was Pythagoras Chinese? An Examination of Right Triangle Theory in Ancient China*. University Park, Penn.: Pennsylvania State Univ. Studies.

Swetz, F., A. Tian-Se. 1984. "A Brief Chronological and Bibliographic Guide to the History of Chinese Mathematics." *Historia Mathematica* 11, pp. 39-56. Section 30.

Szabó, A. 1968. "Wie kamen die Pythagoreer zu dem Satz Eucl., *Elem.* II,5?" *Akademia Athenon, Athens, Praktika*, pp. 203-32. [MoKL.] Section 5,L.i.

Szabó, A. 1969. *Anfänge der griechischen Mathematik*. Vienna: Oldenburg, 1969. French translation *Les débuts des mathématiques grecques*. Translated by M. Federspiel. Paris: Vrin, 1977. [OLU; OOUM.] Section 5,L.i.

Szabó, A. 1974a. "Die Muse der Pythagoreer, zur Frühgeschichte der Geometrie." *Historia Mathematica* 1, pp. 291-316. Section 5,L.i.

Szabó, A. 1974b. "The Origin of the Pythagorean 'Application of Areas'." *Proceedings, No. 2, XIVth International Congress of the History of Science*. Tokyo and Kyoto, Japan, August 1974, pp. 207-13. Tokyo: Science Council of Japan. [InRE.] Section 5,L,i.

Taisbak, C. 1971. *Division and Logos/A Theory of Equivalent Couples and Sets of Integers/Propounded by Euclid in The Arithmetical Books of the "Elements"*. Acta Historia Scientiarum Naturalium et Medicinalium, vol. 25. Copenhagen: Odense Univ. Press.

Taisbak, C. 1982. *Coloured Quadrangles: A Guide to the Tenth Book of Euclid's 'Elements'*. Opuscula Graecolatina, 24. Copenhagen: Museum Tusculaneum Press. Section 1.

Tannery, P. 1882. "De la solution géométrique des problèmes du second degré avant Euclide." *Société des sciences physiques et naturelles de Bordeaux, mémoires*. 2nd ser., 4, pp. 395-416. Sections 5,J,K; 12,iii.

Tannery, P. 1887. *La géométrie grecque*. Gauthier-Villars. Section 20,A,i,ii.

Taylor, A. 1928. *A Commentary on Plato's "Timaeus"*. Oxford: Clarendon Press, 1928. [OOCC.] Section 16,C.

Taylor, A. 1937. *Plato, The Man and His Work*. 4th ed. London: Methuen. [OOCC.] Section 16,intro.

Taylor, C. 1967. "Plato and the Mathematicians: An Examination of Professor Hare's Views." *Philosophical Quarterly* 17, pp. 193-203. [OOCC.] Section 16,A.

Taylor, T. 1979. "Side and Diagonal Numbers Imply the Existence of Incommensurables." *James Madison Journal* 37, pp. 33-37. [ViU.] Section 6.

Terquem, M. 1838. "Notes historiques 1: Sur la locution: diviser une droite en moyenne et extrême raison." *Journal de Mathématiques Pures et Appliquées* 3, pp. 97-98. [OOCU.] Appendix I,C,6.

Terquem, M. 1853. "Sur la locution: diviser une droite en extrême et moyenne raison." *Nouvelles Annales de Mathématiques* 12, pp. 36-40. Appendix I,C,6.

Thabit ibn Qurra–Luckey. "Tabit b. Qurra über den geometrischen Richtigkeitsnachweis der Auflösung der quadratischen Gleichungen." *Saechsische Akademie der Wissenschaften Leipzig, Mathematisch-naturwissenschaftlische Klasse, Sitzungsberichte* 13 (1941), pp. 93-114. Sections 5,G; 28,intro.

Theon of Smyrna–Dupuis. *Théon de Smyrne, Philosophe Platonicien/Exposition des connaisances mathématiques utiles pour la lecture de Platon*. Translated by J. Dupuis. Paris, 1892; Brussels: Culture et Civilization, 1966. Section 6.

Thomas, I. 1939. *Selections Illustrating the History of Greek Mathematics*. 2 vols. London: W. Heinemann, Loeb Classical Library, 1951. [OOCC.]

Thompson, F. 1970. "[Dodecahedrons Again]." *Antiquaries Journal* 50, pp. 92-96.

Thompson, R. 1901*, 1904*. *Cuneiform Texts from Babylonian Tablets, etc., in the British Museum*, vols. 12, 19. London: British Museum. Section 9,A.

Thurlow, B., I. Vecchi. 1979. *Italian Cast Coinage: Italian Aes Grave and Italian Aes Rude, Signatum and the Aes Grave of Sicily*. London: V. C. Vecchi. [British Lending Library.] Section 10.

Toomer, G. 1973. "The Chord Table of Hipparchus and the Early History of Greek Trigonometry." *Centaurus* 18, pp. 6-28. Section 26,B,C.

Toomer, G. 1984. "Lost Greek Mathematical Works in Arabic Translation." *Mathematical Intelligencer* 6, pp. 32-38. Section 28,C.

Toomer, G.–*DSB* 1. "Apollonius of Perga." *DSB*, I, pp. 179-93. Section 24,C.

Toomer, G.–*DSB* 2. "Theon of Alexandria." *DSB*, XIII, pp. 321-25.

Toomer, G.–*DSB* 3. "Hipparchus." *DSB*, XV, Supplement I, pp. 207-24. Section 26B,C.

Toomer, G.–*DSB* 4. "Ptolemy." *DSB*, XI, pp. 186-206. Sections 26; 28,B.

Toomer, G.–*DSB* 5. "al-Khwarizmi." *DSB*, VII, pp. 358-65.

Toomer, G.–*OCD*. "Eudoxus." *OCD*, p. 414. Section 20,intro.

Treves, P.–*OCD*. "Philodemus." *OCD*, p. 818. Section 16,B.

Tropfke, J. 1903. *Geschichte der Elementar-Mathematik*, Band 2. Leipzig: Veil. (mc New York: Readex, Landmarks of Science Series, 1971). [OTY.] Appendix I,C.

Tropfke, J. 1934. "Zur Geschichte der quadratischen Gleichungen über dreieinhalb Jahrtaussend." *Deutsche Mathematiker Vereinigung-Jahresbericht* 44, pp. 26-47, 95-119. (The first part of this article is in 43 [1934], pp. 98-107.)

Unguru, S. 1975. "On the Need to Rewrite the History of Greek Mathematics." *Archive for History of Exact Sciences* 15 (1975-76), pp. 67-114. Section 5,intro.

Unguru, S. 1979. "History of Ancient Mathematics/Some Reflections on the State of the Art." *Isis* 70, pp. 555-65. Section 5,intro.

Valabrega-Gibellato. 1979. "Un'ipotesi sull'origine dell'algebra geometrica di Euclide." *Bolletino Unione Matematica Italiana* A 16, pp. 190-200. Section 5,L,iii.

Van Buren, E. 1939. "The Rosette in Mesopotamian Art." *Zeitschrift für Assyriologie* 45, pp. 99-107. Chapter III,intro.

Van Buren, E. 1945. *Symbols of the Gods in Mesopotamian Art*. Rome: Pontificium Institutum Biblicum. [WU.] Chapter III,intro.; Section 9,B.

Van der Waerden, B. 1954. *Science Awakening I*. Groningen: Noordhoff. [OOCC.] Sections 4; 5; 6; 8; 9; 11; 12; 13; 14; 16; 18; 20.

Van der Waerden, B. 1976. "Defense of a Shocking Point of View." *Archives for History of Exact Sciences* 15, pp. 199-210. Section 5,intro.,F,K.

Verdonk, J. 1966. *Petrus Ramus en de Wiskunde*. Assen: Van Gorcum. [OOUV.] Section 31,H.

Victor, S. 1979. *Practical Geometry in the High Middle Ages*. Philadelphia: American Philosophical Society. [OOCC.] Section 31,A,i.

Vincent, J. 1844. "Sur les deux locutions: 'partager und droite en moyenne et extrême raison' et 'donnée qu'en raison'—Altérations probables dans le texte d'Euclide." *Nouvelles Annales de Mathématiques* 3, pp. 5-14. Appendix I,C,6.

Vogel, C. de. 1963. *Greek Philosophy/A Collection of Texts*, vol. 1. 3rd ed. Leiden: Brill. [OOCC.] Section 11,A.

Vogel, C. de. 1966. *Pythagoras and Early Pythagoreanism*. Assen: Van Gorcum. [OOCC.] Chapter III,intro., Section 11,A.

Vogel, K.–*DSB* 1. "Diophantus of Alexandria." *DSB*, IV, pp. 110-19.

Vogel, K.–*DSB* 2. "Aristaeus." *DSB*, I, pp. 245-46. Section 24,C.

Vogel, K.–*DSB* 3. "Fibonacci." *DSB*, IV, pp. 604-13. Section 31,A.

Wagner, D. 1978. "Lui Hui and Tsu Keng-chih on the Volume of a Sphere." *Chinese Science* 3, pp. 59-79. Section 30.

Wagner, D. 1979. "An Early Chinese Derivation of the Volume of a Pyramid: Lui Hui, Third Century A.D." *Historia Mathematica* 6, pp. 164-88. Section 30.

Wagner, D. 1985. "A Proof of the Pythagorean Theorem by Liu Hui (Third Century A.D.)." *Historia Mathematica* 12, pp. 71-73. Section 30.

Ward-Perkins, J.–*OCD*. "Etruscans." *OCD*, p. 410. Section 10,H.

Waterhouse, W. 1973. "The Discovery of the Regular Solids." *Archive for History of Exact Sciences* 9 (1972-73), pp. 212-21. [NIC.] Sections 11,B; 18,vi.

Wedberg, A. 1955. *Plato's Philosophy of Mathematics*. Stockholm: Almquist and Wiksell, 1975. [OOCC.] Section 16,A,C.

Weil, A. 1978. "Who Betrayed Euclid?" *Archive for History of Exact Sciences* 19, pp. 91-93. Section 5,intro.

Weiss, B. 1972. "Mishnat Ha-middot." In *Encyclopedia Judaica*, vol. 12, p. 110. Jerusalem: Macmillan. Section 28,A,ii.

White, F. 1975. "Plato on Geometry." *Apeiron* 9, pp. 5-14. [OOCC.] Section 16,A.

Wiegand, A. 1847. *Geometrische Lehrsätze und Aufgaben aus des Herrn Professor C. F. A. Jacobi Anhängen zu Van Swinden's Elementen der Geometrie*. Halle: Schmidt. [NIC.] Appendix I,6.

Wiegand, A. 1848. *Die merkwürdigen Punkte des Dreiecks mit Rücksicht auf harmonische Theilung*. 2nd ed. Halle: H. W. Schmidt. [NIC.] Appendix I,6.

Wiegand, A. 1849. *Der allgemeine goldene Schnitt und sein Zusammenhang mit der harmonischen Theilung/Ein neuer Beitrag zum Ausbau der Geometrie*. Halle: Schmidt. [Göttingen.] Appendix I,6.

Wieleitner, H. 1927. *Mathematische Quellenbücher. I, II*. Mathematisch-Naturwissenschaftlich-Technische Bücherei, 3,11. Berlin: Otto Salle. [IU.] Section 31,K; Appendix I,C.

Williams, R. 1972. *The Geometrical Foundation of Natural Structure*. New York: Dover, 1979 (original: *Natural Structures*. Moorpark, Calif.: Eudaemon, 1972). Section 31,C,I.

Wilson, D. 1976. "Herakleia." In *The Princeton Encyclopedia of Classical Sites*, p. 383. Edited by R. Stillwell. Princeton: Princeton Univ. Press. Section 18,A.

Woepcke, M. 1855. "Analyse et extrait d'un recueil de constructions géométriques par Aboul Wafa." *Journal asiatique* 5, pp. 218-56, 309-59. [OTU; also mc Washington, D.C.: Microcard Editions, n.d., OWU.]

Youschkevitch (Juschkewitsch), A. 1961. *Geschichte der Mathematik im Mittelalter*. Leibzig: Teubner, 1964 (Russian edition, 1961). [OOUM.] Section 31,A,i.

Youschkevitch, A. 1976. *Les mathématiques arabes (VIIIe-XVe siècles)*. Paris: Vrin. [DLC.] Section 28,intro.

Zeising, A. 1854. *Neure Lehre von den Proportionen des menschlichen Körpers, aus ein bisher unerkannt....* Leipzig: Weigel. [B.N.] Appendix I,C,6.